Green Chemistry
Environmentally Benign Reaction

Green Chemistry
Environmentally Benign Reaction

V.K. Ahluwalia

Taylor & Francis
Taylor & Francis Group
Boca Raton London New York

CRC is an imprint of the Taylor & Francis Group,
an informa business

Ane Books Pvt. Ltd.

Green Chemistry : Environmentally Benign Reaction

V. K. Ahluwalia

© Author

First Published in 2009

Reprinted : 2010

Ane Books Pvt. Ltd.

4821 Parwana Bhawan, 1st Floor
24 Ansari Road, Darya Ganj, New Delhi -110 002, India
Tel: +91 (011) 2327 6843-44, 2324 6385
Fax: +91 (011) 2327 6863
e-mail: anebooks@vsnl.net
Website: www.anebooks.com

For

CRC Press

Taylor & Francis Group
6000 Broken Sound Parkway, NW, Suite 300
Boca Raton, FL 33487 U.S.A.
Tel : 561 998 2541
Fax : 561 997 7249 or 561 998 2559
Web : www.taylorandfrancis.com

For distribution in rest of the world other than the Indian sub-continent

ISBN-10 : 1 42007 070 3
ISBN-13 : 978 1 42007 0705

British Library Cataloguing in Publication Data
A catalogue record for this book is available from the British Library

Printed at Thomson Press, Delhi

Preface

Green chemistry is basically environmentally benign chemical synthesis and is helpful to reduce environment pollution. A large number of organic reaction were earlier carried out under anhydrous conditions and using volatile organic solvents like benzene, which cause environmental problems and are also potential carcinogenic. Also the bye products are difficult to dispose of.

With the advancements of knowledge and new developments, it is now possible to carry out large number of reactions in aqueous phase, using phase transfer catalysts, using sonication and microwave technologies. Some reactions have also be performed enzymatically and photochemically. It is now possible to carry out a number of reactions using the versatile liquids and also in solid state.

The book is divided into three chapters. Introduction to Green Chemistry is described in Chapter 1. The second chapter deals with those reactions which are now performed under the so called green conditions. Such reactions are now referred to as Green Reactions.. Finally in chapter 3 are described a number of preparations in aqueous phase, using phase transfer catalysis using sonication and microwave technologies. Also some preparation carried out enzymatically and photochemically. It is now possible to perform by using ionic liquids as solvents are also described.

The author expresses his sincere thanks to Dr. Pooja Bhagat, Dr. Madhu Chopra for all the help they have rendered.

Grateful thanks are due to Prof. Sukh Dev FNA, INSA Professor, New Delhi, Prof. J. M. Khurana, Department of Chemistry and Dr. R. K. Suri, Additional Director, Ministry of Forests, Government of India.

Finally I take the opportunity to thank Prof. Ramesh Chandra, Director, Dr. B. R. Ambedkar Centre for Biomedical Research University of Delhi, Delhi for all the help rendered.

V. K. Ahluwalia

Foreword

I feel happy to congratulate Prof. V. K. Ahluwalia on his writing a book on "Green Chemistry - Environmentally Benign Reactions". The book is replete with basic principles of Green Chemistry and requisite details that are necessary to obtain a desirable organic reactions (which earlier needed anhydrous conditions and used volatile organic solvents) under green condition. It is hoped that this development will go a long way in reducing not only environmental pollution but also effecting atom economy.

The book has been very well written and presented in a lucid manner. The book is so comprehensive that it can serve as a practical guide to the researchers (including M.Sc., M.Phil. and Ph.D) in various Industries, Universities and College Laboratories.

Dr. R. K. Suri

Additional Director
Government of India
Ministry of Environment & Forests
Paryavaran Bhavan, C.G.O. Complex
Lodi Road, New Delhi-110003

Contents

Preface   v
Foreword   vii

1. GREEN CHEMISTRY   1

    Introduction   1
    1.1 The Need of Green Chemistry   1
    1.2 Principles of Green Chemistry   2
    1.3 Planning a Green Synthesis in a Chemical Laboratory   10

2. GREEN REACTIONS   17

    Introduction   17
    2.1 Acyloin Condensation   17
         Mechanism   18
         2.1.1 Acyloin Condensation using coenzyme, Thiamine   20
         2.1.2 Applications   21
    2.2 Aldol Condensation   22
         Mechanism   22
         The Aldol   23
         2.2.1 Acid-Catalysed Aldol Condensation   23
         2.2.2 Crossed Aldol Condensation   24
         2.2.3 Aldol Type Condensations of Aldehydes with
                NitroAlkanes and Nitriles   27
         2.2.4 Vinylogous Aldol Reaction   28
         2.2.5 Aldol Condensation of Silyl Enol Ethers in
                Aqueous Media   28
         2.2.6 Aldol Condensation in Solid Phase   29
         2.2.7 Applications   30
    2.3 Arndt-Eistert Synthesis   33
         Mechanism   34
         2.3.1 Applications   35
    2.4 Baeyer–Villiger Oxidation   38

Mechanism		39
Migratory Aptitude		40
2.4.1	Baeyer–Villiger Oxidation in Aqueous Phase	41
2.4.2	Baeyer–Villiger Oxidation in Solid State	42
2.4.3	Enzymatic Baeyer–Villiger Oxidation	42
2.4.4	Applications	43
2.5	Barbier Reaction	47
2.5.1	Barbier Reaction under sonication	47
2.5.2	Applications	48
2.6	Barton Reaction	50
Mechanism		50
2.6.1	Applications	50
2.7	Benzoin Condensation	52
Mechanism		53
2.7.1	Benzoin condensation under catalytic conditions	53
2.7.2	Applications	54
2.8	Baker–VenkatAraman Rearrangement	56
2.8.1	PTC catalysed synthesis of Flavones	56
Mechanism		57
2.8.2	Application	57
2.9	Bouveault Reaction	57
2.9.1	Bouveault Reactions under sonication	58
2.10	Cannizzaro Reaction	58
Mechanism		59
2.10.1	Crossed Cannizzaro Reaction	60
2.10.2	Intramolecular Cannizzaro Reaction	60
2.10.3	Cannizzaro reactions under sonication	61
2.10.4	Applications	61
2.11	Claisen Rearrangement	63
Mechanism		63
2.11.1	Claisen Rearrangement in Water	66
2.11.2	Applications (Classical Claisen Condensation)	66
2.11.3	Applications (Aqueous Phase Claisen Rearrangement)	67
2.12	Claisen–Schmidt Reaction	69
Mechanism		70
2.12.1	Claisen schmidt reaction in aqueous phase	70
2.12.2	Applications	71
2.13	Clemmensen Reduction	73
Mechanism		73
Limitations		73
2.13.1	Applications	74
2.14	Dakin Reaction	76
Mechanism		77
2.14.1	Applications	77

2.15 Darzen Reaction   79
   2.15.1 Darzen reaction in presence of phase transfer catalyst   79
   Mechanism   80
   2.15.2 Applications   81
2.16 Dieckmann Condensation   82
   Mechanism   83
   2.16.1 Dieckmann Condensation in Solid State   83
   2.16.2 Dieckmann Condensation Under Sonication   83
   2.16.3 Dieckmann Condensation Using Polymer
        Support Technique   84
   2.16.4 Applications   84
2.17 Diels-Alder Reaction   86
   Mechanism   89
   2.17.1 Diels–Alder Reactions under microwave Irradiation   89
   2.17.2 Diels–Alder reactions in aqueous phase   89
   2.17.3 Diels–Alder reaction in Ionic Liquids   89
2.18 Grignard Reaction   90
   Grignard Reagent   90
   2.18.1 Grignard reaction under Sonication   90
   Structure of grignard Reagent   91
   Grignard Reaction   91
   Reaction Mechanism   92
   Limitations   92
   2.18.2 Grignard Reaction in Solid State   93
   2.18.3 Applications   94
2.19 Heck Reaction   103
   Mechanism   103
   2.19.1 Heck reaction in aqueous phase   103
   2.19.2 Heck reaction under PTC conditions   104
   2.19.3 Heck reaction in Ionic Liquids   105
   2.19.4 Applications   105
2.20 Knoevenagel Condensation   106
   Mechanism   107
   2.20.1 Knoevenagel Reaction in Water   108
   2.20.2 Knoevenagel Reaction in Solid State   109
   2.20.3 Knoevenagel Reaction in Ionic Liquids   109
   2.20.4 Applications   109
2.21 Michael Addition   111
   Mechanism   111
   2.21.1 Michael Addition Under PTC Conditions   113
   2.21.2 Michael Addition in Aqueous Medium   113
   2.21.3 Michael Addition in Solid State   116
   2.21.4 Michael Addition in Ionic liquids   118
   2.21.5 Applications   119

2.22 Mukaiyama Reaction   123
   2.22.1 Mukaiyama reaction in Aqueous Phase   124
2.23 Reformatsky Reaction   125
   Mechanism   125
   2.23.1 Reformatsky Reaction Using Sonication   126
   2.23.2 Reformatsky Reaction in Solid State   127
   2.23.3 Applications   127
2.24 Simmons–Smith Reaction   130
   Mechanism   130
   2.24.1 Simmons–Smith Reaction Under Sonication   131
   2.24.2 Applications   132
2.25 Strecker Synthesis   133
   Mechanism   134
   2.25.1 Strecker Synthesis Under Sonication   134
   2.25.2 Applications   135
2.26 Ullmann Reaction   136
   Mechanism   136
   2.26.1 Ullmann coupling under sonication   137
   2.26.2 Applications   138
2.27 Weiss—Cook Reaction   140
2.28 Williamsons Ether Synthesis   140
   Mechanism   141
   2.28.1 Phase Transfer Catalysed Williamson Ether Synthesis   141
   2.28.2 Applications   142
2.29 Wittig Reaction   143
   The Phosphorus Ylides   144
   Mechanism   144
   2.29.1 The Wittig Reaction with Aqueous Sodium Hydroxide   145
           Modifications of Wittig Reagent   146
   2.29.2 Wittig Reaction in Solid Phase   148
   2.29.3 Wittig Reaction in Ionic Liquids   149
   2.29.4 Applications   149
2.30 Wurtz Reaction   152
   Mechanism   153
   2.30.1 Wurtz Reaction under Sonication   154
   2.30.2 Wurtz Reaction in Water   154
   2.30.3 Applications   154

3   **GREEN PREPARATION**   **155**

3.1 Aquous Phase Reactions   155
   3.1.1   Hydrolysis of Methyl Salicylate with Alkali   155
   3.1.2   Chalcone   156
   3.1.3   6-Ethoxycarbonyl-3,5-diphenyl-2-cyclohexenone   157
   3.1.4 $\Delta^{1,9}$-Octalone   159

3.1.5   p-Ethoxyacetanilide (Phenacetin)   160
3.1.6   p-Acetamidophenol (Tylenol)   161
3.1.7   Vanillideneactone   162
3.1.8   2,4-dihydroxybenzoic aicd (β–resorcylic acid)   163
3.1.9   Iodoform   164
3.1.10   Endo-cis-1,4-endoxo–Δ^5-cyclohexene-2,
         3-dicarboxylic acid   164
3.1.11   Trans stilbene   165
3.1.12   2-Methyl-2-(3-oxobutyl)-1, 3-cyclopentanedione.   166
3.1.13   Hetero Dields-Alder Adduct   167
3.2   Solid state (solventless) reactions   168
3.2.1   3-Pyridyl-4(3H) quinazolone   168
3.2.2   Diphenylcarbinol   168
3.2.3   Phenyl benzoate   169
3.3   Photochemical Reactions   169
3.3.1   Benzopionacol   169
3.3.2   Conversion of trans azobenzene to cis azobenzene   170
3.3.3   Conversionn of trans stilbene into cis stilbene   172
3.4   PTC catalysed reactions   172
3.4.1   Phenylisocyanide ($C_6H_5N \equiv C$)   172
3.4.2   1-Cyano Octane ($CH_3(CH_2)_6CH_2CN$)   173
3.4.3   1-Oxaspiro-[2,5]-octane-2-carbonitrile   174
3.4.4   3,4-Diphenyl-7-hydroxycoumarin   174
3.4.5   Flavone   176
3.4.6   Dichloronorcarane [2,2-Dichlorobicyclo (4.1.0) heptane]   177
3.4.7   Oxidation of toluene to benzoic acid   178
3.4.8   Benzonitrile from benzamide   179
3.4.9   n-Butyl benzyl ether   180
3.4.10   Salicylaldehyde   181
3.5   Rearrangement Reactions   182
3.5.1   Benzopinacolone   182
3.5.2   2-Allyl phenol   183
3.6   Microwave induced reactions   184
3.6.1   9,10-Dihydroanthracene-endo-α,β-succinic anhydride
         (Anthracene-maleic anhydride adduct)   184
3.6.2   3-methyl-1-phenyl-5-pyrazolone   185
3.6.3   Preparation of derivatives of some organic compounds   186
3.7   Enzymatic Transformations   187
3.7.1   Ethanol   187
         Analysis of the distillate   188
3.7.2   (S)-(+)-Ethyl 3-hydroxybutanoate   189
         Calculation of optical purity or Enantiomeric excess   190
3.7.3   Benzoin   191

      3.7.4   1-Phenyl-(1S) ethan-1-ol from acetophenone   193
      3.7.5   Deoximation of oximes by ultrasonically
              stimulated Bakers yeast   194
  3.8 Sonication reactions   195
      3.8.1   Butyraldehyde   195
      3.8.2   2-Chloro-N-aryl anthranilic acid   196
  3.9 Esterification   196
      3.9.1   Benzocaine (Ethyl p-aminobenzoate)   196
      3.9.2   Isopentyl acetate (Banana oil)   198
      3.9.3   Methyl salicylate (oil of wintergreen)   198
  3.10 Enamine reaction   200
      3.10.1 2–Acetyl cyclohexanone   200
  3.11 Reactions in ionic liquids   201
      3.11.1 1-Acetylnaphthalene   201
      3.11.2 Ethyl 4-methyl-3-cyclohexene carboxylate   202
  Index   205

1

Green Chemistry

INTRODUCTION

Green chemistry is defined as environmentally benign chemical synthesis. It focusses on a process (whether carried out in industry or chemical laboratory) that reduce the use and generation of hazardous substances or byproducts. Strict laws have been passed by various governments particularly in advanced countries like USA to strictly follow the procedures for various synthesis so as to reduce or eliminate the products (or by products) that are responsible for the pollution of the environment. The chemists all over the globe are motivated not only for the environmentally benign synthesis of new products but also to develop green synthesis for existing chemicals. This has been possible by the replacement of the organic solvents, which are hazardous by water or eliminate the use of solvent altogether.

There is absolutely no doubt that green chemistry has brought about medical revolution (e.g., synthesis of drugs etc.). The world's food supply has increased many fold due to the discovery of hybrid varieties, improved methods of farming, better seeds and use of agro chemicals like fertilizers, insecticides and herbicides etc. Also the quality of life has improved due to the discovery of dyes, plastics, cosmetics and other materials. All these developments increased the average life expectancy from 47 years in 1900 to 75 years in 1990's. However, the ill effects of all the development became pronounced. The most important effect is the release of harzardous by products of chemical industries and the release of agro chemicals in the atmosphere, land and water bodies; all these are responsible for polluting the environment including atmosphere, land and water bodies. Due to all these green chemistry assumed special importance.

1.1 THE NEED OF GREEN CHEMISTRY

It has already been stated that various scientific developments in the 20th century brought about various benefits to the mankind, but all this was responsible for a number of environmental problems at the local and global levels. It is, of course, important to formulate guidelines and pass strict rules for the practicising chemists. But the most important is to bring about changes at the grass root level. And this can be achieved by bringing about necessary changes in the chemistry curriculum in the colleges and the universities and also in the secondary schools. A concerted and pervasive effort is needed to reach the widest audience. Bringing green chemistry to the class room and the laboratory will have the desired effect in educating the students at various levels about green chemistry.

1.2 PRINCIPLES OF GREEN CHEMISTRY

Green chemistry deals with environmentally benign chemical synthesis with a view to devise pathways for the prevention of pollution. According to Paul T. Anastas,[1] the following twelve basic principles of green chemistry have been formulated.

1. **It is better to prevent waste than to treat or clean up waste after it is formed.**

   It is best to carry out a synthesis by following a pathway so that formation of waste (by products) is minimum or absent. It must be kept in mind that in most of the cases, the cost involved in the treatment and disposal of waste adds to the overall cost of production. The unreacted starting materials (which may or may not be hazardous) form part of the waste. The basic principle 'prevention is better than cure' should be followed. The waste if discharged in the atmosphere, sea or land not only causes pollution but also requires expenditure for cleaning up.

2. **Synthetic materials should be designed to maximize the incorporation of all materials used in the process into the final product.**

   It has so far been believed that if the yield in a particular reaction is about 90%, it is considered to be good. The percentage yield is calculated by

   $$\% \text{ yield} = \frac{\text{Actual yield of the product}}{\text{Theoretical yield of the product}} \times 100$$

   The above calculation implies that if one mole of a starting material produces one mole of the product, the yield is 100%. However, such a synthesis may generate significant amount of waste or by products which is not visible in the above calculation. Such a synthesis, even though is 100% (by above calculation) is not considered to be a green synthesis. For example, reactions like Grignard reactions and Wittig reaction may proceed with 100% yield but they do not take into account the large amount of by products obtained.

   A reaction or a synthesis is considered to be green if there is maximum incorporation of the starting materials or reagents in the final product. One should take into account the percentage atom utilization, which is determined by the following equation

   $$\% \text{ atom utilization} = \frac{\text{MW of desired product}}{\text{MW of desired product} + \text{MW of waste products}} \times 100$$

   This concept of atom economy was developed by B.M. Trost[2] in a consideration of how much of the reactants end up in the final product.

The same concept was also determined by R.A. Sheldon[3] as given below.

$$\% \text{ atom economy} = \frac{\text{FW of atoms utilized}}{\text{FW of the reactants used in the reaction}} \times 100$$

The most common reaction we generally come across in organic synthesis are rearrangement, addition, substitution and elimination reactions. Let us find out which of the above reactions is more atom economical.

(a) Rearrangement Reactions

These reactions involve rearrangement of atoms that make up a molecule. For example, allyl phenyl ether on heating at 200°C gives o-allyl phenol (Scheme-1).

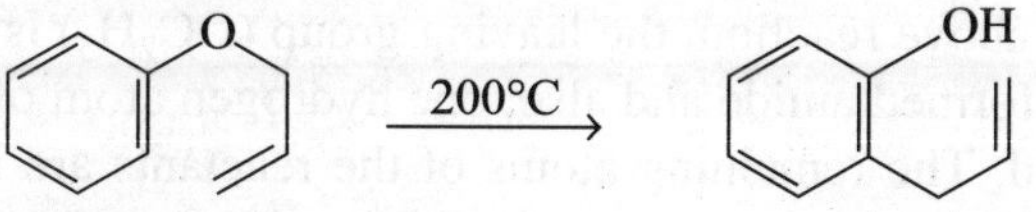

Allyl phenyl ether   o - allyl phenol

(Scheme-1)

The rearrangement reaction (in fact all rearrangement reactions) is 100% atom economical reaction, since all the reactants are incorporated into the product.

(b) Addition Reactions

Consider the bromination of propene (Scheme-2).

$$H_3C\ CH = CH_2 + Br_2 \xrightarrow{CCl_4} H_3C\ CH\ Br\ CH_2Br$$

Propene   1,2-dibromopropane

(Scheme-2)

Here also all elements of the reactants (propene and bromine) are incorporated into the final product (1,2-dibromopropane). So this reaction is also 100% atom economical reaction.

In a similar way cycloaddition reaction of butadiene and ethene (Scheme-3) and addition of hydrogen to an olefin (Scheme-4) is 100% atom economical reaction.

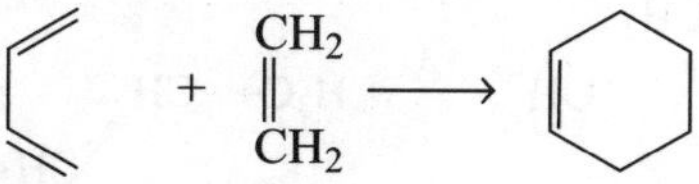

Butadiene   ethene   cyclohexene

(Scheme-3)

$$H_3C - CH = CH_2 + H_2 \xrightarrow{Ni} H_3C - CH_2 - CH_3$$

Propene   Propane

(Scheme-4)

(c) Substitution Reactions

In substitution reactions, one atom (or group of atoms) is replaced by another atom (or group of atoms). The atom or group that is replaced is not utilised in the final product. So the substitution reactions are less atom economical than rearrangement or addition reactions.

Let us consider the reaction of ethyl propionate with methyl amine (Scheme-5).

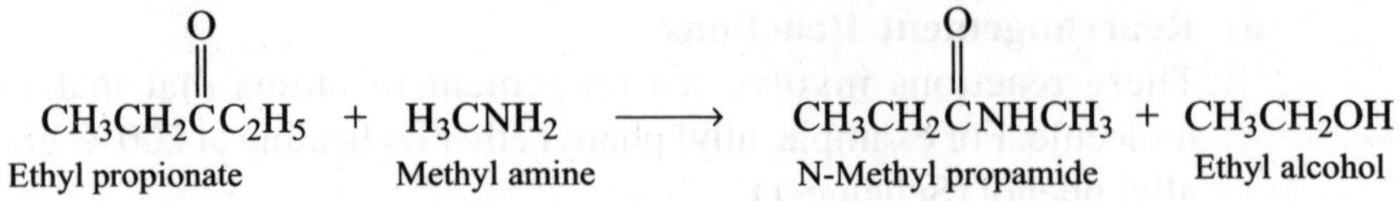

$$CH_3CH_2\overset{\overset{\displaystyle O}{\|}}{C}C_2H_5 \; + \; H_3CNH_2 \; \longrightarrow \; CH_3CH_2\overset{\overset{\displaystyle O}{\|}}{C}NHCH_3 \; + \; CH_3CH_2OH$$

Ethyl propionate   Methyl amine   N-Methyl propamide   Ethyl alcohol

(Scheme-5)

In the above reaction, the leaving group (OC_2H_5) is not incorporated in the formed amide and also, one hydrogen atom of the amine is not utilized. The remaining atoms of the reactants are incorporated into the final product.

The total of atomic weights of the atoms in reactants that are utilized is 87.106 g/mole, while the total molecular weight including the reagent used is 133.189 g/mole. Thus a molecular weight of 46.069 g/mole remains unutilized in the reaction.

	Reactants		Utilized		Unutilized	
	Formula	FW	Formula	FW	Formula	FW
	$C_5H_{10}O_2$	102.132	C_3H_5O	57.057	C_2H_5O	45.061
	CH_5N	31.057	CH_4N	30.049	H	1.008
Total	$C_6H_{15}NO_2$	133.189	C_4H_9NO	87.106	C_2H_5OH	46.069

Therefore, the atom economy (%) = $\dfrac{87 \cdot 106}{133 \cdot 189} \times 100 = 65.40\%$

(d) Elimination Reactions

In an elimination reaction, two atoms or groups of atoms are lost from the reactant to form a π bond. Consider the following Hofmann elimination reaction (Scheme-6).

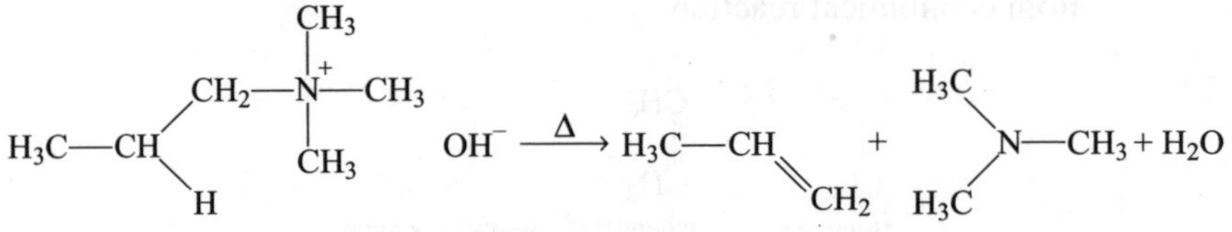

$$H_3C-CH\underset{H}{\overset{\displaystyle CH_2-\overset{+}{\underset{CH_3}{N}}-CH_3}{}} \; OH^- \xrightarrow{\Delta} H_3C-CH\overset{\displaystyle CH_2}{} \; + \; \underset{H_3C}{\overset{\displaystyle H_3C}{}}N-CH_3 + H_2O$$

(Scheme-6)

The above elimination reaction is not very atom economical. The percentage atom economy is 35.30% and is the least atom economical of all the above reactions.

Consider another elimination reaction involving dehydrohalogenation of 2-bromo-2-methylpropane with base to give 2-methylpropene (Scheme-7).

$$\underset{\substack{\text{Br}\quad\text{H}\\ \text{2-Bromo-2-methylpropane}}}{H_3C-\overset{\overset{\displaystyle CH_3}{|}}{\underset{|}{C}}-CH_2}\ \xrightarrow{NaOC_2H_5}\ \underset{\text{2-Methyl propene}}{H_3C-\overset{\overset{\displaystyle CH_3}{|}}{C}=CH_2}+C_2H_5OH+NaBr$$

(Scheme-7)

The above dehydrohalogenation reaction (an elimination reaction) is also not very atom economical. The percentage atom economy is 27% which is even less than the Hofmann elimination reaction.

3. **Wherever practicable, synthetic methodologies should be designed to use and generate substances that possess little or no toxicity to human health and the environment.**

   One of the most important principle of green chemistry is to prevent or at least minimize the formation of hazardous products which may be toxic and or environmentally harmful. In case hazardous products are formed, their effects on the workers must be minimized by the use of protective clothing, respirator etc. This, of course, will add to the cost of production. At times, it is found that the controls may fail and there may be more risk involved. Green chemistry, in fact, offers a scientific option to deal with such situations.

4. **Chemical products should be designed to preserve efficacy of function while reducing toxicity.**

   It is extremely important that the chemicals synthesised or developed (e.g., dyes, paints, cosmetics, pharmaceuticals etc.) should be safe to use. A typical example of an unsafe drug is thalidomide (introduced in 1961) for reducing the effects of nausea and vomitting during pregnancy (morning sickness). The children born to women taking thalidomide suffered birth defects. Subsequently, the use of thalidomide was banned, the drug withdrawn and strict regulations passed for testing all new drugs.

   With the advancement of technology, the designing and production of safer chemicals has become possible. In fact, it is possible to manipulate the molecular structure to achieve this goal.

5. **The use of auxiliary substances (solvents, separation agents, etc.) should be made unnecessary whenever possible and, when used, innocuous.**

   A number of solvents like methylene chloride, chloroform, perchloroethylene, carbon tetrachloride, benzene and other aromatic hydrocarbons have been used (in a large number of reactions) due to their excellent solvent properties. However, the halogenated solvents (mentioned above) have been identified as suspected human carcinogens. Also, benzene and other aromatic hydrocarbons are believed to promote cancer in humans and other animals.

The solvent selected for a particular reaction should not cause any environmental pollution and health hazard. The use of liquid carbon dioxide should be explored. If possible, the reaction should be carried out in aqueous phase or without the use of a solvent in solid phase.

A lot of concern has been expressed about the use of solvents which have direct hazardous effect on the environment. One such example is chlorofluocarbons (CFCs) which have been widely used as cleaning agents, blowing agent and as refrigerants. These CFCs are responsible for depleting the ozone layer, which in turn has disastrous effect on human survival. Even the volatile organic compounds (VOCs) like carbon tetrachloride, methylene chloride, chloroform etc, which have been used as solvents in a number of applications have disasterous effects in the atmosphere. In view of all these effects, regulations have been made under the clear air act in some of the advanced countries like USA to control many classes of chemical used as solvents.

It has already been stated that a major problem with many solvents is their volatility that may damage human health and the environment. To avoid this, a lot of work has been carried out on the use of immobilised solvents. These solvents maintain the solvency of the material, but are non-volatile and do not expose humans or the environment to the hazards of that substance.

As far as possible the pathway for a reaction should be such that there is no need for separation or purification. By this procedure, the energy requirement is kept to a minimum.

6. **Energy requirement should be recognized for their environmental and economic impacts and should be minimized.**

In any chemical synthesis, the requirement of energy should be kept to a minimum. For example, if the starting materials and reagents are soluble in a particular solvent, the reaction mixture has to be heated to reflux for completing the reaction. In such cases, the time required for completion of the reaction should be minimum, so that least amount of energy is required. The use of a catalyst has the great advantage of lowering the requirement of energy of a reaction.

In case the final product is not pure, it has to be purified by distillation, recrystallisation or ultrafiltration. All these steps require energy. The process should be designed in such a way that there is no need for separation or purification.

It is possible, the energy to a reaction can be supplied photochemically, by using microwave or sonication.

7. **A raw material or feedstock should be renewable rather than depleting, whenever technically and economically practicable.**

The starting materials can be obtained from renewable or non-renewable material. For example, petrochemicals are mostly obtained from petroleum oil, which is a non-renewable source since its formation take million of years from animal and vegetable remains. The starting materials which can be obtained from agricultural or biological processes are referred to as renewable starting materials; however, these cannot be obtained in continuous supply due to factors like failure of crops etc.

Substances like carbon dioxide (generated from natural sources or synthetic routes) and methane gas (obtained from natural sources such as marsh gas, natural gas etc.) are available in abundance; these are considered as renewable starting materials.

8. **Unnecessary derivatization (blocking group, protection/deprotection, temporary modification of physical/chemical processes) should be avoided whenever possible.**

   A commonly used technique in organic synthesis is the use of protecting or blocking group. These groups are used to protect a sensitive moiety from the conditions of the reaction, which may make the reaction to go in an unwanted way if it is left unprotected. A typical example of this type of transformation would be protection of amine by making benzyl ether in order to carry out an transformation of another group present in the molecule. After the reaction is complete, the NH_2 group can be regenerated through cleavage of the benzyl ether (Scheme-8).

(Scheme-8)

Derivatization of this type is quite common in the synthesis of fine chemicals, pharmaceuticals, pesticides and certain dyes. In the above example, benzyl chloride (a known hazard) needs to be handled with care and used in the preparation of the desired material and then regenerated as waste upon deprotection.

In the above procedure, the protecting group is not incorporated into the final product, their use makes a reaction less atom economical. Thus, as far as possible, the use of protecting groups be avoided. Though atom economy is a valuable criteria in evaluating a particular synthesis as 'green', other aspects of efficiency must be considered.

9. **Catalytic reagents (as selective as possible) are superior to stoichiometric reagents.**

In some reactions the reactants (A and B) react to form a product (C), in which all the atoms contained within A and B are incorporated in the product (C). In such cases, stoichiometric reactions are equally environmentally benign from the point of material usage as any other type of reactions. However if one of the starting material (A or B) is a limiting reagent; in such cases even if the yield is 100%, some unreacted starting material will be left over as waste. In other cases, if the reagents A and B do not give 100% yield of the product (C), both the excess of unreacted reagents will form part of waste. It is found that due to the reason mentioned above, catalysts, where available offer distinct advantages over typical stoichiometric reagents. The catalyst facilitates the transformations without being consumed or without being incorporated into the final product.

Catalysts are selective in their action in that the degree of reaction that takes place is controlled, e.g., monoaddition vs multiple addition; also the stereochemistry is controlled (e.g., R vs S enantiomer). By using catalysts, both starting-material utilization is enhanced and formation of waste reduced. An additional advantage of the use of catalyst is that the activation energy of a reaction is reduced and so the temperature necessary for the reaction is also lowered. This results in saving the energy.

It should be understood that in stoichiometric processes the product obtained is one mole for every mole of the reagent used. However, a catalyst will carry out thousands of transformations before being exhausted.

Following are given some of the applications of the use of catalysts.

(i) Hydrogenation of olefins in presence of nickel catalyst gives much better yields (Scheme-9)

$$H_3C-CH=CH_2 + H_2 \xrightarrow{Ni} H_3C-CH_2-CH_3$$
Propene   Propane

(Scheme-9)

(ii) Conversion of benzyl chloride into benzyl cyanide in much better yields using phase transfer catalysts. (Scheme-10).

$$C_6H_5CH_2Cl + aq.\ KCN \xrightarrow{PTC} C_6H_5CH_2CN$$
Benzyl chloride   > 90% yield
                                      Benzyl cyanide

(Scheme-10)

(iii) Oxidation of toluene with $KMnO_4$ in presence of crown ether gives much better yields (Scheme-11).

Toluene   >85% yield
                                     Benzoic acid

(Scheme-11)

(iv) Even in those cases where no reaction occurs usually, the reaction becomes feasible. As example is the hydration of alkynes to give aldehydes or ketones (Scheme-12).

$$HC \equiv CH + H_2O \xrightarrow[H_2SO_4]{HgSO_4} CH_3CHO$$

Acetylene　　　　　　　　　　　　acetaldehyde

$$CH_3C \equiv CH + CO + CH_3OH \xrightarrow{Pd} CH_3 - \underset{\underset{CH_2}{\|}}{\overset{O}{\overset{\|}{C}}} - OCH_3$$

Propyne

Methyl methacrylate
(shell corporation)

(Scheme-12)

(v) The selectivity enhancement takes place as shown by reduction of a triple bond to double bond (Scheme-13).

$$H_3C - C \equiv CH + H_2 \xrightarrow[\text{mono addition}]{Pd\text{-}BaSO_4} H_3C - CH = CH_2$$

Propyne　　　　　　　　　　　　　Propene

(Scheme-13)

(vi) Selectivity in C-methylation versus O-methylation (Scheme-14).

$$C_6H_5 - \underset{\underset{O}{\|}}{C} CH_2COCH_3 \xrightarrow[CH_3I]{NaOEt} C_6H_5 - \underset{\underset{O}{\|}}{C} - \overset{CH_3}{\underset{}{\overset{|}{CH}}} - COCH_3$$

Benzoyl acetone　　　　　　　　　α-benzoyl-α-methylacetone

(Scheme-14)

10. **Chemical products should be so designed that at the end of their function they do not persist in the environment and break down into innocuous degradation products.**

It is of utmost importance that the products that are synthesised should be biodegradable; they should not be 'persistant chemicals' or 'persistant bioaccumulators'. Such chemicals (which are non-biodegradable) remain in the same form in the environment or are taken up by various plants and animal species and accumulate in their systems; this is deterimental to the concerned species. The problem of non-biodegradability is generally associated with pesticides, and plastics and a host of other organic molecules.

Most of the pesticides in use are organohalogen-based compounds. These pesticides though effective tend to bioaccumulate in plants and animals. The pesticide DDT was one of the first pesticide which bioaccumulated in plants and animals. Whenever a chemical is being designed, it should be made sure that it will be biodegradable. It is now possible to place functional groups and other features in the molecule which will facilitate its degradation. Functional groups which are susceptable to hydrolysis, photolysis or other cleavage have been used to ensure that products will biodegrade. It is equally important to make sure that the degradation products should not possess any toxicity and be detrimental to the environment.

Plastics are known to remain persistant and are not biodegrable. The waste plastics were mostly used for landfills etc. However, it has now been possible to make plastics (particularly for garbage bags etc.) which are biodegradable.

11. **Analytical methodologies need to be further developed to allow for real-time, in-process monitoring, and control prior to the formation of hazardous substances.**

Analytical methodologies and technology have been developed which allow the prevention and minimization of the generation of hazardous substances in chemical processes. One need to have accurate and reliable sensors, monitors, and analytical techniques to assess the hazards that are present in the process stream. Using various techniques, a chemical process can be monitored for generation of hazardous by products and side reactions. These procedures can prevent any accident which may occur in chemical plants.

12. **Substances and the forms of a substance used in a chemical process should be chosen so as to minimize the potential for chemical accidents, including releases, explosions and fires.**

The occurrence of accidents in chemical industry must be avoided. The accidents in Bhopal (India) and Seveso (Italy) and many others have resulted in the loss of thousands of life.

At times it is possible to increase accident potential inadvertently in an attempt to minimize the generation of waste in order to prevent pollution. It has been noticed that in an attempt to recycle solvents from a process (in order to be economical and also prevent escape of solvent to the atmosphere) increases the potential for a chemical accident or fire. In fact, a process must balance the accident prevention with a desire for preventing pollution. A possible course is not to use volatile solvents, instead solids or low vapour pressure substance can be used.

1.3 PLANNING A GREEN SYNTHESIS IN A CHEMICAL LABORATORY

Following are given some of the points, which should be kept in mind for carrying out a synthesis in a chemical laboratory.

1. **Percentage atom utilization**

   There should be maximum incorporation of the starting materials and reagents into the final product (also see section 1.3).

2. Evaluating the type of the reaction involved.

The reaction involved must be evaluated with regard to its environmental impact or consequences. For this purpose, the nature of the starting material and the by products (if formed) must be examined. Following are given the different type of reaction which may be involved in a particular synthesis.

(a) Rearrangements

These reactions, as the name indicates in which the atoms that comprise a starting molecule change its orientation relative to one another including their connectivity and bonding pattern. Such reactions can be performed using a variety of procedures including thermal, photo and chemical means. From the point of green chemistry, in such reactions, both the starting material and the end products contain the same atoms and so there is no waste generated. In fact, a rearrangement reaction is 100% atom economical and fully efficient.

(b) Addition reactions

In these reaction a reagent adds to a substrate, all reagent and the substrate are consumed during the reaction. No additional by products are generated and such reactions are very efficient and like rearrangement reaction are also 100% atom economical. Some typical addition reactions include the addition of bromine to an olefin, grignard reagent to a carbonyl compound and hydrogen cyanide to an α, β-unsaturated carbonyl compounds. (Scheme-15).

Cyclohexene   + Br$_2$ Br / Br

Carbonyl compoind   Grignard reagent   + R″ MgBr

ethyl crotonate   hydrogen cyanide   + HCN β– cyanoethyl butyrate

(Scheme-15)

(c) Substitution reactions

In these reactions the functional group of a substrate is replaced with another functional group. Typical examples include the well known S_N1 and S_N2 reactions. In these reactions, nucleophilic reagents displace a leaving group in an aliphatic carbon atom; the product formed incorporates the nucleophile with removal of the leaving group. Typical examples are given below (Scheme-16).

$$C_6H_5CH_2Cl \quad + \quad KCN \longrightarrow \quad C_6H_5CH_2CN \quad + \quad KCl$$

Benzyl chloride   Benzyl cyanide

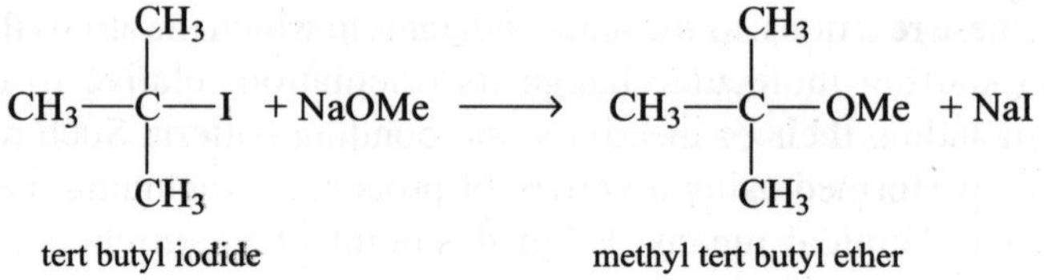

tert butyl iodide   methyl tert butyl ether

(Scheme-16)

In some cases, the leaving group is the desired product. For example potassium iodide demethylation of a carboxylic acid methyl ester to give free carboxylate salt and methyl iodide. (Scheme-17)

$$C_6H_5COOCH_3 \quad + \quad KI \longrightarrow \quad C_6H_5COOK \quad + \quad CH_3I$$

Methyl benzoate   Pot-benzoate   Methyl iodide

(Scheme-17)

The usefulness of the methods depends on the nature of the leaving group generated. This pathway can be convient and efficient if a substitution reaction sequence can be designed where the leaving group has been carefully selected.

(d) Elimination reactions

These are reverse of addition reactions and are procedures to generate unsaturation in the molecule. Examples of this type include dehydration of an alcohol to generate an olefin and loss of an alcohol from a hemiacetal to give an aldehyde (Scheme-18).

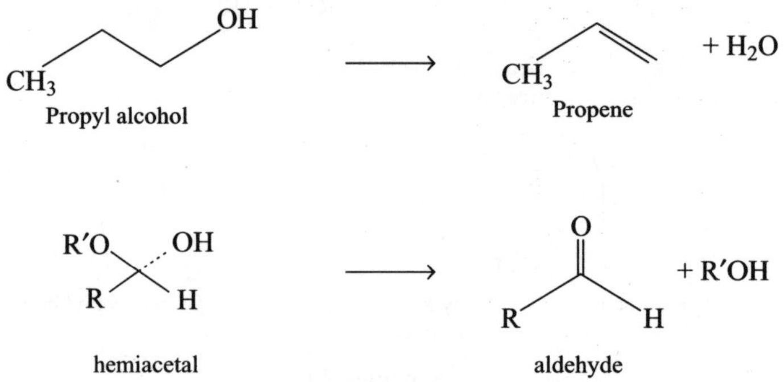

Propyl alcohol   Propene

hemiacetal   aldehyde

(Scheme-18)

As in the case of substitution reactions, the environmental implications of the leaving group should be examined, evaluated and controlled.

(e) Pericyclic reactions

These are concerted reactions and are characterised by the making or breaking of bonds in a single concerted step through a cyclic transition state involving π or σ electrons. Energy of activation for pericyclic reaction is supplied by heat in a thermally induced reaction and by ultraviolet light in a photo-induced reaction. Pericyclic reactions are highly stereospecific, often the thermal and photochemical processes yield products with different but specific stereochemistry. Since pericyclic reactions do not involve ionic or free radical intermediates; solvents, and nucleophiles or electrophilic reagents have no effect on the course of the reaction. We normally come across three type of pericyclic reaction.

(i) **Cycloaddition reactions.** In these reactions two molecules combine to form a ring; two π bonds being converted to two signal bonds in the process. The most common example of a cycloaddition reaction is the Diels-Alder reaction (Scheme-19).

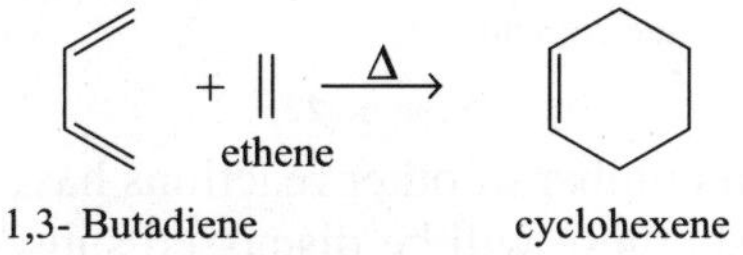

1,3- Butadiene   ethene   cyclohexene

(Scheme-19)

(ii) **Electrocyclic reaction.** These are reversible reactions in which a compound with two π electrons are used to form a sigma bond (Scheme-20).

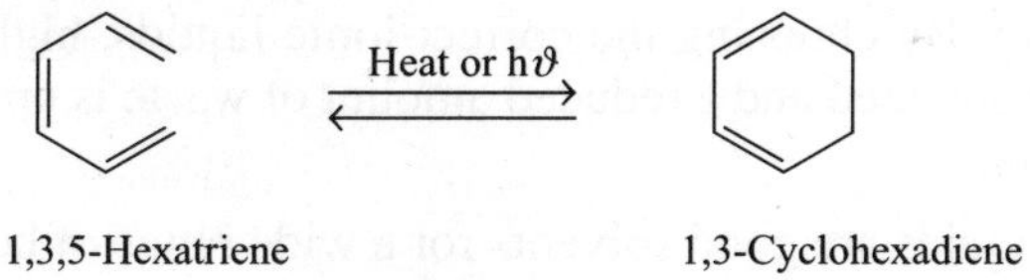

1,3,5-Hexatriene   Heat or $h\vartheta$ 1,3-Cyclohexadiene

(Scheme-20)

(iii) **Sigmatropic rearrangements.** These are concerted intramolecular rearrangements in which an atom or a group of atoms shift from one position to another (Scheme-21).

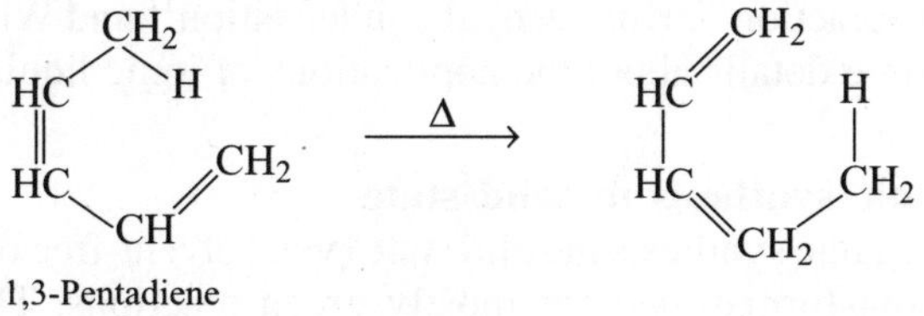

1,3-Pentadiene

(Scheme-21)

Pericylic reactions are very convenient from the point of view of environmental problems, since there is no by product obtained in these reactions.

3. Selection of appropriate solvent.

The solvent selected for a particular reaction should not have any environmental pollution and health hazard. As far as possible the reactions should be performed in aqueous phase or without the use of a solvent (solventless reaction). Recently a novel type of solvent known as "Ionic liquids" have been developed and are used for various synthesis.

(i) Aqueous phase reaction

A typical reaction which has been carried out in aqueous phase is the Diels-Alder reaction (Scheme-22).

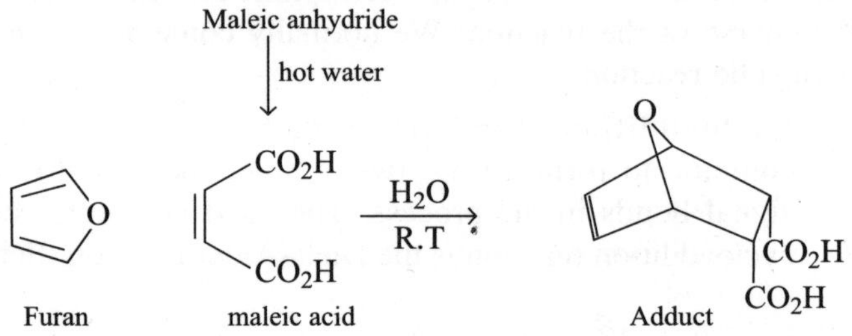

Maleic anhydride

| hot water
↓

Furan   maleic acid   Adduct CO_2H

(Scheme-22)

Besides this a number of other reactions have also been performed in aqueous phase; these will be discussed subsequently.

(ii) Reactions in ionic-liquid

Ionic liquids are made up of at least two components in which the cation and anion can be varied. In these cases, the properties such as melting point, viscosity, density and hydrophobicity can be varied by simple changes to the structure of ions. The ionic liquids are immiscible in water. By choosing the correct ionic liquids, higher product yield can be obtained and a reduced amount of waste is produced in a given reaction.

Ionic liquids are good solvents for a wide range of both inorganic and organic materials. They are also immiscible with a number of organic solvents and provide a non-aqueous, polar alternative for two phase system.

Ionic liquids find application in alkylations[4], allylations[5], hydroformyla-tions[6], epoxidations[7], synthesis of ethers[8], Friedel craft reaction[9], Diel-Alder Reaction[10], Knoevengal condensation[11] and Wittig reaction[12]. For more details about the applications of ionic liquids see reference 13.

(iii) Organic synthesis in solid state

The organic synthesis in solid state (viz. solvent-free organic synthesis and transformations) are mostly green reactions. These are of two types:

(a) Solid phase organic synthesis without any solvent

The earliest record of an organic reaction in dry state is the claisen rearrangement of allyl phenyl ether to o-allyl phenol (Scheme-23)

allyl phenyl ether   O-allyl phenyl

(Scheme-23)

A large number of reactions have now been performed without any solvent. These will be discussed subsequently.

(b) **Solid supported organic synthesis**

In these reactions, the reactants are stirred in a suitable solvent (for examples water, alcohol, methylene chloride etc) with a suitable adsorbent or solid support like silica get, alumina, phyllosilicate (m^{n+}–montomorillonite etc). After stirring, the solvent is removed in vacuo and the dried solid support on which the reactants have been adsorbed are used for carrying out the reaction under microwave irradiation.

4. **Selection of starting materials.**

As far as possible the starting materials selected should be obtainable from renewable sources. The starting materials should not cause any harm (eg allergy etc.) to the person handling these.

5. **Use of protecting group** should be avoided as far as possible, since these generate wastes.

6. **Use of catalyst**

We know that catalysts facilitates transformation and the conversions can be effected in short duration of time and consumes less energy. Such reactions should be preferred. Use of phase transfer catalysts and crown ethers is very well known.

7. **Use of microwaves, sonication** saves lots of energy and time and gives much better yields.

REFERENCES

1. Paul T. Anastas and John C. Warner, Green Chemistry Theory and Practice, Oxford University Press, New York 1998.
2. Barry M. Trost, Science, 1991, *254*, 1471.
3. Roser A. Sheldon, Chem. Ind. (London), 1992, 903.
4. Chauvin, Y., Hirschauer, H., Olivier, H., J. Mol. Catal., 1994, *92*, 155.
5. Chen, W., Xu L., Chatterton, C., Xiao, Chem. Commun, 1999, 1247.
6. Fuller, A., Breda, A.C., Carlin, R.T., J. Electrochem. Soc., 1997, *144*, 67; Favrc, F., Olivier, H. *et al.*, Chem. Commun., 2001, 1360.
7. Song, C.E., Roh, E.J., Chem. Commun., 2000, 837.

8. Fraga-Dubreeuil, J., Bourahla, K., Rahmouni, M., Bazureau, J.P., Hamelin, J., Catal. Commun. 2002, *3*, 185.

9. Boon, J.A., Levisky, J.A., Pflug, J.I., Wilkes, J.S., J. Org. Chem., 1986, *51*, 480.

10. Earle, M.S., McMormae, P.B., Sedden, K.R., Green Chem., 1999, *1*., 23.

11. Harjani, J.R., Orara, S.J., Salunkhe, M.M., Tetrahedron Lett., 2002, *43*, 1127.

12. Wheeler, C., West, K.N., Liottda, C.L. Eckert, C.A., Chem. Commun., 2001, *88*.

13. Ahluwalia V.K., Aggarwal, K. Organic Synthesis, Special Techniques, Narosa Publishing House, 2005, 195-210.

2

Green Reactions

INTRODUCTION

A majory of organic reactions require the use of volatile solvents, dry conditions and produce a number of by products which are harmful to the environment. With the advancement of knowledge and development of new and better techniques, a large number of organic reactions could be carried out in eco friendly conditions. It has now been possible to carry out those reactions, which earlier needed anhydrous conditions, in aqueous phase. Also the yields in a number of reaction has been increased by the use of microwaves, sonication and catalysis. Such reactions are now designated as Green Reaction. Following are given some of such reaction. For the shake of understanding, the reaction conditions used earlier and their mechanism and the new conditions which make them green are given.

2.1 ACYLOIN CONDENSATION

It consists in treating carboxylic ester with metallic sodium in large volume of benzene or toluene followed by protic solvents. The product obtained in α-hydroxy ketone, called acyloin. This condensation is called acyloin condensation. In this reaction, large dilution is required to ensure intramolecular condensation as against intermolecular reaction. A typical acyloin condensation is given in Scheme-1.

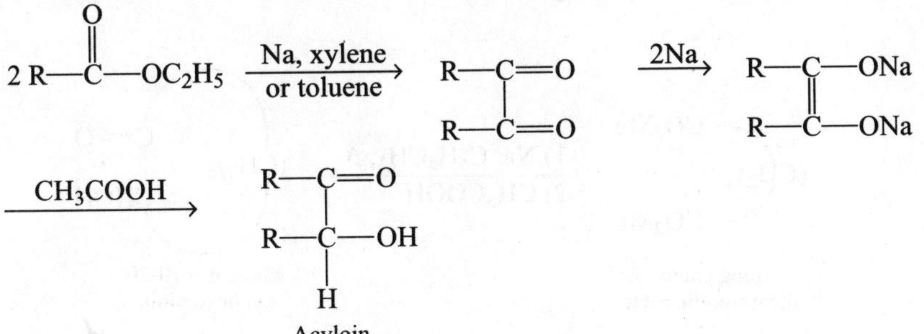

$$2\ R-\overset{\overset{\displaystyle O}{\|}}{C}-OC_2H_5 \xrightarrow[\text{or toluene}]{\text{Na, xylene}} \begin{array}{c} R-C=O \\ R-C=O \end{array} \xrightarrow{2Na} \begin{array}{c} R-\overset{\|}{C}-ONa \\ R-\overset{\|}{C}-ONa \end{array}$$

$$\xrightarrow{CH_3COOH} \begin{array}{c} R-C=O \\ R-\overset{|}{\underset{|}{C}}-OH \\ H \end{array}$$

Acyloin

(Scheme-1)

MECHANISM

A radical mechanism is believed to occur. The metallic sodium donates its electron to the carbonyl carbon to give the species (1), which dimerises to give (2). Subsequent loss of alkoxy group produces 1,2-diketone (3). Further reduction gives sodium salt of ene diol (4). Finally addition of acids yields 1,2-diol which tautomerises to acyloin (5).

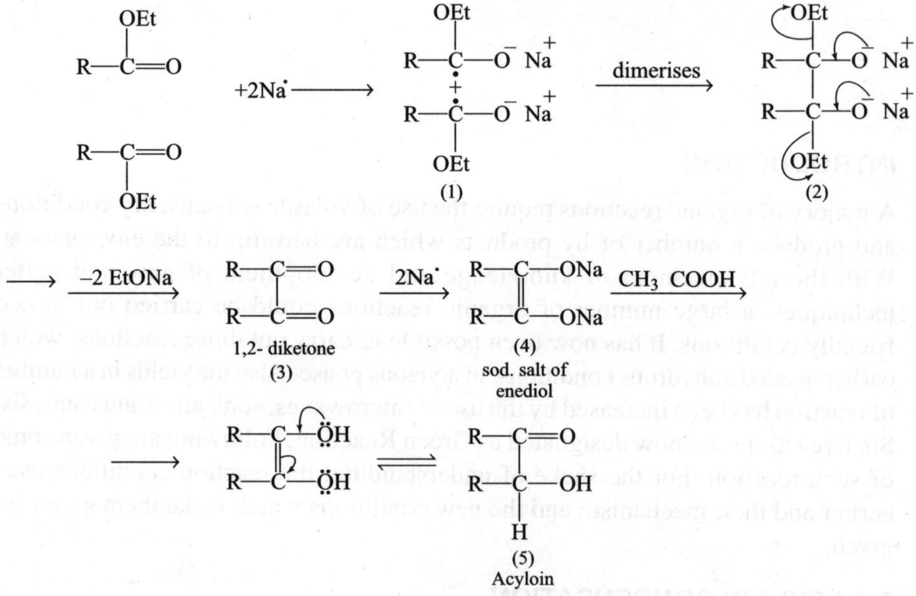

(Scheme-2)

Large-ring compounds (cyclic acyloins) are obtained by using long-chain dicarboxylic esters. The method is best suited for closing rings of ten members or more. In cyclic acyloins, dilution technique is not required. The yield is 60-95% for 10-20 membered ring (Scheme-3).

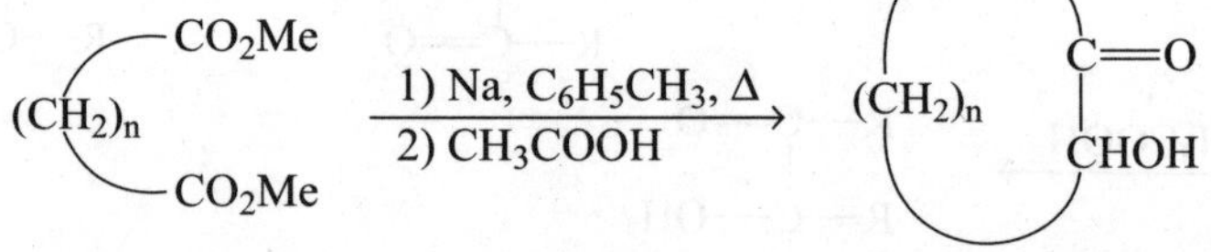

long chain
dicarboxylic ester

where n = 10-20
cyclic acyloin

(Scheme-3)

In case of cyclic acyloins, the mechanism suggested is given in Scheme-4.

$$\underset{(CH_2)_n}{\overset{\displaystyle CH_2-CO_2Et}{\diagdown CH_2-CO_2Et}} \quad \xrightarrow{2Na^{\cdot}} \quad \underset{(CH_2)_n}{\overset{\displaystyle CH_2-\overset{OEt}{\underset{\cdot}{C}}-O^-}{\diagdown CH_2-\underset{OEt}{\overset{\cdot}{C}}-O^-}} \quad \longrightarrow$$

$$\longrightarrow \quad \underset{(CH_2)_n}{\overset{\displaystyle CH_2-\overset{OEt}{C}-O^-}{\diagdown CH_2-\underset{OEt}{C}-O^-}} \quad \xrightarrow{2OEt^-} \quad \underset{(CH_2)_n}{\overset{\displaystyle CH_2-C=O}{\diagdown CH_2-C=O}} \quad \longrightarrow$$

$$\xrightarrow{2Na^{\cdot}} \quad \underset{(CH_2)_n}{\overset{\displaystyle CH_2-C-O^-\ Na^+}{\diagdown CH_2-C-O^-\ Na^+}} \overset{\|}{\underset{\|}{}} \quad \xrightarrow{H^+} \quad \underset{(CH_2)_n}{\overset{\displaystyle CH_2-C-OH}{\diagdown CH_2-C-OH}} \overset{\|}{\underset{\|}{}}$$

$$\underset{(CH_2)_n}{\overset{\displaystyle CH_2-C=O}{\diagdown CH_2-CHOH}} \quad \xleftarrow[\text{tautomerism}]{\text{keto - enol}}$$

acyloin

(Scheme-4)

In acyloin condensation, much better results are obtained if the reactions and conducted in trimethyl silane; even four membered rings can be prepared easily under these conditions. The chlorotrimethyl silane acts as a scavenger for the alkoxide ion liberated during the reaction (see Scheme-4), so that the reaction medium is kept neutral and the wasteful base-catalysed side reactions such as β-elimination and Clasien or Dickmann condensations are avoided. Also, the oxygen sensitive ene-diol is protected as the bis-trimethyl silyl ether which may be isolated and purified before hydrolysis to acyloin (Scheme-5).

$$\underset{(CH_2)_n}{\overset{\displaystyle CH_2-C-OH}{\diagdown CH_2-C-OH}} \overset{\|}{\underset{\|}{}} \quad \xrightarrow{(CH_3)_3SiCl} \quad \underset{(CH_2)_n}{\overset{\displaystyle CH_2-C-O\ SiMe_3}{\diagdown CH_2-C-O\ SiMe_3}} \overset{\|}{\underset{\|}{}}$$

ene diol

$$\downarrow HCl$$

$$\underset{(CH_2)_n}{\overset{\displaystyle CH_2-C=O}{\diagdown CH_2-CHOH}}$$

acyloin

(Scheme-5)

Using the above methodology diethyl suberate was converted into seven membered cyclic acyloin in 75% yield and the four membered ring compound in 90% yield (Scheme-6).

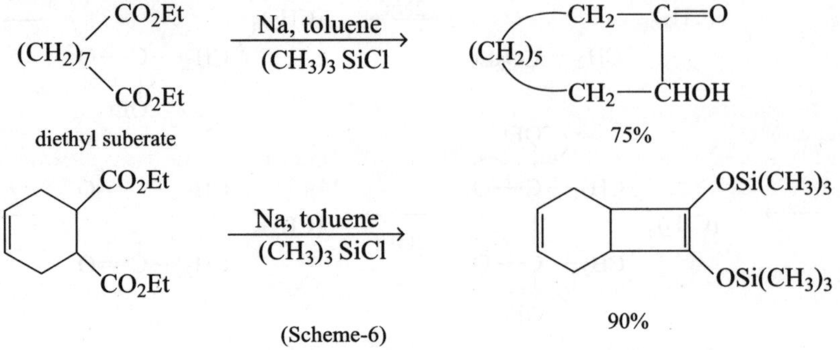

diethyl suberate

75%

90%

(Scheme-6)

2.1.1 Acyloin Condensation using Coenzyme, Thiamine

The acyloin condensation can also be catalysed by the coenzyme 'Thiamine'. Thus acetaldehyde on reaction with the air gives acyloin.

$$CH_3\ CHO \xrightarrow{\text{Thiamine}} CH_3\text{---}CO\text{---}CH\text{---}CH_3$$

acyloin OH

The mechanism of the reaction is given below.

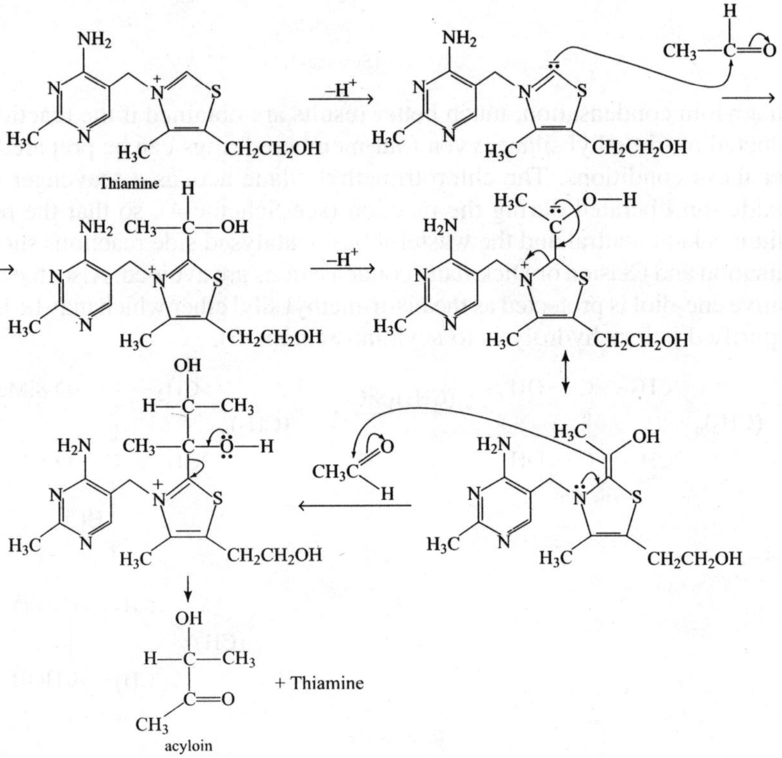

The above reaction is similar to benzoin condensation.

2.1.2 Applications

Acyloin condensation is a very useful synthetic reaction.

1. Cyclic acylions can be conveniently prepared (see Schemes 3, 4, 5 and 6).

2. Preparation of catenane.

Formation of catenanes has been observed by carrying out ring closure with the ester of the 34-carbon dicarboxylic acid.

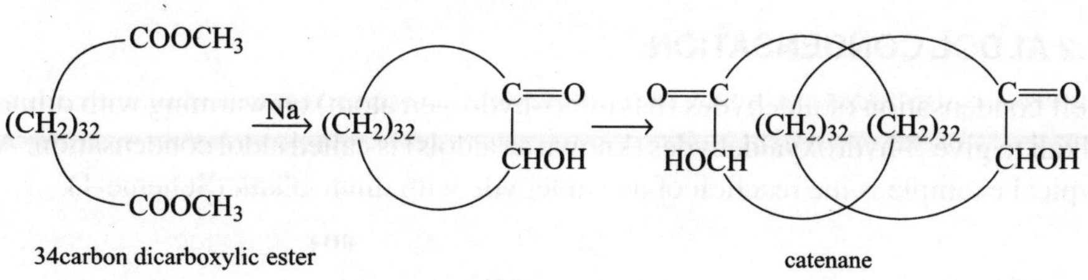

34carbon dicarboxylic ester   catenane

Some other applications are given below.

(i)   76%   Ref 1

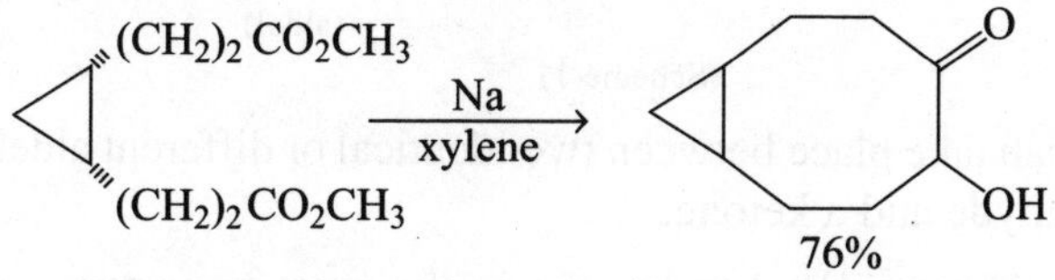

(ii)   11%   Ref 3

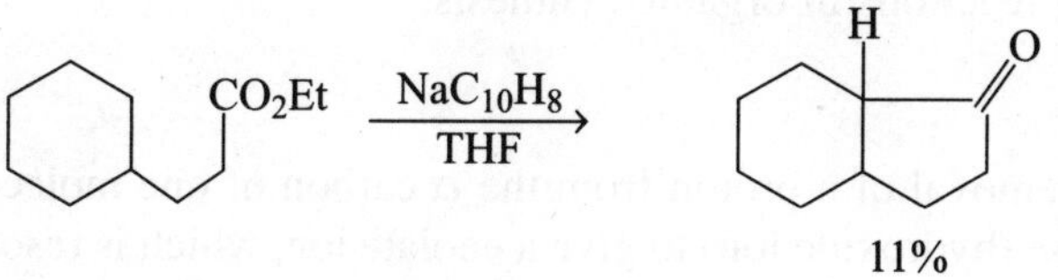

(iii)   70%   Ref 4

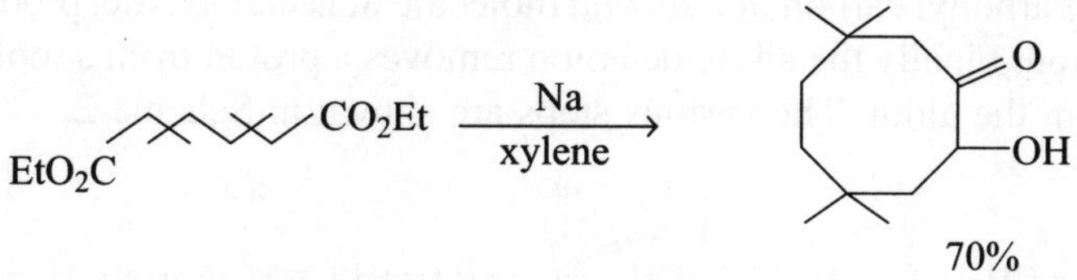

(iv)   78%   70%   Ref 5

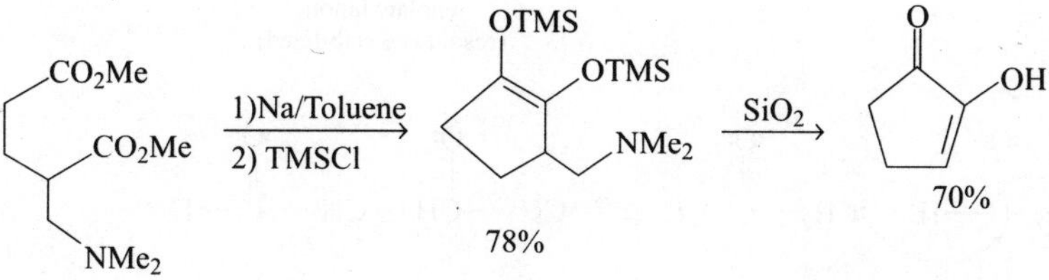

REFERENCES

1. P.G. Gassman, J. Seter and F.J. Williams, J.Am. Chem. Soc., 1971, *93,* 1673.
   J.M. Bloomfield and D.C. Owsley, J. Org. Chem., 1975, *40*, 393.
2. K. Ruhlann, H. Seefluth and H. Becker, Chem. Ber., 1967, *100*, 3820.
   J.J. Bloomfield, Tetrahedron Lett., 1968, 587.
3. G.D. Gutsche, I.Y.C. Tao and J. Kozma, J. Org. Chem., 1967, *32*, 1782.
4. A. Krebs, Tetrahedron Lett., 1968, 4511.
5. R.C. Cookson and S.A. Smith, J. Chem. Soc. Perkin I, 1979, 2447.

2.2 ALDOL CONDENSATION

Self condensation of aldehydes (having α-hydrogen atom) on warming with dilute alkali to give β-hydroxyaldehydes (known as aldols) is called aldol condensation. A typical example is the reaction of acetaldehyde with dilute alkali (Scheme-1).

$$CH_3CHO + CH_3CHO \underset{}{\overset{^-OH}{\rightleftharpoons}} CH_3\overset{\overset{\displaystyle OH}{|}}{C}HCH_2CHO$$

acetaldehyde

β- hydroxy butyraldehyde

(aldol)

(Scheme-1)

The aldol reaction can take place between two identical or different aldehydes or ketones and an aldehyde and a ketone.

Aldol condensation is considered to be one of the most important carbon-carbon bond-forming reactions in organic synthesis.

MECHANISM

The first step is the removal of a proton from the α carbon of one molecule of acetaldehyde by a base (hydroxide ion) to give a enolate ion, which is resonance stabilized. The formed enolate ion then acts as a nucleophile—as a carbanion—and attacks the carbonyl carbon of a second molecule of acetaldehyde, producing an alkoxide anion. Finally the alkoxide anion removes a proton from a molecule of water to form the aldol. The various steps are shown in Scheme-2.

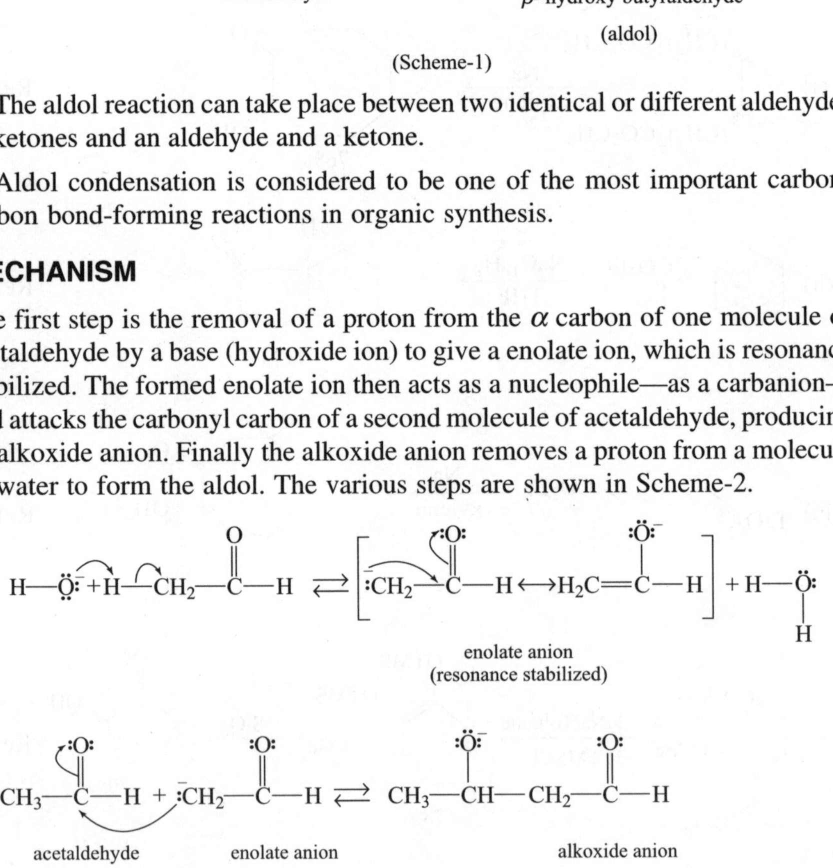

enolate anion
(resonance stabilized)

acetaldehyde   enolate anion   alkoxide anion alkoxide anion   aldol

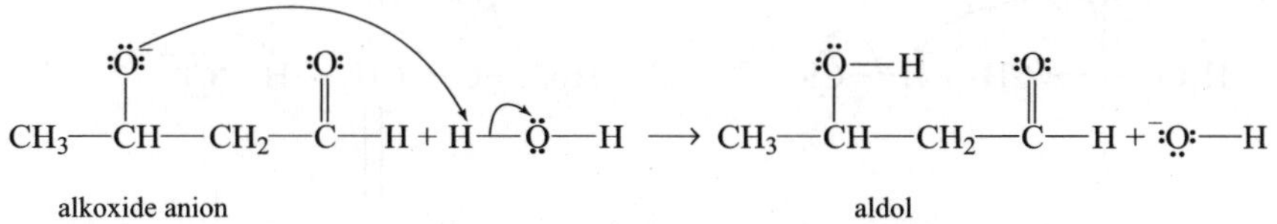

(Scheme-2)

THE ALDOL

The aldol obtained under basic conditions (in the example cited above) on heating, dehydration takes place to give crotonaldehyde (2-butenal). The dehydration is prompted due to acidity of the remaining α-hydrogen and also because the product is stabilized by having conjugated double bonds (Scheme-3).

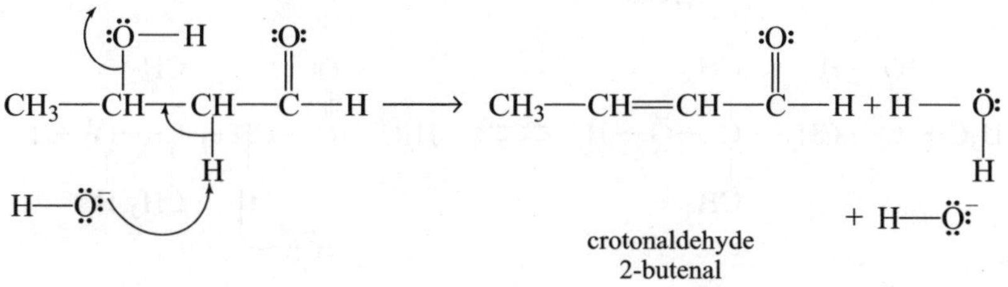

crotonaldehyde
2-butenal

(Scheme-3)

In some cases, the dehydration in the aldol occurs readily and it is not possible to isolate the aldol.

2.2.1 Acid-Catalysed Aldol Condensation

In the aldol condensation cited above the condensation takes place in presence of a base. However, aldol condensations can also be brought about with acid catalysts. For example, treatment of acetone with hydrogen chloride gives the aldol condensation product, viz., 4-methyl-3-penten-2-one. In general, in acid-catalysed aldol reactions there is simultaneous dehydration of the initially formed aldol (Scheme-4).

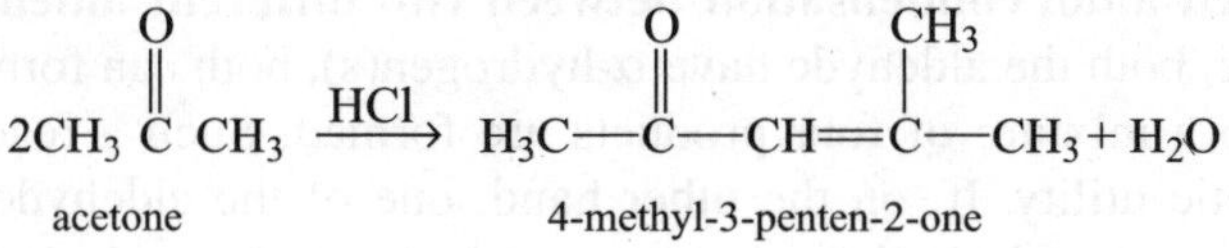

acetone   4-methyl-3-penten-2-one

(Scheme-4)

The mechanism of the acid-catalysed aldol condensation starts with the acid-catalysed formation of enol, which adds to the protonated carbonyl group of another molecule of acetone. The final step is proton transfer and dehydration leading to the final end product (Scheme-5).

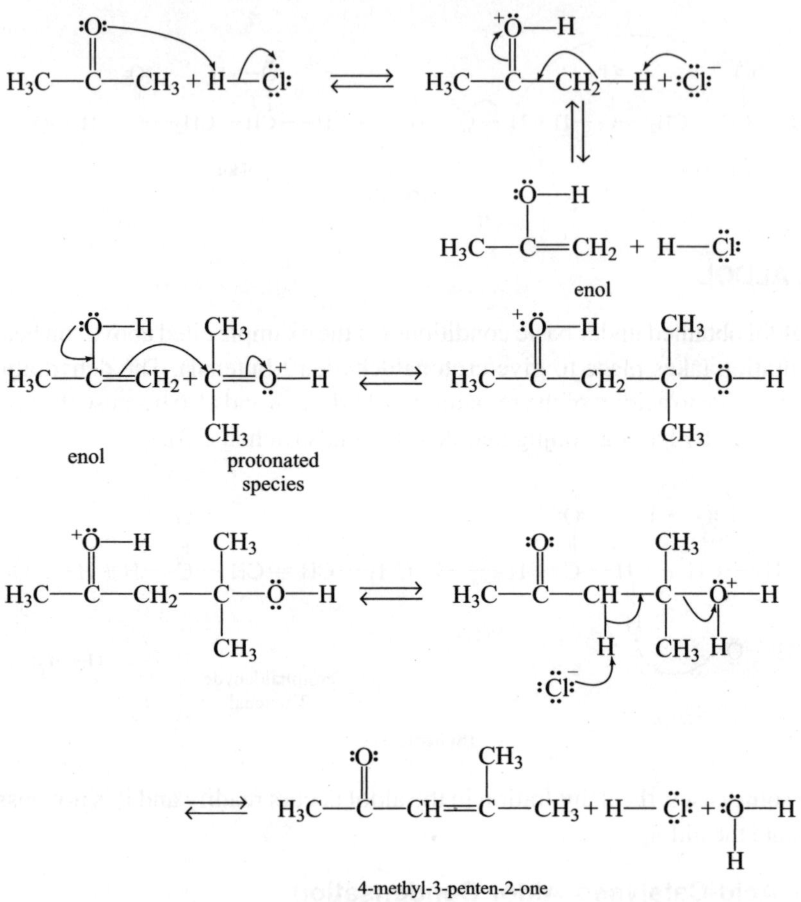

(Scheme-5)

2.2.2 Crossed Aldol Condensation

An aldol condensation that uses two different carbonyl compounds is called a crossed aldol condensation. In such a situation the following three situations may be there:

(i) **Crossed aldol condensation between two different aldehydes.**

In case, both the aldehyde have α-hydrogen(s), both can form carbanions and so a mixture of four products are formed. Such a reaction has no synthetic utility. If, on the other hand, one of the aldehydes has no α-hydrogen then in such a case two products are formed as shown below (Scheme-6).

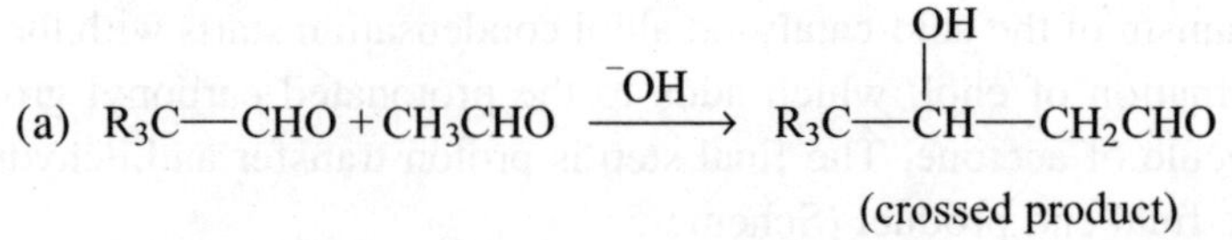

(a) $R_3C—CHO + CH_3CHO \xrightarrow{^-OH} R_3C—\overset{\overset{\displaystyle OH}{|}}{CH}—CH_2CHO$

(crossed product)

(b) $CH_3CHO + CH_3CHO \xrightarrow{\ ^-OH\ } H_3C-\overset{\overset{\displaystyle OH}{|}}{CH}-CH_2CHO$

(normal simple product)

(Scheme-6)

The formation of the crossed product can be achieved (and the formation of normal simple product can be avoided) by placing the aldehyde which has no α-hydrogen along with sodium hydroxide in a flask and then slowly adding (dropwise) the aldehyde with an α-hydrogen, to the mixture. Under these conditions, the concentration of the reactant with an α-hydrogen is always low and much of the reactant is present in as an enolate anion. So the main reaction that takes place is one between this enolate anion and the component that has no α-hydrogen. In other words, major amount of crossed aldol condensation product will be obtained. Following table gives some of the typical crossed aldol condensation.

Table 2.1: Crossed Aldol Condensation

Aldehyde without α-hydrogen	Aldehyde with α-hydrogen	Reaction conditions	Product	Yield %
$\overset{\overset{\displaystyle O}{\|\|}}{C_6H_5\,CH}$ benzaldehyde	$\overset{\overset{\displaystyle O}{\|\|}}{CH_3CH_2CH}$ propanal	$^-OH/10°C$	$C_6H_5\,CH{=}\overset{\overset{\displaystyle CH_3}{\|}}{C}-\overset{\overset{\displaystyle O}{\|\|}}{CH}$ 2-methyl-3-phenyl -2-propenal (α-methylcinnamaldehyde)	68
$\overset{\overset{\displaystyle O}{\|\|}}{C_6H_5C}-H$ benzaldehyde	$C_6H_5\,CH_2\overset{\overset{\displaystyle O}{\|\|}}{CH}$ phenyl acetaldehyde	$^-OH/20°C$	$C_6H_5CH{=}\overset{\underset{\displaystyle C_6H_5}{\|}}{C}\,\overset{\overset{\displaystyle O}{\|\|}}{CH}$ 2,3-diphenyl-2-propenal	65
$\overset{\overset{\displaystyle O}{\|\|}}{H-C}-H$ formaldehyde	$CH_3\,\overset{\overset{\displaystyle CH_3}{\|}}{CH}-\overset{\overset{\displaystyle O}{\|\|}}{CH}$ 2-methyl propanal	$40\,°C$ dil. Na_2CO_3	$CH_3-\overset{\overset{\displaystyle CH_3}{\|}}{\underset{\underset{\displaystyle CH_2OH}{\|}}{C}}-\overset{\overset{\displaystyle O}{\|\|}}{CH}$ 2-hydroxy methyl-2-methyl propanal	>64%

(ii) **Crossed aldol condensation between two different ketones.**

In such cases poor yield is obtained and so is not much useful. The poor reactivity of carbonyl carbons of ketones is responsible for low yields.

(iii) **Crossed aldol condensation between an aldehyde and a ketone.**

(a) When an aldehyde and a ketone, both having α-hydrogens are condensed, only two products are obtained (Scheme-7). This is because ketones are poor carbanion acceptors and do not undergo self condensation.

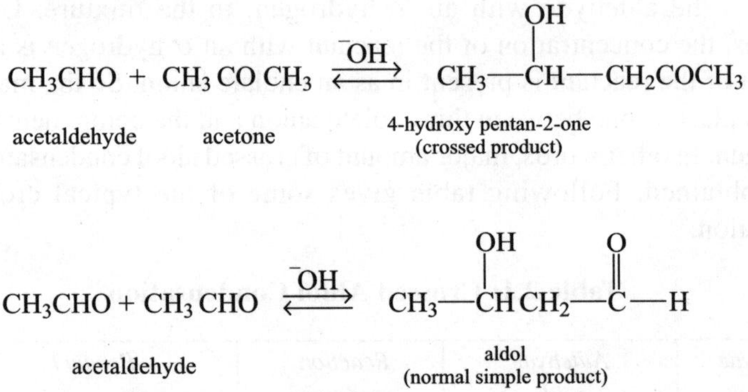

$$CH_3CHO + CH_3\,CO\,CH_3 \xrightleftharpoons{\text{}^-OH} CH_3 \overset{\overset{\text{OH}}{|}}{-}CH-CH_2COCH_3$$

acetaldehyde   acetone   4-hydroxy pentan-2-one
(crossed product)

$$CH_3CHO + CH_3\,CHO \xrightleftharpoons{\text{}^-OH} CH_3 \overset{\overset{\text{OH}}{|}}{-}CH\,CH_2 \overset{\overset{\text{O}}{\|}}{-}C-H$$

acetaldehyde   aldol
(normal simple product)

(Scheme-7)

Normally the crossed product predominates. The formation of aldol can be minimized by slow addition of the aldehyde to the mixture of ketone and alkali.

(b) When the reaction is between a ketone and an aldehyde with no α-hydrogen, only one product is obtained. Such a condensation is called **claisen-schmidt** reaction.

A reaction, worth mentioning as a crossed aldol condensation between an aldehyde and a ketone is the mixed aldol condensation of acetone with formaldehyde. In this case, formaldehyde cannot form an enolate since it lacks α-hydrogen. However, it is a good electron pair acceptor because of freedom from steric hinderance and because it has an usually weak carbonyl bond. Acetone forms an enolate easily, but is a relatively poor acceptor. So the following reaction occurs (Scheme-8).

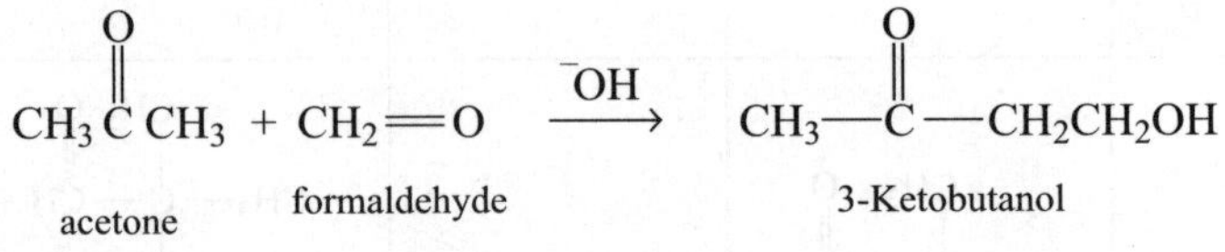

$$CH_3 \overset{\overset{\text{O}}{\|}}{C}\,CH_3 + CH_2{=}O \xrightarrow{\text{}^-OH} CH_3 \overset{\overset{\text{O}}{\|}}{-}C-CH_2CH_2OH$$

acetone   formaldehyde   3-Ketobutanol

(Scheme-8)

The reaction does not stop as indicated (Scheme-8). In fact, all six α-hydrogens can be replaced by —CH_2OH groups (Scheme-9).

$$\text{CH}_3\overset{\overset{\displaystyle O}{\|}}{\text{C}}\!-\!\text{CH}_3 + 6\text{CH}_2\!=\!\text{O} \xrightarrow{\ ^-\text{OH}\ } (\text{HOH}_2\text{C})_3\text{C}\!-\!\overset{\overset{\displaystyle O}{\|}}{\text{C}}\!-\!\text{C}(\text{CH}_2\text{OH})_3$$

acetone   formaldehyde

(Scheme-9)

2.2.3 Aldol Type Condensations of Aldehydes with NitroAlkanes and Nitriles

(i) Condensation with nitroalkanes.

The α-hydrogens of nitroalkanes are considerably acidic, much more than those of aldehydes and ketones. The acidity of nitroalkanes is attributed to electron withdrawing effect of NO_2 group and the resonance stabilization of the formed anion.

$$R\!-\!CH_2\!-\!\overset{+}{N}\!\!\underset{O^-}{\overset{O}{<}} \ +\!:B \longrightarrow R\!-\!CH\!-\!\overset{+}{N}\!\!\underset{O^-}{\overset{O}{<}} \longleftrightarrow R\!-\!CH\!=\!\overset{+}{N}\!\!\underset{O^-}{\overset{O^-}{<}}$$

nitroalkane   resonance stabilized anion

Nitroalkanes having α-hydrogens undergo base catalysed aldol type condensations with aldehydes and ketones. As example is the condensation of benzaldehyde and nitromethane (Scheme-10).

$$\text{C}_6\text{H}_5\overset{\overset{\displaystyle O}{\|}}{\text{CH}} + \text{CH}_3\text{NO}_2 \xrightarrow{\ ^-\text{OH}\ } \text{C}_6\text{H}_5\text{CH}\!=\!\text{CHNO}_2$$

benzaldehyde nitromethare

(Scheme-10)

The rate of the above condensation is increased by using basic alumina catalyst and **sonication.**

The reaction of a nitroalkane with an aldehyde in presence of base is called **Henry Reaction.**

(ii) Condensation with nitriles.

Like nitroalkanes, the α-hydrogens of nitriles are also acidic (but less so than those of aldehydes and ketones) and so undergo aldol type condensations. One such example is the condensation of benzaldehyde with phenylacetonitrile (Scheme-11).

$$\text{C}_6\text{H}_5\overset{\overset{\displaystyle O}{\|}}{\text{CH}} + \text{C}_6\text{H}_5\text{CH}_2\text{CN} \xrightarrow[\text{EtOH}]{\ ^-\text{OEt}\ } \text{C}_6\text{H}_5\text{CH}\!=\!\underset{\underset{\displaystyle \text{C}_6\text{H}_5}{|}}{\text{C}}\!-\!\text{CN}$$

benzaldehyde phenylacetonitrile

(Scheme-11)

2.2.4 Vinylogous Aldol Reaction

The γ-hydrogen of an α, β-unsaturated ketones, nitriles and esters is 'active' hydrogen and so electrophylic addition takes place at γ-position. This is known as **vinylogous aldol addition**, when the electrophile is an aldehyde. Thus, the reaction of isophorone with benzaldehyde gives only vinylogous aldol addition in low yields. However, in the presence of CTACl, the condensation product, (E) benzylidene isophorone, is obtained in excellent yield. By using tetrabutylammonium chloride (TBACl), a mixture of addition and condensation product is obtained[3] (Scheme-12).

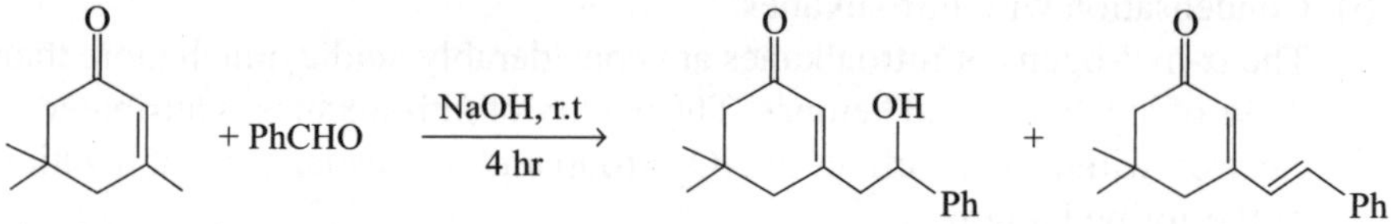

isophorone   benzaldehye

(E)

water	24 %	—
CTACl	—	80%
TBACl	27%	58%

(Scheme-12)

2.2.5 Aldol Condensation of Silyl Enol Ethers in Aqueous Media

The aldol condensation of silyl enol ethers with benzaldehydes, catalysed by titanium tetrachloride was first reported[4] in 1973. However, these reactions are carried out in anhydrous solvents[5]. It has now been possible to perform aldol condensation of silyl enol ethers with aldehydes in aquous phase[6]. (Scheme-13).

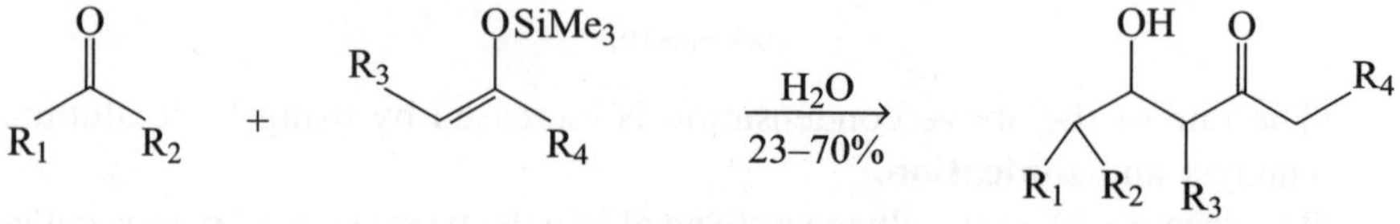

(Scheme-13)

The aqueous phase reaction (Scheme-13) was carried out without any catalyst, but it took several days for completion, since water serves as a weak lewis acid. The addition of stronger lewis acid (e.g., ytterbium triflate) greatly improved[7] the yield and the rate. (Scheme-14).

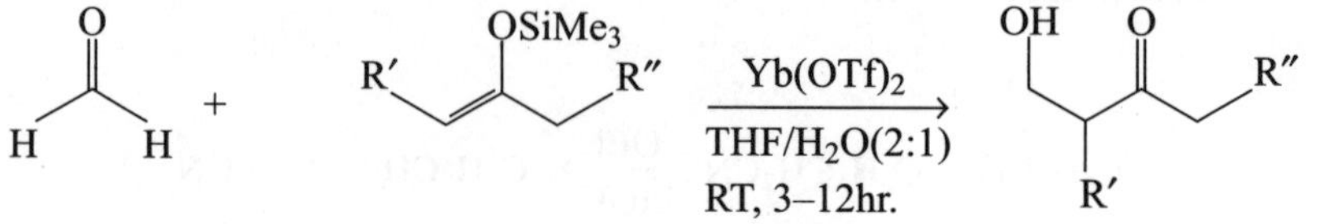

(Scheme-14)

The reaction of silyl enol ether of propiophenone with commercial formaldehyde in presence of ytterbium triflate gave the adduct (Scheme-15).

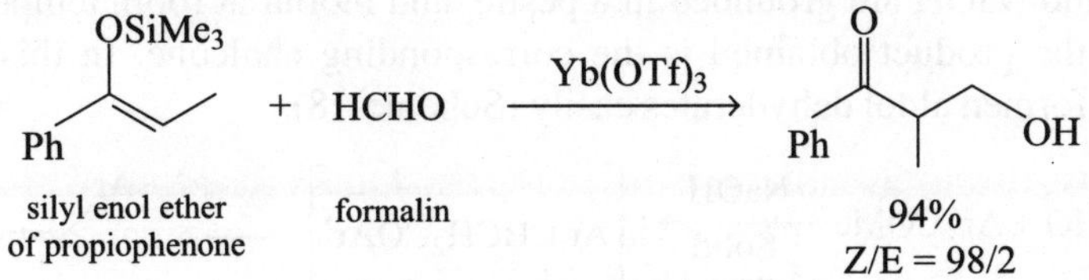

silyl enol ether
of propiophenone   formalin   94%
                                                                Z/E = 98/2

(Scheme-15)

Aldehydes other than formaldehydes can also be used. Thus, the reaction of 1-trimethylsilyloxycyclohexene with benzaldehyde in presence of catalytic amount of $\mathrm{Yb(OTf)_3}$ (10 mol %) in H_2O-THF (1 : 4) give 91% yield of the adduct[9,7b] (Scheme-16).

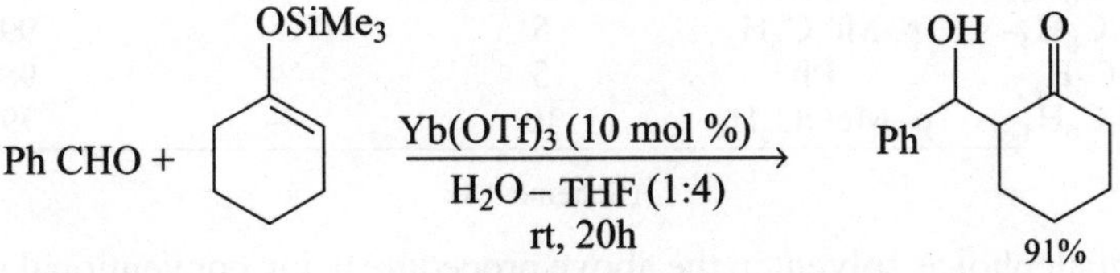

(Scheme-16)

2.2.6 Aldol Condensation in Solid Phase

The aldol condensation of the lithium enolate of methyl 3,3-dimethylbutanoate with aromatic aldehydes gave[10] 8:92 mixture of the syn and anti products in 70% yields. (Scheme 17).

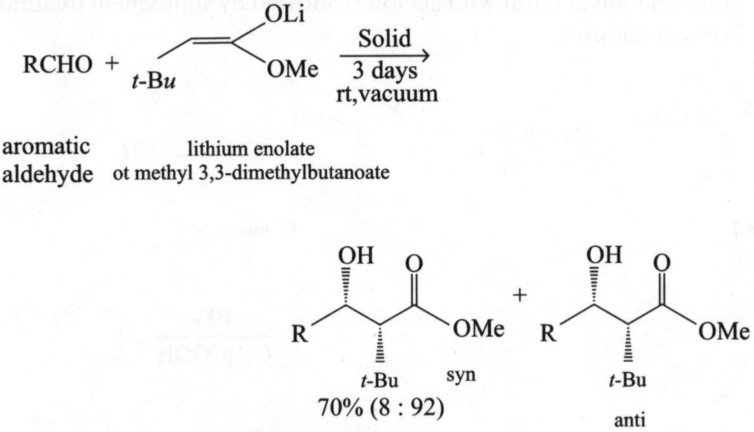

aromatic   lithium enolate
aldehyde   ot methyl 3,3-dimethylbutanoate

t-Bu   syn t-Bu
70% (8 : 92)   anti

R = 2 – OCH$_3$ C$_6$ H$_4$–, 4 – Cl C$_6$ H$_4$–,
4 – NO$_2$ C$_6$ H$_4$–, 3 NO$_2$ C$_6$ H$_4$–,
2 – NO$_2$ C$_6$ H$_4$–, 4 NO$_2$ –2– thenyl
(Scheme-17)

The above reaction (Scheme-17) is carried out by mixing freshly ground mixture of the starting materials in vacuum for 3 days at room temp.

In the absence of any solvent, some aldol condensations proceed[11] more efficiently and stereoselectively. In this procedure the appropriate aldehyde and ketone and NaOH are grounded in a pestle and mortar at room temperature for 5 min., the product obtained is the corresponding chalcone. In this case, the initially formed aldol dehyderates easily (Scheme-18).

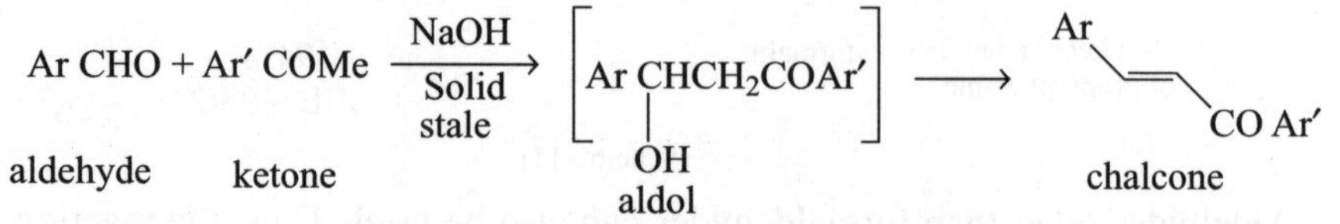

$$Ar\ CHO + Ar'\ COMe \xrightarrow[\substack{Solid \\ stale}]{NaOH} \left[Ar\ \underset{\underset{aldol}{OH}}{CHCH_2COAr'} \right] \longrightarrow \underset{chalcone}{Ar\diagdown\diagup CO\ Ar'}$$

aldehyde   ketone

Ar	Ar′	Reaction time	Yield aldol	Yield Chalcone
Ph	Ph	30	10	–
p–Me C$_6$H$_4$–	Ph	5	–	97
p–Me C$_6$H$_4$–	p–Me C$_6$H$_4$–	5	–	99
p–Cl C$_6$H$_4$–	Ph	5	–	98
p–Cl C$_6$H$_4$	p–MeOC$_6$H$_4$–	10	–	79

(Scheme-18)

Use of alcohol as solvent in the above procedure using conventional procedure gave only aldol in poor yield (10–25%).

2.2.7 Applications

A number of synthetic procedures based on aldol condensation of aldehydes and ketones are known.

(i) β-Ionone required for the synthesis of Vitamin A is prepared by the condensation of citral with acetone followed by subsequent treatment with boron trifluoride.

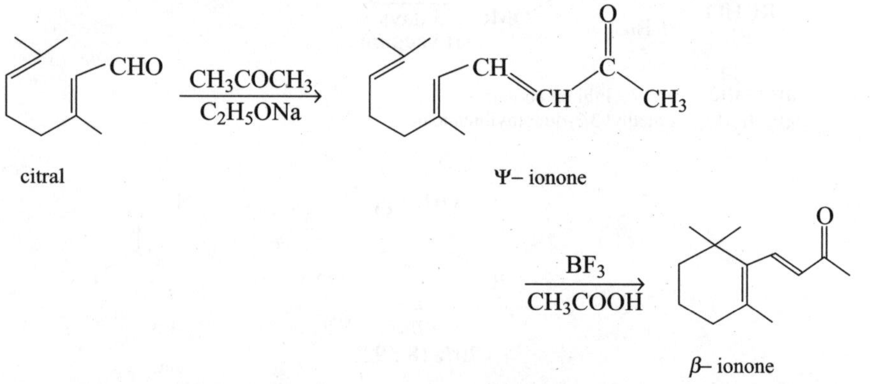

citral   Ψ– ionone

β– ionone

(ii) A commercially important mixed condensation involve the reaction of acetaldehyde and excess formaldehyde in presence of calcium hydroxide to give trihydroxymethyleneacetaldehyde (which has no α-hydrogen) and undergoes **crossed cannizzaro reaction** with formaldehyde to give a tetrahydroxy alcohol, known as pentaerythritol.

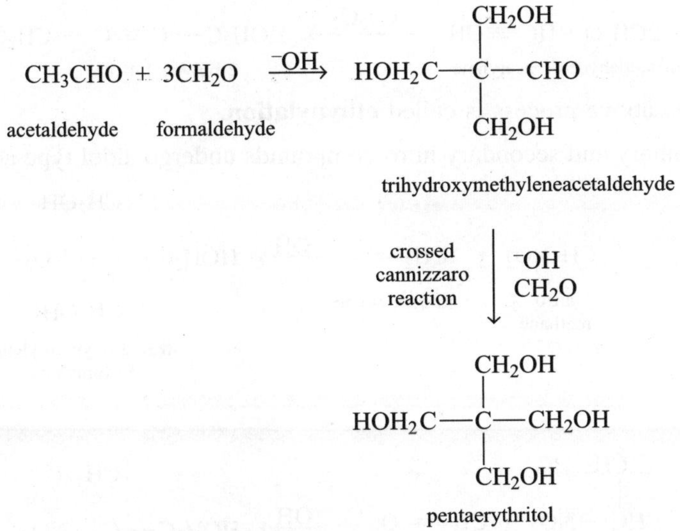

$$CH_3CHO + 3CH_2O \xrightarrow{^-OH} HOH_2C-\underset{\underset{CH_2OH}{|}}{\overset{\overset{CH_2OH}{|}}{C}}-CHO$$

acetaldehyde   formaldehyde

trihydroxymethyleneacetaldehyde

crossed cannizzaro reaction $\quad\Big\downarrow\begin{array}{l}^-OH\\CH_2O\end{array}$

$$HOH_2C-\underset{\underset{CH_2OH}{|}}{\overset{\overset{CH_2OH}{|}}{C}}-CH_2OH$$

pentaerythritol

(iii) Aldol condensation of acetone in presence of acids (dry hydrogen Chloride gas) gives mesityl oxide and phorone.

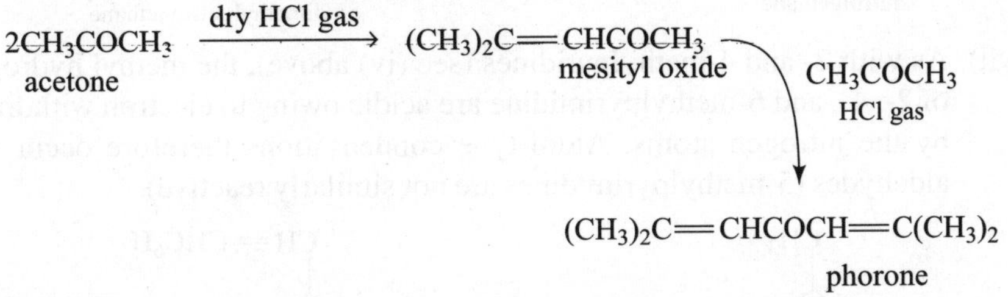

$$2CH_3COCH_3 \xrightarrow{\text{dry HCl gas}} (CH_3)_2C=CHCOCH_3$$
acetone   mesityl oxide

$$\xrightarrow[\text{HCl gas}]{CH_3COCH_3}$$

$$(CH_3)_2C=CHCOCH=C(CH_3)_2$$
phorone

(iv) An important class of cyanine dyes (photographic sensitisers) are obtained by the aldol condensation of pyridine and quinoline (having methyl groups at positions 2 and 4) with aldehydes.

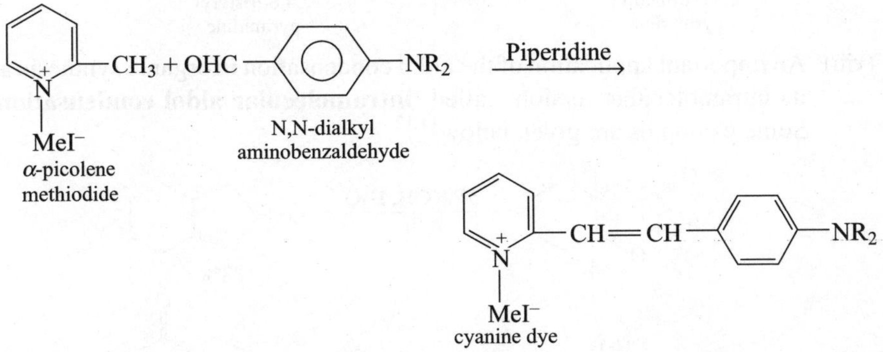

MeI⁻
α-picolene
methiodide

N,N-dialkyl
aminobenzaldehyde

cyanine dye

(v) Acetylenic alcohol, viz. 2-butyne-1,4-diol, a valuable commercial product is obtained by aldol type condensation of acetylene (compounds having acidic C—H bonds) with formaldehyde in presence of Cu_2C_2.

$$2CH_2O + HC \equiv CH \xrightarrow{Cu_2C_2} HOH_2C—C \equiv C—CH_2OH$$

formaldehyde   acetylene   2-butyne-1,4-diol

The above process is called **ethynylation**.

(vi) Primary and secondary nitro compounds undergo aldol type additions.

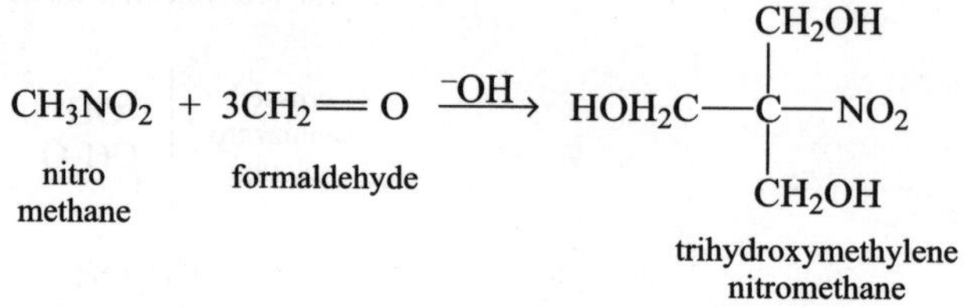

$$CH_3NO_2 + 3CH_2 = O \xrightarrow{^-OH} HOH_2C—\overset{\displaystyle CH_2OH}{\underset{\displaystyle CH_2OH}{\overset{|}{\underset{|}{C}}}}—NO_2$$

nitro   formaldehyde
methane

trihydroxymethylene
nitromethane

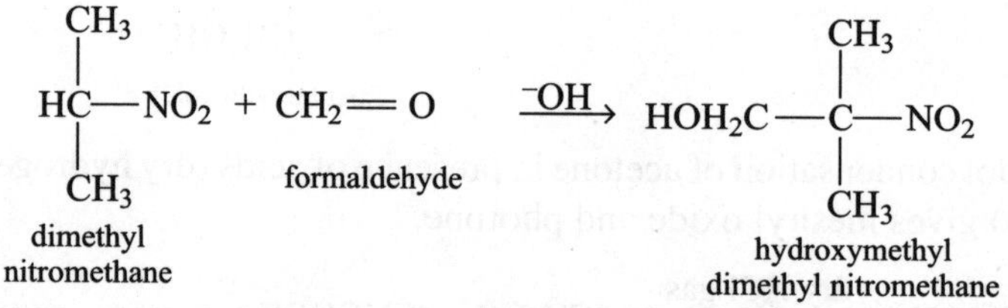

$$\overset{\displaystyle CH_3}{\underset{\displaystyle CH_3}{\overset{|}{\underset{|}{HC}}}}—NO_2 + CH_2 = O \xrightarrow{^-OH} HOH_2C—\overset{\displaystyle CH_3}{\underset{\displaystyle CH_3}{\overset{|}{\underset{|}{C}}}}—NO_2$$

dimethyl   formaldehyde
nitromethane

hydroxymethyl
dimethyl nitromethane

(vii) As with 2- and 4-methylpyridines (see (iv) above), the methyl hydrogens of 2-, 4-, and 6-methylpyrimidine are acidic owing to electron withdrawal by the nitrogen atoms. Aldol-type condensations therefore occur with aldehydes (5-methylpyrimidines are not similarly reactive).

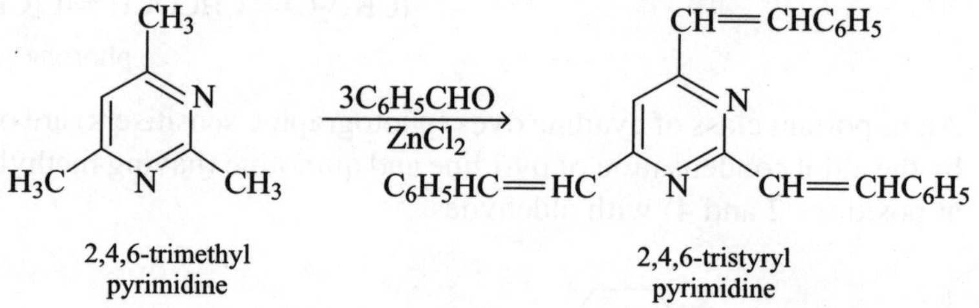

$$\xrightarrow[ZnCl_2]{3C_6H_5CHO}$$

2,4,6-trimethyl
pyrimidine

2,4,6-tristyryl
pyrimidine

(viii) An important application of the aldol condensation in organic synthesis are its intramolecular version, called **'intramolecular aldol condensation'**. Some examples are given below[11,12].

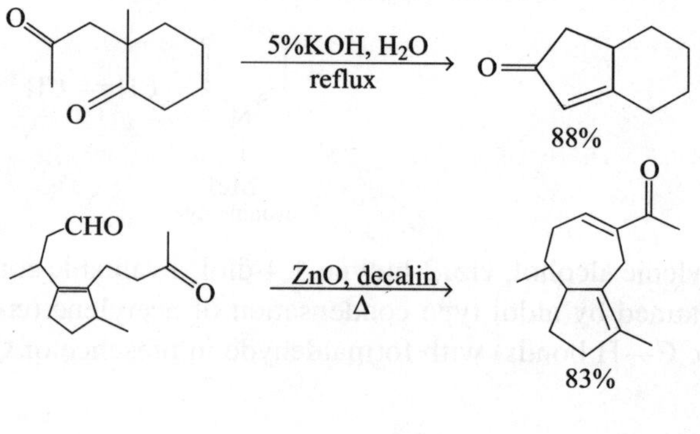

$$\xrightarrow[reflux]{5\%KOH, H_2O}$$

88%

$$\xrightarrow[\Delta]{ZnO, decalin}$$

83%

$$\text{(structure: methyl ketone cyclohexanone)} \xrightarrow[-H_2O]{\text{KOH, } \Delta} \text{(octalone structure)} \quad 85\%$$

$$\underset{CH_3}{O} \underset{C}{\overset{O}{\parallel}} CH_2\,CH_2CH_2\,CH_2\,\overset{O}{\underset{CH}{\parallel}} \xrightarrow{\ ^{-}OH\ } \text{(cyclopentene-}\overset{O}{\underset{C}{\parallel}}\text{—}CH_3\text{)} \quad 73\%$$

REFERENCES

1. A.T. Nielsen, W.J. Houlihan, Organic Reactions, 1968, *16*, 1; W. Forest Ed., Newer method of Preparative Organic Chemistry, 1971, *6*, 48, H.O. House, Modern Synthetic Reactions (W.A. Benzman, California, 2nd. ed.), 1972, pp. 629–682.
2. R.S. Verma and G.W. Kabalka, Heterocycles, 1985, *23*, 139.
3. F. Fringuelli, G. Pani, O. Piermalti and F. Pizzo, Tetrahedron, 1994, *50*, 11499; F. Fringuelli, G. Pani, O. Piermatti and F. Pizzo, Life Chem. Rep. 1995, *13*, 133.
4. T. Mukaiyama, K. Narasaka and T. Banno, Chem. Lett., 1973, 1011; T. Mukaiyama, K. Banno and K. Narasaka, J. Am. Chem. Soc., 1974, *96*, 7503.
   T. Mukaiyama, Org. React., 1982 *28*, 2303.
5. K. Takai, C.M. Heathcock, J. Org. Chem., 1985, *50*, 3247; A.E. Vougioukas and H.B. Kagan Tetrahedron Lett., 1987, *28*, 5513.
6. A. Lubineau, J. Org. Chem., 1986, *51*, 2142.
   A. Lubineau, E. Mayer, Tetrahedron, 1988, *44*, 6065.
7. (a) S. Kobayashi and I. Hachiya, J. Org. Chem., 1994, *59*, 3590,
   (b) For a review on lanthanide catalysed organic reaction in aqueous media, S. Kobayashi, Synlett., 1994, 589.
8. S. Murata, M. Suzuki and R. Nojori, Tetrahedron Lett, 1980, *21*, 2527.
9. S. Kobayashi and I. Hachiya, Tetrahedron Lett, 1992, 1625.
10. Y. Wef and R. Bakthavatechalan, Tetrahedron Lett., 1991, *32*, 1535.
11. S.C. Welch, J. M. Assercq and J. P. Loh, Tetrahedron Lett., 1986, 1115.
12. W. Hoffmann and H. Siegel, Tetrahedron Lett., 1975, 533.

2.3 ARNDT-EISTERT SYNTHESIS

It is a convenient method of converting an acid (RCOOH) to the next homologous acid (RCH$_2$COOH). In this procedure, the acid is first converted to its acid chloride, which is then treated with diazomethane (CH$_2$N$_2$) resulting in the formation of diazoketone (RCOCHN$_2$). The diazoketone on treatment with silver oxide gets converted to a ketene (RCH=C=O), which gets esterified under the conditions of the reaction. Subsequent hydrolysis of the ester provides the homologous acid (RCH$_2$COOH). Various steps are shown in (Scheme-1).

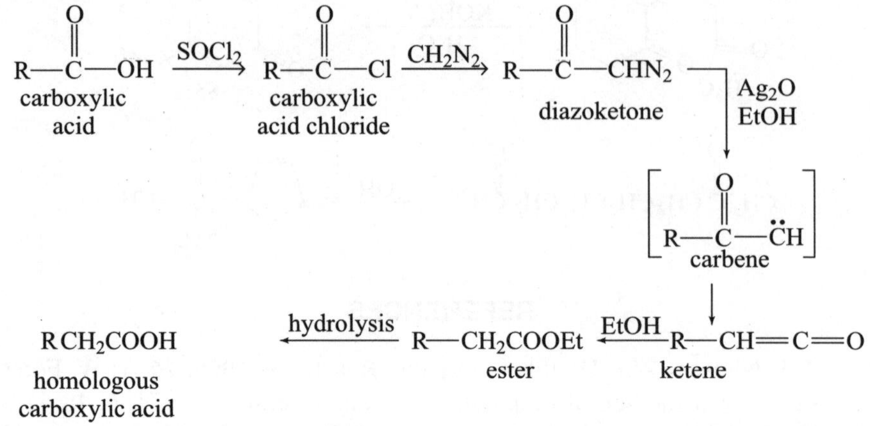

(Scheme-1)

As seen (Scheme-1), the conversion of the diazoketone into the ketene involves a rearrangement known as **Wolff rearrangement** under the catalytic influence of silver oxide (see mechanism given below).

MECHANISM

The various steps involved in the mechanism are

  (i) nucleophilic attack of diazomethane on the carbonyl carbon of the acid chloride to give the diazoketone.
 (ii) diazoketone eliminates a molecule of nitrogen to form a carbene.
(iii) rearrangement of the carbene to the ketene (**Wolff's rearrangement**)
 (iv) the reactive ketene reacts with the nucleophile present (H_2O) to form the higher homologue of the acid.

The various steps of mechanism are depicted in Scheme-2.

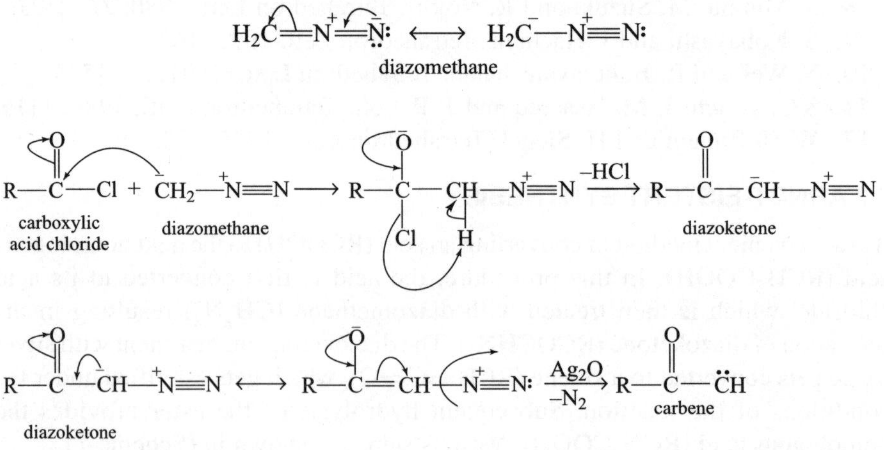

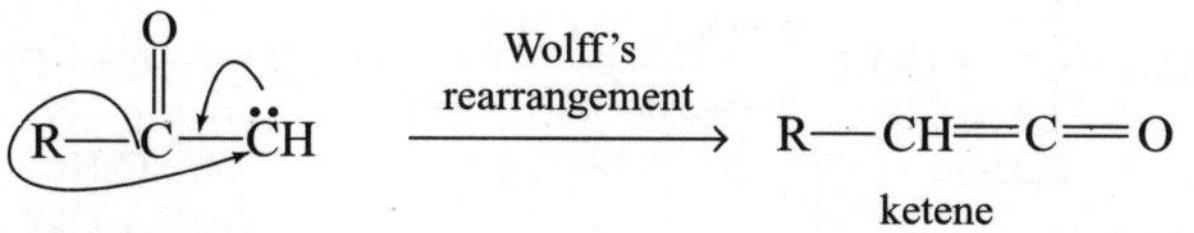

(Scheme-2)

The above mechanism is supported by the fact that the formed intermediate ketene can be trapped. Also, the isotopic labelling experiment has shown that the carbonyl of the acid chloride or the diazoketone is present in the resulting acid as the carbonyl carbon.

2.3.1 Applications

(i) Synthesis of higher homologues of carboxylic acids, amides and esters. The ketene obtained by the Wolff's rearrangement of the carbene (obtained as an intermediate during Arndt-Eistert Synthesis) on reaction with H_2O, NH_3, or R^1OH give the corresponding carboxylic acid, amide or ester resps., all of which have one carbon atom more than the starting carboxylic acid.

(ii) Synthesis of various carboxylic acids

(a)

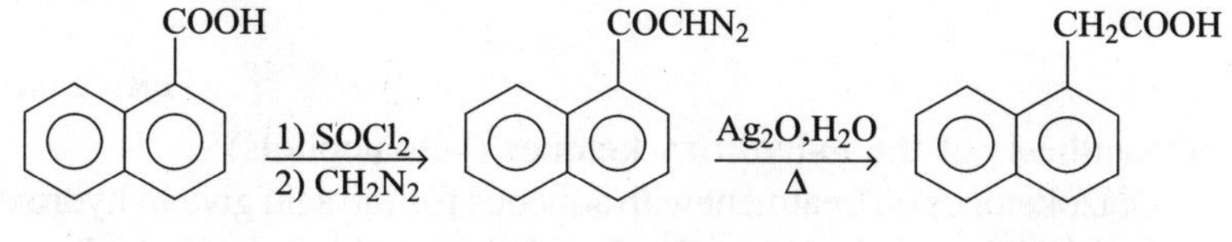

α-naphthoic acid   diazo-α-acetonaphthone   α-naphthyl acetic acid (80%)

(b) CH₃CH₂—C(CH₃)(C₆H₅)—COOH $\xrightarrow[\text{3) Ag}_2\text{O, H}_2\text{O(twice)}]{\text{1) SOCl}_2, \text{2) CH}_2\text{N}_2}$ CH₃CH₂—C(CH₃)(C₆H₅)—CH₂CH₂COOH

2-methyl-2-phenyl
butyric acid

4-methyl-4-phenyl
caproic acid

(c) *o*-nitrobenzoic acid $\xrightarrow[\text{3) Ag}_2\text{O, H}_2\text{O}]{\substack{\text{1) SOCl}_2 \\ \text{2) CH}_2\text{N}_2}}$ 2-nitrophenyl acetic acid

(iii) **Synthesis of homoveratroyl chloride, an intermediate in the synthesis of papaverine**

3,4-dimethoxy
benzoic acid $\xrightarrow[\text{3) Ag}_2\text{O, H}_2\text{O}]{\substack{\text{1) SOCl}_2 \\ \text{2) CH}_2\text{N}_2}}$ 3,4-dimethoxy
phenylacetic acid $\xrightarrow{\text{SOCl}_2}$ homoveratroyl
chloride

(iv) **Synthesis of mescaline**

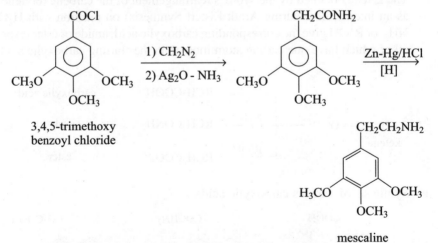

3,4,5-trimethoxy
benzoyl chloride $\xrightarrow[\text{2) Ag}_2\text{O - NH}_3]{\text{1) CH}_2\text{N}_2}$ (CH₂CONH₂ compound) $\xrightarrow[\text{[H]}]{\text{Zn-Hg/HCl}}$

mescaline

(v) **Synthesis of the ω-hydroxy ketones (ketoalcohols)**
Diazoketones on treatment with aqueous formic acid give ω-hydroxy ketones.
In the absence of catalyst the diazoketone is hydrolyed to a keto alcohol.

$$\text{RCOCHN}_2 \xrightarrow{\text{HCOOH + H}_2\text{O}} \text{RCOCH}_2\text{OH}$$

(vi) Trimethylsilyl diazomethane can also be used[2] for homologation in Arndt-Eistert synthesis

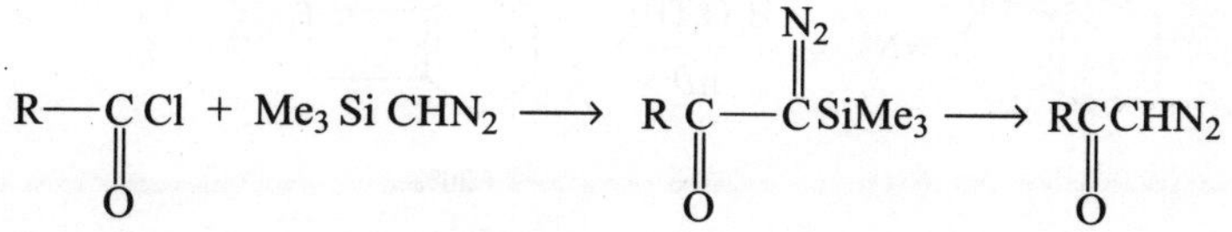

$$R-\underset{O}{\overset{}{C}}Cl + Me_3 Si CHN_2 \longrightarrow R \underset{O}{\overset{N_2}{C}}-\overset{\parallel}{C} SiMe_3 \longrightarrow RCCHN_2$$

(vii) A **photochemical Arndt-Eistert** reaction can also be performed. Some examples are

(a) Synthesis of methyl γ-cyclohexyl butyrate[3]

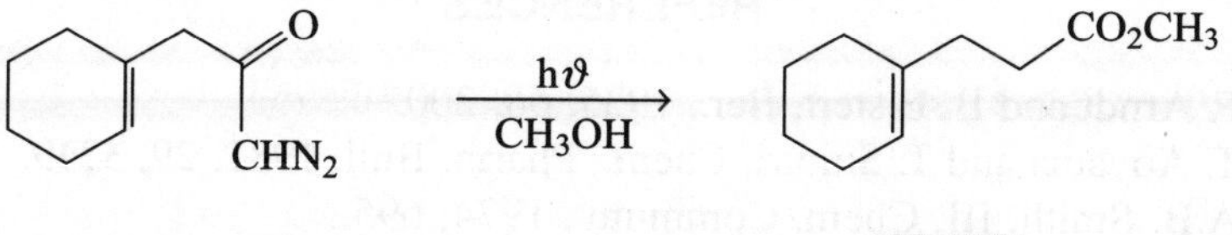

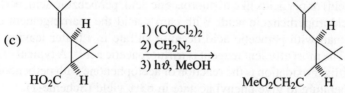

80–95%

(b) Synthesis of methyl δ-furanyl glutarate

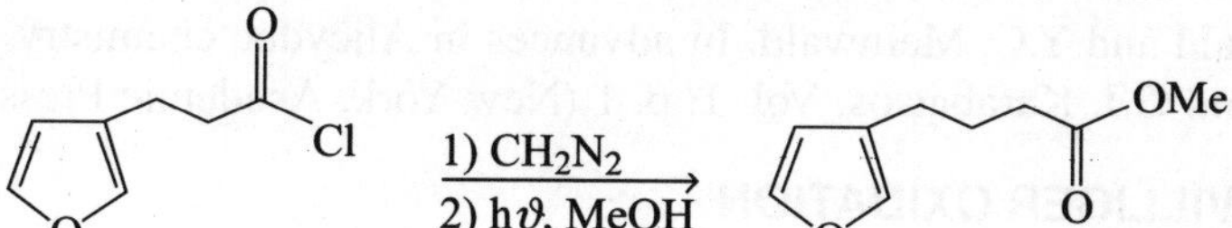

methyl δ-furanyl glutarate
(90%)

(c)

(viii) A variation of Arndt-Eistert reaction is that in cyclic diazoketones, the rearrangement leads to ring contraction. This reaction has been widely used[6] for the preparation of strained small ring compounds such as bicyclo [2, 1, 1]-hexane and benzocyclobutene.

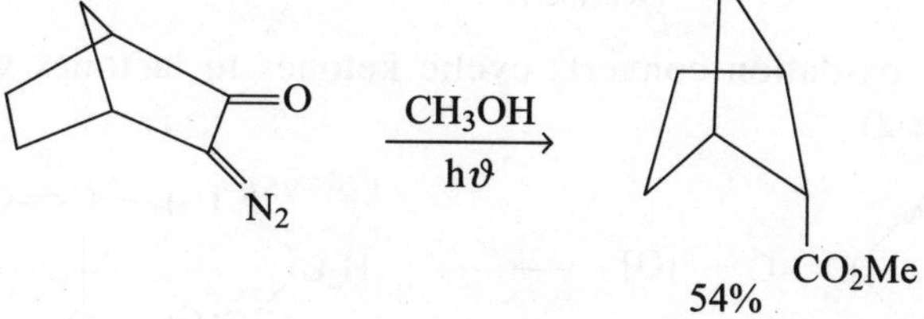

54% CO₂Me

methylbicyclo [2, 1, 1] hexane carboxylate

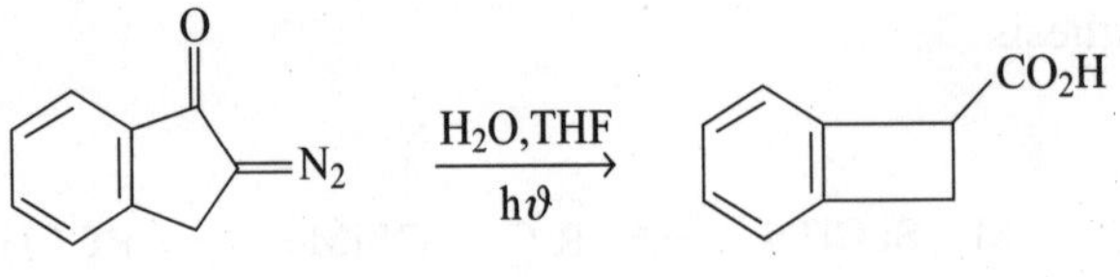

20%

bicyclobutene carboxylic
acid

REFERENCES

1. F. Arndt and B. Eistert, Ber., 1935, *68*, 200.
2. T. Aoyama and T. Shioiri, Chem., Pharm. Bull. 1981, *29*, 3249.
3. A.B. Smith, III, Chem. Commum., 1974, 695.
4. E.J. Walsh, Jr. and G.B. Stone, Tetrahedron Lett., 1986, 1127.
5. A.B. Smith, III, B.D. Dorsey, M. Visnick, T. Maeda and M.S. Malamas, J. Am. Chem. Soc., 1986, *108*, 3110.
6. J. Meinwald and Y.C. Meinwald. In advances in Alicydoc chemistry, ed. H. Hart and G.J. Karabatsos, Vol. 1, p. 1 (New York, Academic Press).

2.4 BAEYER–VILLIGER OXIDATION

The oxidation of ketones to esters with hydrogen peroxide or with peracids (RCO_3H) is known as Baeyer–Villiger oxidation[1]. The reaction can be brought about conveniently by hydrogen peroxide in weakly basic solution, peroxy sulfuric acid (caro's acid) or per acids like trifluoroacetic acid, per benzoic acid, performic acid and m-chloroperbenzoic acid. With caro's acid the rearrangement step is much faster than with peracetic acid because sulfate is a better leaving grcup than acetate. The most efficient reagent is trifluoroacetic acid[2]. A typical example of Baeyer–Villiger oxidation is the reaction of acetophenone with perbenzoic acid at room temperature to give phenyl acetate in 63% yield (Scheme-1).

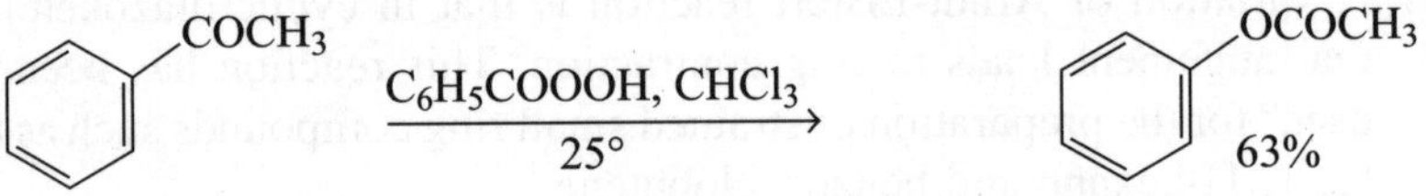

acetophenone

phenyl acetate

(Scheme-1)

Baeyer–Villiger oxidation converts cyclic ketones to lactones with ring expansion (Scheme-2).

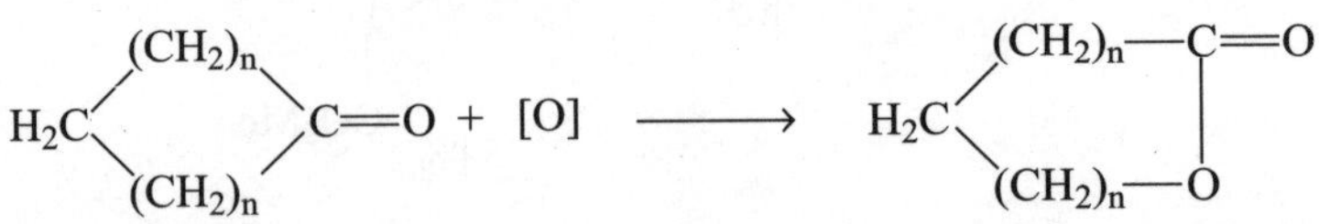

cyclic ketone

lactone

cyclopentanone → (CF₃CO₃H) → δ-valero lactone

$$\text{cyclopentanone} \xrightarrow{\text{CF}_3\text{CO}_3\text{H}} \delta\text{-valero lactone}$$

(Scheme-2)

This is a convenient method for the synthesis of lactones. The overall reaction is an insersion of oxygen atom between the carbonyl carbon and the adjacent carbon (in ketone). Organic solvents, which are inert under the conditions of the reaction may be used. Of course, the choice of the solvent depends on the solubility of the reactants. Solvents like acetic acid and chloroform are commonly used. An important modification of the Baeyer–Villiger oxidation is **Dakins reaction**.

Baeyer–Villiger oxidation cannot be used in case of ketone which contain

$$C=C \left(\text{which gets converted into epoxide} -\underset{\underset{O}{\diagdown}}{C}-\underset{}{C}- \right),$$

—S— (which gets converted into $-\underset{\underset{O}{\parallel}}{S}-$) or $\diagup N—R$ (which gets

converted into $-\underset{\underset{R}{\mid}}{N}\rightarrow O$ groups). In case, an alkene is present, bis [trimethylsilyl] peroxide is used to carry out the Baeyer–Villiger oxidation[3].

MECHANISM

The mechanism of Baeyer–Villiger oxidation is not clear. However, it is understood that the reaction takes the following course.

 (i) The carbonyl reactant removes a proton from the acid (H—A) to give the protonated carbonyl reactant (1).
 (ii) The peroxy acid attacks the protonated carbonyl reactant (1) to give the oxonium ion (2).
(iii) A proton is removed from the oxonium ion (2) to give the species (3).
(iv) The species (3) abstracts a proton from the acid (H—A) to give the species (4).
 (v) The phenyl group migration with an electron pair takes place (from the species 4) to the adjacent oxygen, simultaneous with the departure of RCO₂H as a leaving group to give 5.
(vi) Final step is the removal of a proton which results in the formation of ester (Scheme-3).

acetophenone (1) Peracid

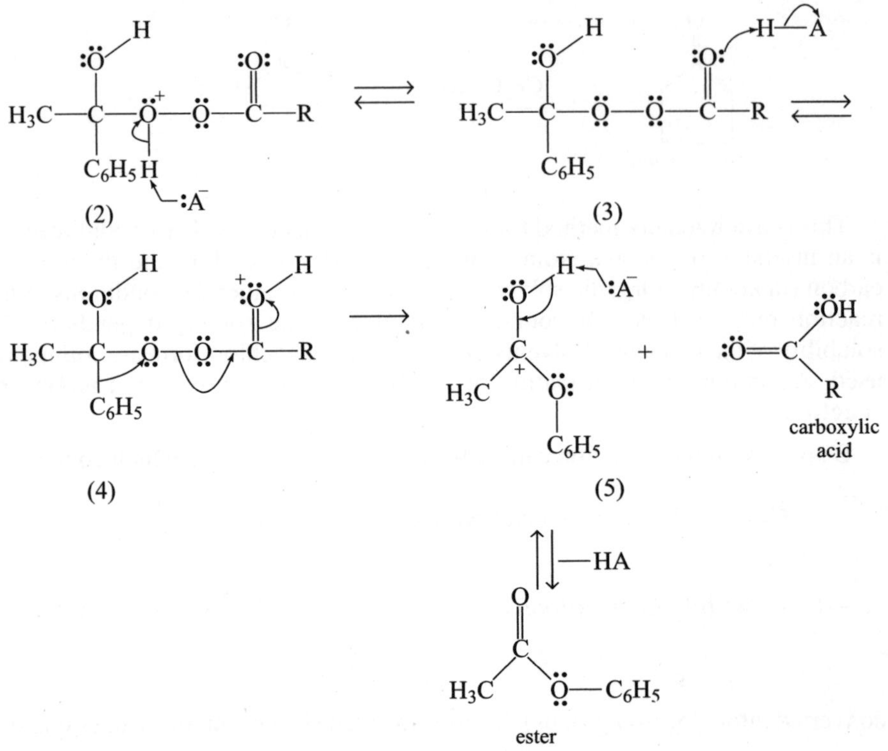

(2)   (3)

(4)   (5)

carboxylic
acid

ester

(Scheme-3)

The products of Baeyer–Villiger oxidation (Scheme-3) show that a phenyl group has a greater tendency to migrate than a methyl group. Had this not been the case, the product would have been $C_6H_5COOCH_3$, and not $CH_3COOC_6H_5$. The above mechanism is supported by the observation that the labelled carbonyl oxygen atom of the ketone becomes the carbonyl oxygen atom of the ester (the ester has the same ^{18}O content as the ketone) (Scheme-4).

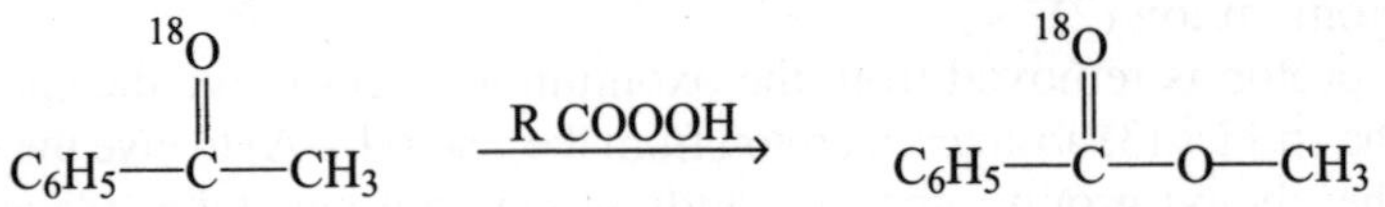

(Scheme-4)

MIGRATORY APTITUDE

The tendency of a group to migrate is called its migratory aptitude. Studies have shown that in the Baeyer–Villiger oxidation (and also other reactions), the migratory aptitude of the groups is H > phenyl > 3° alkyl > 2° alkyl > 1° alkyl > methyl. In all cases, this order is for groups migrating with their electron pairs, i.e., as anions. The aryl group migrates in preference to methyl and primary alkyl groups. In the aryl series, migration is facilitated by electron-releasing para substituents. Thus, migratory aptitude among aryl group is $p\text{-}CH_3OC_6H_4 > C_6H_5 > p\text{-}O_2NC_6H_4$.

For example, phenyl p-nitrophenyl ketone yields only phenyl p-nitrobenzoate by the migration of phenyl group (Scheme-5).

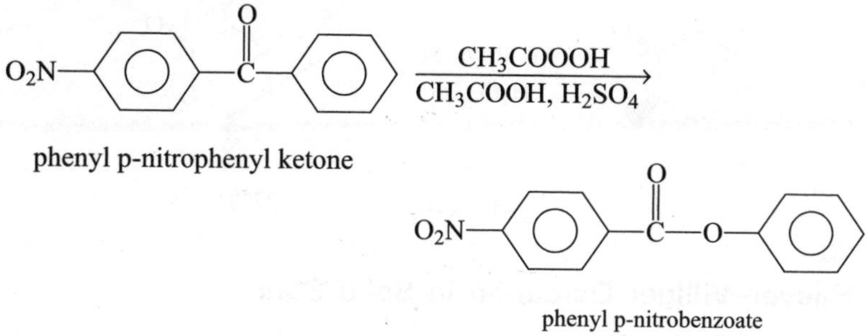

phenyl p-nitrophenyl ketone

phenyl p-nitrobenzoate

(Scheme-5)

The Baeyer–Villiger oxidation takes place with retention of configuration of the migrating group. Two such examples are given in Scheme-6.

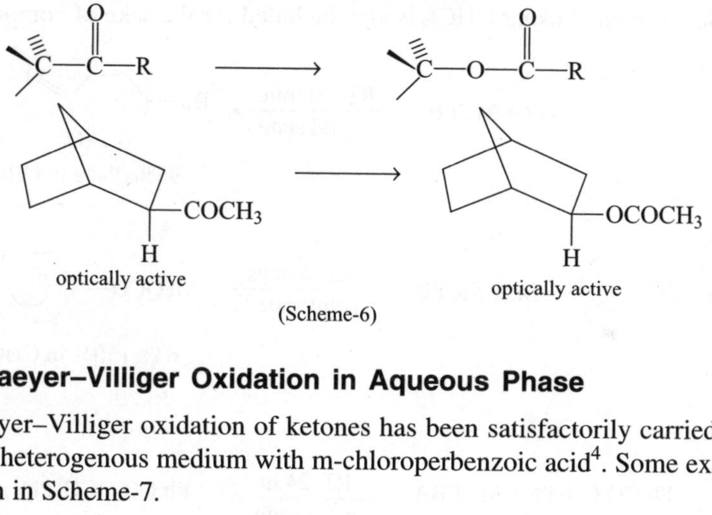

optically active

optically active

(Scheme-6)

2.4.1 Baeyer–Villiger Oxidation in Aqueous Phase

The Baeyer–Villiger oxidation of ketones has been satisfactorily carried out in aqueous heterogenous medium with m-chloroperbenzoic acid[4]. Some examples are given in Scheme-7.

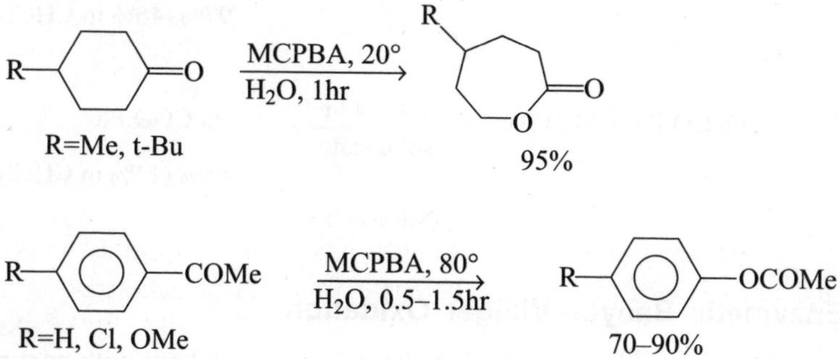

R=Me, t-Bu

95%

R=H, Cl, OMe

70–90%

(Scheme-7)

The above procedure can also be used for reactive ketones (e.g., anthrone, which usually gives anthraquinone) and ketones which are unreactive or give expected lactones in organic solvents with difficulty[5] can also be oxidised (Scheme-8).

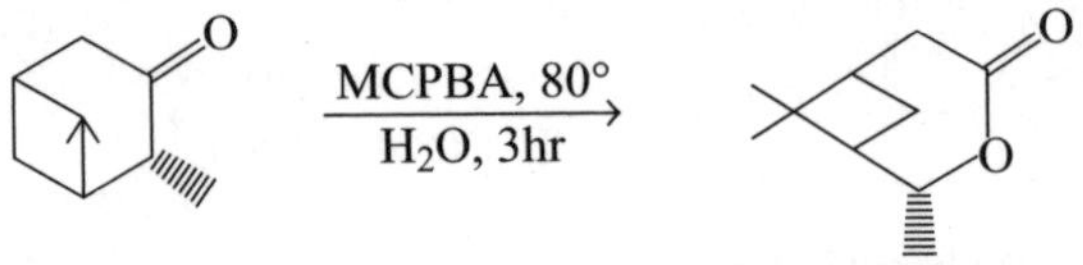

$$\xrightarrow[\text{H}_2\text{O, 3hr}]{\text{MCPBA, 80°}}$$

(Scheme-8)   27%

2.4.2 Baeyer–Villiger Oxidation in Solid State

Some Baeyer–Villiger oxidation of ketones with m-chloroperbenzoic acid proceed much faster in the solid state than in solution. In this procedure, a mixture of powdered ketone and 2 mole equivalent of m-chloroperbenzoic acid is kept at room temperature to give the product[6]. Some examples are given below in Scheme-9 (the yield obtained using $CHCl_3$ is also included for the sake of comparison).

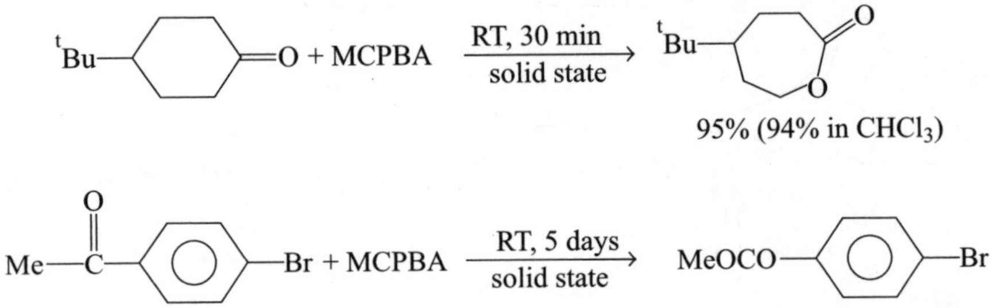

$$\text{'Bu}\text{—}\langle\rangle\text{=O} + \text{MCPBA} \xrightarrow[\text{solid state}]{\text{RT, 30 min}}$$

95% (94% in $CHCl_3$)

$$\xrightarrow[\text{solid state}]{\text{RT, 5 days}} \text{MeOCO}\text{—}\langle\rangle\text{—Br}$$

64% (50% in $CHCl_3$)

$$\text{Ph CO CH}_2\text{Ph} + \text{MCPBA} \xrightarrow[\text{solid state}]{\text{RT, 24 hr}} \text{Ph CO OCH}_2\text{Ph}$$

97% (46% in $CHCl_3$)

$$\text{Ph CO Ph} + \text{MCPBA} \xrightarrow[\text{solid state}]{\text{RT, 24 hr}} \text{Ph COO Ph}$$

85% (13% in $CHCl_3$)

(Scheme-9)

2.4.3 Enzymatic Baeyer–Villiger Oxidation

A number of Baeyer–Villiger oxdiations have been carried out with enzymes. A typical transformation is the enzymatic Baeyer–Villiger oxidation, which gives lactone from cyclohexanone using[7,8] a purified cyclohexanone oxygenase enzyme (Scheme-10).

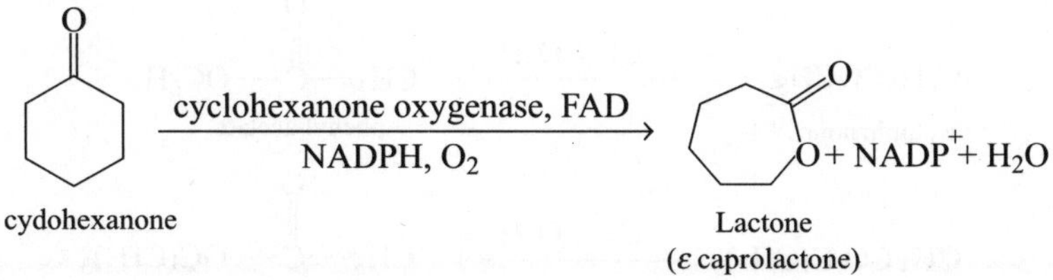

(Scheme-10)

Similarly, 4-methyl cyclohexanone can be converted into the corresponding lactone in 80% yield[9] with > 98% ee with cyclohexanone oxygenase obtained from Acineto bacter (Scheme 11).

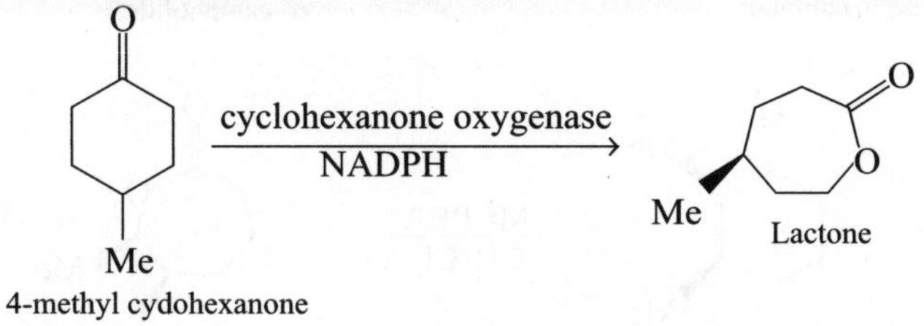

(Scheme-11)

Cyclohexanone oxygenase in presence of NADH (reduced nicotinamide adenine dinucleotide) converts[9] phenyl acetone into benzyl acetate (Scheme-12).

$$C_6H_5CH_2CO\,CH_3 \xrightarrow[\text{O}_2,\ \text{ENZ-FAD, NADPH, H}^+]{\text{cyclohexanone oxygenase}} C_6H_5CH_2OCOCH_3$$

phenylacetone   benzylacetate

(Scheme-12)

A number of enzymatic Baeyer–Villiger oxidations have been reported in steroids (see Applications).

2.4.4 Applications

Baeyer–Villiger oxidation has great synthetic utility. Some important applications are:

(i) Transformation of ketones into esters. An oxygen atom is introduced between the carbon of the carbonyl and the adjacent carbon. This reaction is applicable to both acyclic and cyclic ketones. Oxidation of cyclic ketones results in ring expansion and forms lactones as illustrated by the conversion of cydopentanone to δ-valerolactone (Scheme-2). Similarly, camphor on Baeyer–Villiger oxidation gives α-compholide in 30% yield by using caro's acid, and 2,3-dimethylcyclohexanone is converted[10] into lactone by this method.

$$C_6H_5COCH_3 \xrightarrow{CF_3CO_3H} CH_3-\overset{\overset{\displaystyle O}{\|}}{C}-OC_6H_5$$

acetophenone   phenyl acetate

$$CH_3\,CO\,C(CH_3)_3 \xrightarrow{CF_3CO_3H} CH_3-\overset{\overset{\displaystyle O}{\|}}{C}-OC(CH_3)_3$$

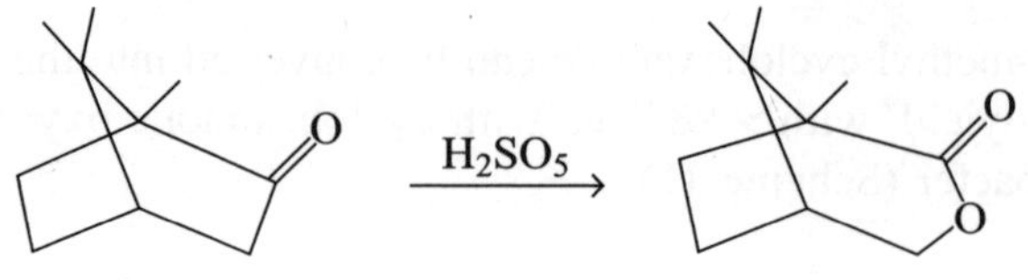

pinacolone   t-butylacetate

camphor $\xrightarrow{H_2SO_5}$ α- campholide

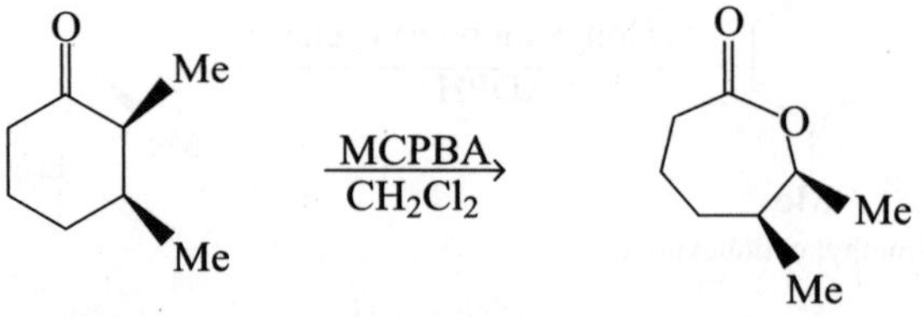

2,3-dimethyl cyclohexanone $\xrightarrow[CH_2Cl_2]{MCPBA}$ lactone

(ii) Synthesis of carboxylic acids from ketones or aldehydes. Baeyer–Villiger oxidation of aldehydes or ketones give esters, which on saponification give carboxylic acids.

$$RCOR + R'-\overset{\overset{\displaystyle O}{\|}}{C}-O-O-H \longrightarrow RCO_2R + R'CO_2H$$
$$\underset{\overset{\displaystyle \quad}{\xrightarrow{H^+}}}{} RCO_2H + ROH$$

(iii) Synthesis of anhydrides. α-Diketones on oxidation under Baeyer–Villiger conditions give anhydrides.

$$CH_3\overset{\overset{\displaystyle O}{\|}}{C}-\overset{\overset{\displaystyle O}{\|}}{C}-CH_3 \xrightarrow{RCO_3H} CH_3-\overset{\overset{\displaystyle O}{\|}}{C}-O-\overset{\overset{\displaystyle O}{\|}}{C}-CH_3$$

diacetyl   acetic anhydride

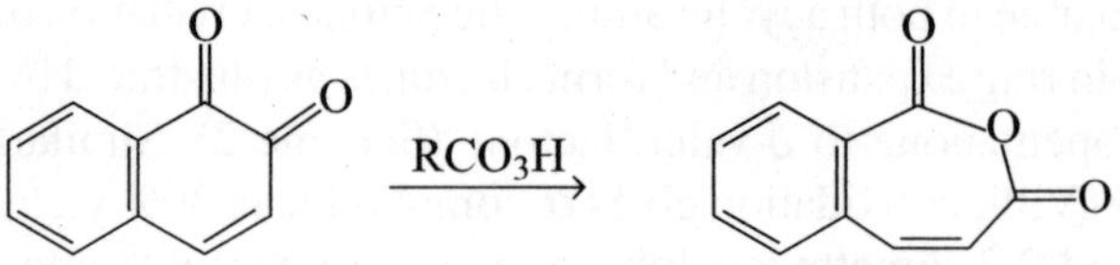

α-napthaquinone $\xrightarrow{RCO_3H}$

(iv) Synthesis of large ring lactones (which are difficult to prepare by other methods)

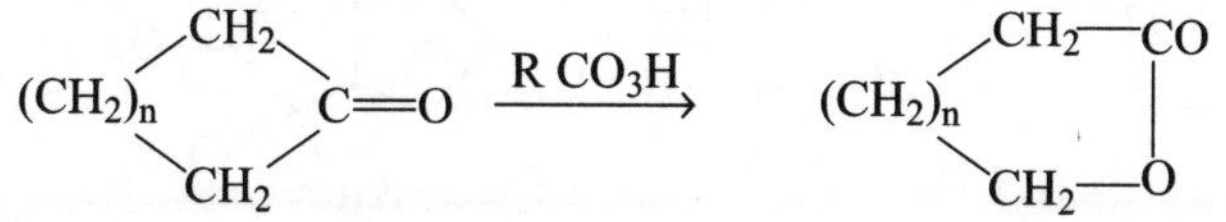

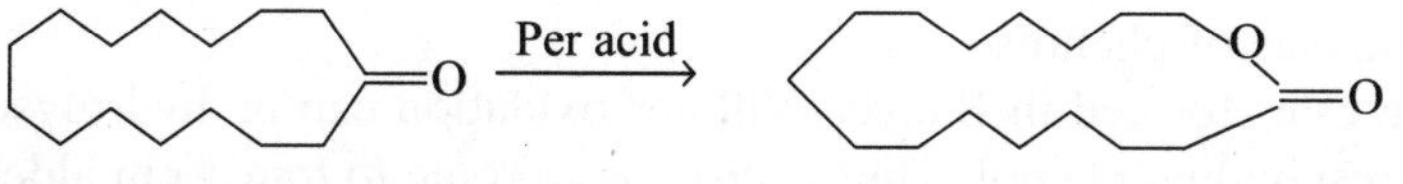

(v) Synthesis of long chain hydroxyesters. The long chain lactones obtained by Baeyer–Villiger oxidation of ketone (see iv above) on treatment with conc. H_2SO_4/C_2H_5OH give long chain α, ω-hydroxyesters.

$$HOCH_2 \text{---}(CH_2)_n \text{---} CH_2 \text{---} CO_2Et$$

α,ω-hydroxy esters

lactones

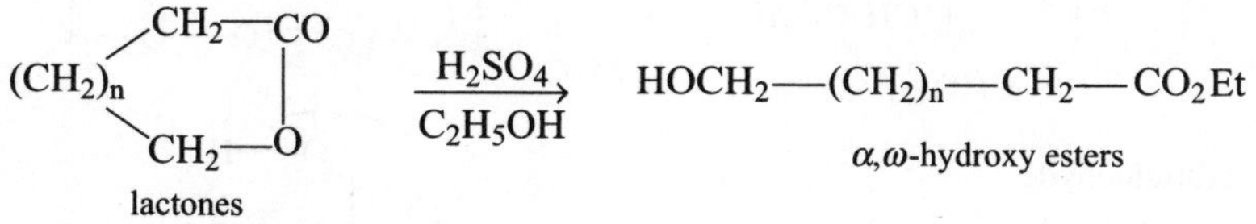

(vi) Biochemical Baeyer–Villiger oxidations of steroids.

19-Nortestosterone on treatment with *Aspergillus tamarii* gives 70% yield of 19-nortestololactone[11]. Progresterone and testosterone are converted into Δ^1-dehydrotestolo-lactone by fermentation with *cylindrocarpon radicicola*[12].

Testololactone is obtained from progesterone by oxidation with *Penicillium chrysogenum* and from 4-androstene-3,17-dione by treatment with *Penicillium lilacinum*[13].

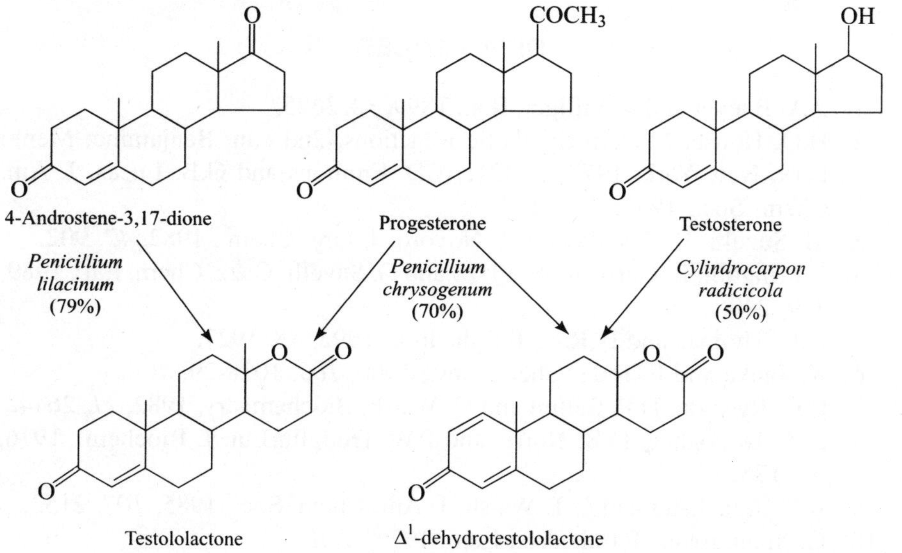

4-Androstene-3,17-dione

Progesterone

Testosterone

Penicillium lilacinum (79%)

Penicillium chrysogenum (70%)

Cylindrocarpon radicicola (50%)

Testololactone

Δ^1-dehydrotestololactone

(vi) Synthesis of bicyclic lactone with retention of configuration[14].

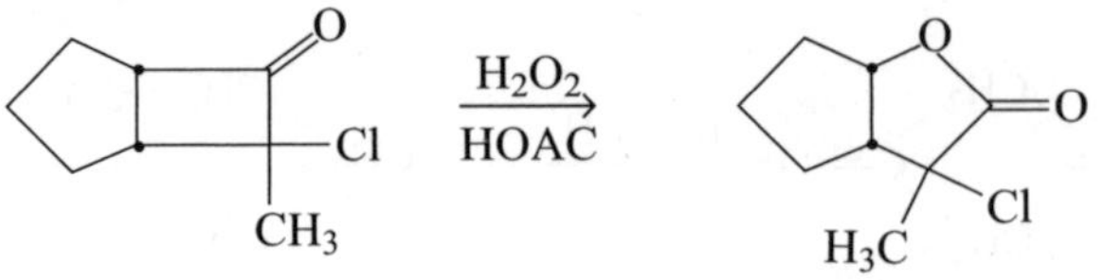

(vii) Synthesis of phenols
   The ester formed in Baeyer–Villiger oxidation can be hydrolysed to the corresponding phenol. Thus it provides a route to transform aldehydes or ketones to phenols. For example veratraldehyde is converted into 3,4-dimethoxyphenol (see also **Dakins reaction**).

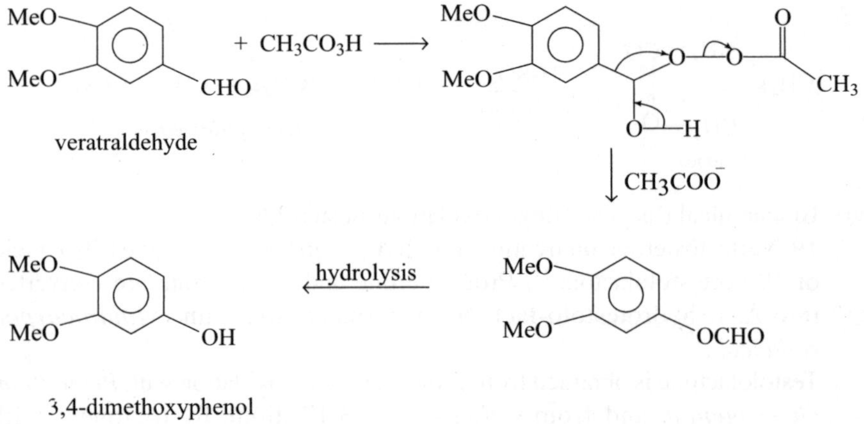

veratraldehyde

3,4-dimethoxyphenol

REFERENCES

1. A.V. Baeyer and V. Villiger, Ber., 1899, *32*, 3625.
2. H.O. House, Modern Synthetic Reactions, 2nd edn. Benjamine; Menlo Park, New York, 1972, p. 321; W.D. Emmons and G.B. Lucas, J. Am. Chem. Soc., 1955, *77*, 2287.
3. M. Suzuki, H. Takada and R. Noyoro, J. Org. Chem., 1982, *47*, 902.
4. F. Fringulli, R. Germani, E. Pizzo and G. Savelli, Gazz. Chem. Ital., 1989, *119*, 249.
5. A.E. Thomas and F. Ray, Tetrahedron, 1992, *48*, 1927.
6. K. Tanka and F. Toda, Chem. Rev., 2000, *100*, 1028-29.
7. C.C. Ryerson, D.P. Ballon and C. Walsh, Biochemistry, 1982, *21*, 2644.
8. N.A. Donoghue, D.B. Norris and P.W. Trudgill, Eur. J. Biochem., 1976, *63*, 175.
9. B.P. Branchaud and C.T. Walsh, J. Am. Chem. Soc., 1985, *107*, 2153.
10. G. Magnusson, Tetrahedron Lett., 1977, 2713.

11. J.T. McCurdy and R.D. Garrett, J. Org. Chem., 1968, *33*, 660.
12. F.J. Fried, R.W. Thoma and A. Klingsberg, J. Am. Chem. Soc., 1953, *75*, 5764.
13. R.L. Prairie and P. Talalay, Biochemistry, 1963, *2*, 203.
14. A. Hassner. J. Org. Chem., 1978, *43*, 1774.

2.5 BARBIER REACTION

The reaction of a ketone with an organometallic reagent to give the corresponding alcohol is known as Barbier reaction (Scheme-1).

(Scheme-1)

The original process discovered by Barbier offers distinct advantages but has a number of drawbacks. It is very important to prepare the organo metallic reagent in situ (in presence of the substrate). As the organometallic intermediate is generally very reactive with water, Barbier type reactions were earlier carried out in anhydrous solvents. Barbier reaction can only be performed with reactive alkyl halides.

An important Barbier reaction[1a] is useful in the synthesis of cyclopentanone.

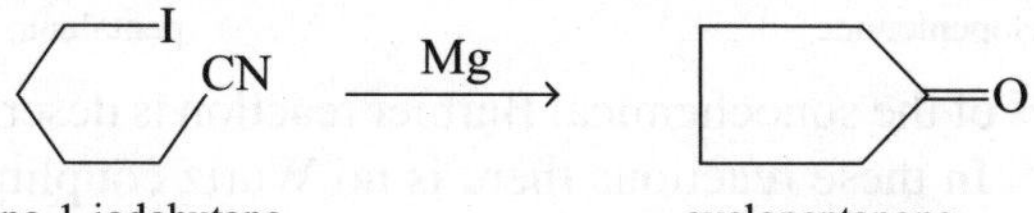

5-cyano-1-iodobutane   cyclopentanone
                                   (61-79%)

2.5.1 Barbier Reaction under sonication

A first improvement was obtained by the replacement of magnesium by lithium[2], but the most crucial step was made by **sonication**[3]. There are significant advantage in this method in that the reactions can be carried out in commercial THF and are largely free from side reactions such as reduction and enolization, which are common in conventional procedure. Even allyl or benzyl halides give much better yields (> 95%) and very little Wurtz coupling, which predominates in non-ultrasonic conventional method, has been observed.

The modified Barbier reaction using Li/THF/sonication is represented as in Scheme 2.

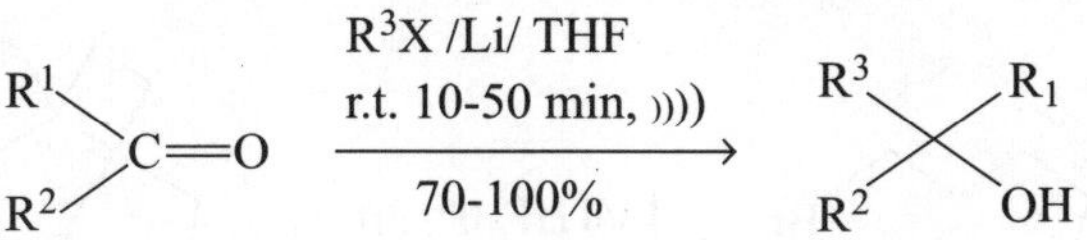

(Scheme-2)

2.5.2 Applications

(i) In Barbier reaction, α, β-unsaturated ketones can be used with good results.

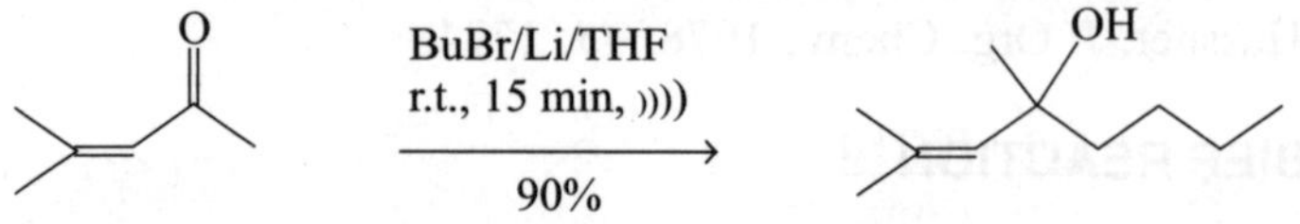
O
BuBr/Li/THF
r.t., 15 min,)))
90%
OH

(ii) A intermediate required for the synthesis of sesquicarene can be obtained by a sonochemical Barbier step[4].

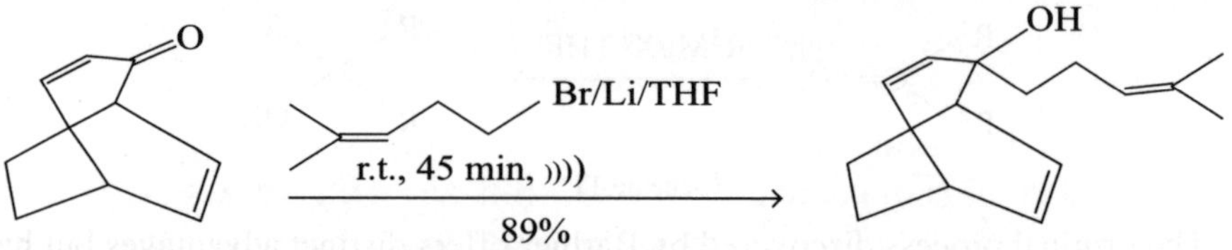
O
Br/Li/THF
r.t., 45 min,)))
89%
OH

(iii) Synthesis of pentalenic acid[5] starting from dimethylcyclopentenone.

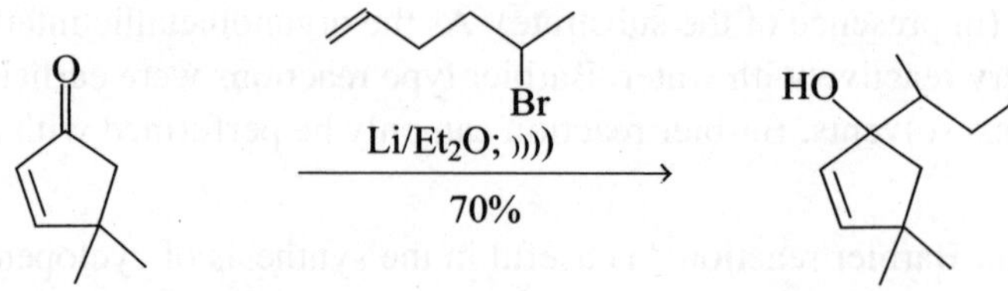
O
Br
Li/Et2O;)))
70%
HO

diethylcyclopentenone

pentalenic acid

(iv) Examples of the sonochemical Barbier reaction is described using benzylic halides[6,7]. In these reactions there is no Wurtz coupling.

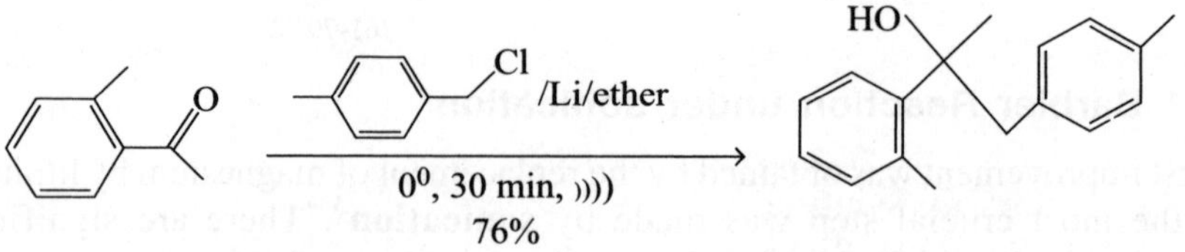
O
Cl
/Li/ether
0⁰, 30 min,)))
76%
HO

(v) An allylic phosphate can be used in place of the corresponding halide[8].

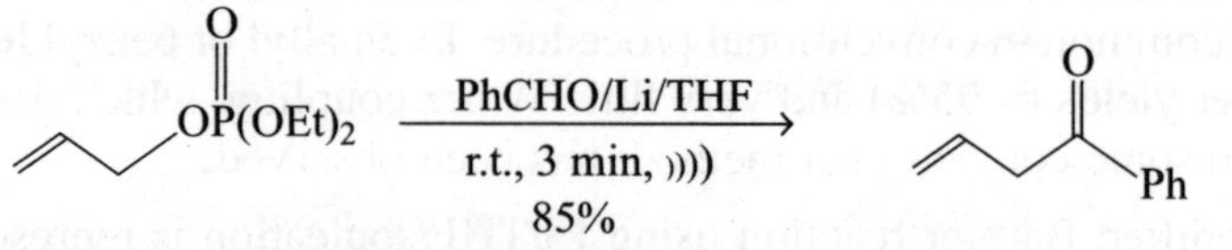
O
OP(OEt)2
PhCHO/Li/THF
r.t., 3 min,)))
85%
O
Ph

(vi) An intramolecular reaction from the following substrate is the key step in the synthesis of trichodiene[9].

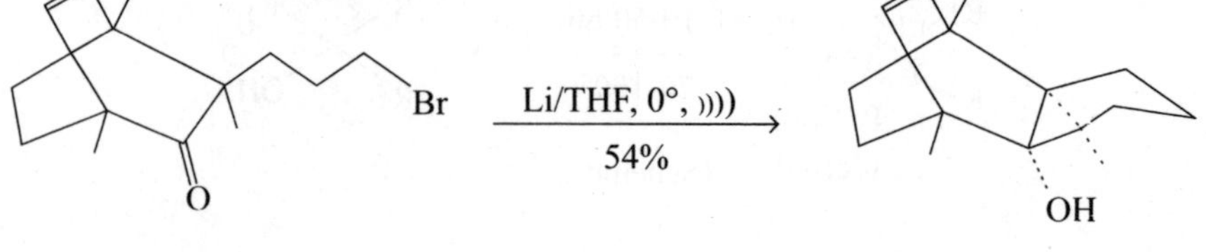
Br
O
Li/THF, 0°,)))
54%
OH

(vi) Barbier procedure can be applied to amides. The usual reaction of organometallic compounds with amides give carbonyl compounds along with formation of many by products. However, under Barbier sonochemical procedure, several improvements are there and good yields are obtained[10,11].

$$RX \xrightarrow[\substack{r.t.,\ 10\text{-}10\ min,\ ))))\\70\text{-}80\%}]{Li/DMF/THF} \left[\begin{array}{c} LiO \quad NMe_2 \\ \diagdown / \\ / \diagdown \\ R \quad X \end{array} \right] \longrightarrow RCHO$$

R = aryl, aryl, benzyl

(viii) Barbier procedure can be applied to isocyanates. Sonochemical Barbier procedure for the reaction with isocyates give much better result. However, in this case, better results are obtained[12] with sodium or magnesium. A comparisation of the result with different metals is given below.

$$t\,BuN == C == O \xrightarrow[r.\,t]{metal\ |PhBr|\ THF} PhCONHBu(t)$$

Metal	Conditions	Time (h)	% yield
Na	↘	48h	53
Li	))))	15 min.	51
Na	))))	45 min.	78
Mg	))))	15 min.	91

Besides what have been stated above, a number of other applicating of sonochemical Barbier procedure have been recorded[13].

REFERENCES

1. P. Barbier, C.R. Acad. Sci., 1898, *128*, 110.
1a. C. Blomberg, Synthesis, 1977, 18.
2. P.J. Pearce, D.H. Richards and N.F. Scilly, J. Chem. Soc. Perkin Trans. I, 1972, 1655.
3. J.-L. Luche and J.C. Damanio, J. Am. Chem. Soc., 1980, *102*, 7926.
4. T. Uychara, J. Yamada, K. Ogata and T. Kato, Bull. Chem. Soc. Japan, 1985, *58*, 211.
5. M. Ihara, M. Katogi, K. Fukumoto and T. Kametani, J. Chem. Soc., Chem. Commun., 1987, 721.
6. I.C. Burkow, L.K. Sydnes and D.C.N. Ubeda, Acta. Chem. Secand. Ser. B. 1987, *B41*, 235
7. S.B. Singh and G.R. Pettit, Syn. Commun., 1987, *17*, 877.
8. S. Araki and Y. Butsugan, Chemistry Lett. 1988, 457.
9. R.L. Snowden, P. Sonnay, J. Org. Chem. 1984, *49*, 1465.
10. C. Petrier, A.L. Gemal, J.-L. Luche, Tetrahedron Lett., 1982, *23*, 3361.

11. J. Einhorn, J.-L. Luchi, Tetrahedron Lett., 1986, *27*, 1791.
12. J. Einhorn, J.-L. Luchi, Tetrahedron Lett., 1986, *27*, 501.
13. C. Einhron, J. Einhorn and J.L. Luche, Synthesis, Review, 1989, 784.

2.6 BARTON REACTION

In Barton reaction[1], a methyl group in the δ-position to an OH group is converted into an Oxime group, which can be oxidised to a CHO group. In this procedure, the alcohol is first converted to the nitrite ester. Photolysis of the nitrite results in the conversion of the nitrite group to the OH group and nitrosation of the methyl group. Hydrolysis of the oxime tautomer gives the aldehyde. The overall reaction is shown in Scheme-1.

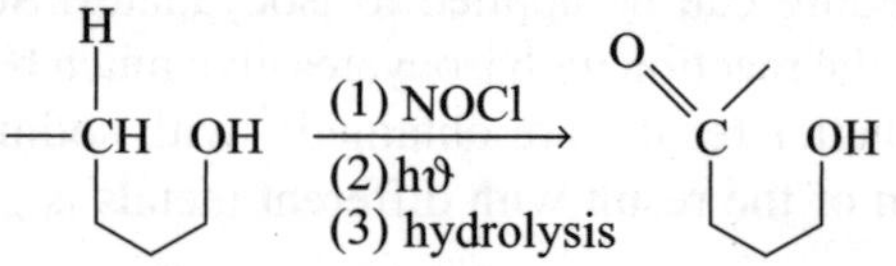

(Scheme-1)

MECHANISM

The alcohol on reaction with nitrosyl chloride (NOCl) gives the nitrite, which on photolysis undergoes homolytic cleavage to give alkoxy radical, subsequently a hydrogen atom is abstracted from a carbon atom in a δ-position to the original hydroxyl group to give nitroso alcohol, which tautomerises to the oxime. The transfer of hydrogen to alkoxy free radical takes place via a six-member transition state[2]. Finally, the oxime group can be hydrolysed to the aldehyde group. Various steps involved are shown in Scheme-2.

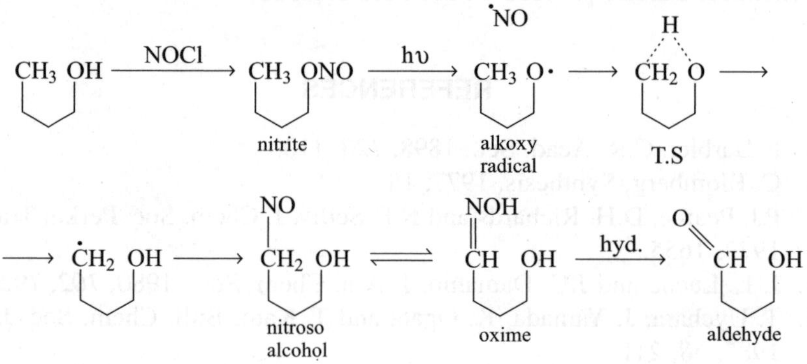

(Scheme-2)

Barton reaction provides a procedure to oxidise a carbon atom separated from an OH group by three other carbon atoms.

2.6.1 Applications

1. The most important synthetic applications of the Barton reaction has been in the steroid series, particularly in the functionalisation of the ten non-activated C-18 and C-19 angular methyl group by photolysis of the nitrites

of suitably disposed hydroxyl group. In principle (see structure below) C-18 methyl group can be attacked by an alkoxy radical at C-8, C-11, C-15 or C-20. The C-19 methyl group can be attacked by an alkoxy radical at C-2, C-4, C-6 and C-11.

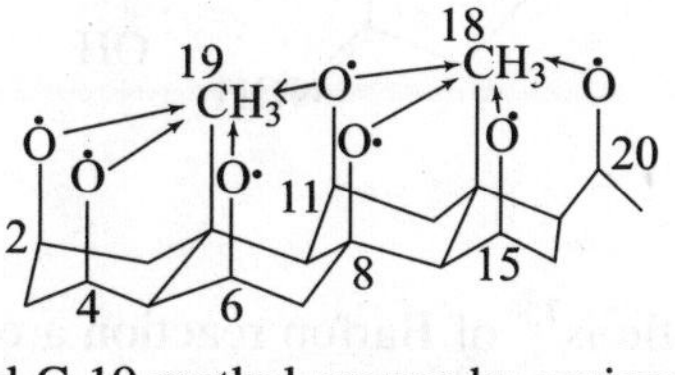

attack of C-18 and C-19 methyl groups by various alkoxy radicals.

Most of the approaches have been realised either through the Barton reaction[1,2] or by the related reaction[3-5]. The reactions are facilitated by the conformational rigidity of the steroid skeleton and by the 1,3-diaxial relationship of the interacting groups; this allows easy conformationally favoured six-membered cyclic transition states. Due to this, attack on the primary H atoms of the methyl groups is much easier than in the aliphatic series.

2. Barton *et al.* were successfull in the synthesis of aldosterone, a biologically important hormone of the adrenal cortex by photolysis of the 11 β-nitrite in toluene solution (which in turn was obtained from corticosterone acetate by reaction with NOCl). The separated oxime on hydrolysis with HNO_2 afforded aldosterone-21-acetate directly.

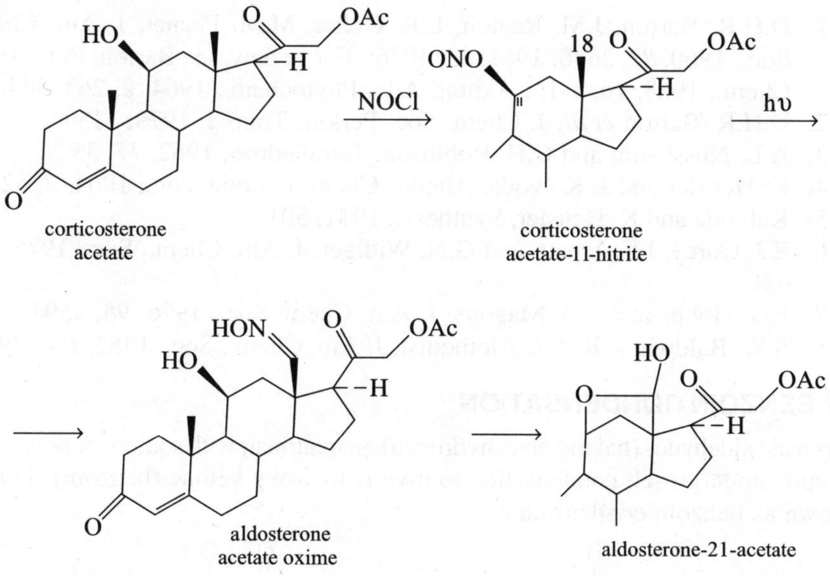

corticosterone acetate

corticosterone acetate-11-nitrite

aldosterone acetate oxime

aldosterone-21-acetate

3. Synthesis of perhydrohistrionicotin was effected[6] by photolysis of the appropriate nitrite; the formed oxime, on Beckmann rearrangement yielded the bicyclic lactam.

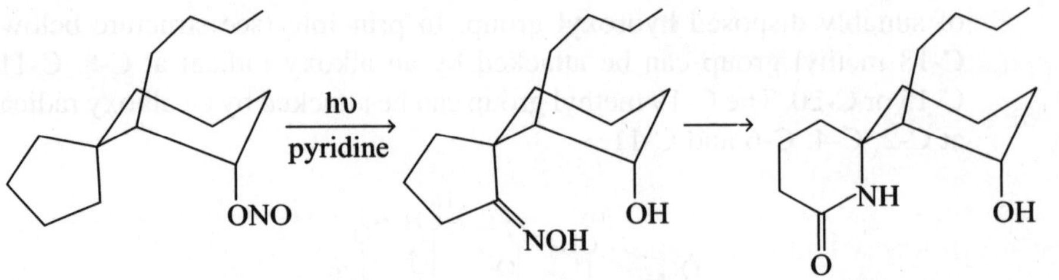

$$\xrightarrow[\text{pyridine}]{h\upsilon}$$

perhydro-histrionicotin

4. Two other applications[7,8] of Barton reaction are given below.

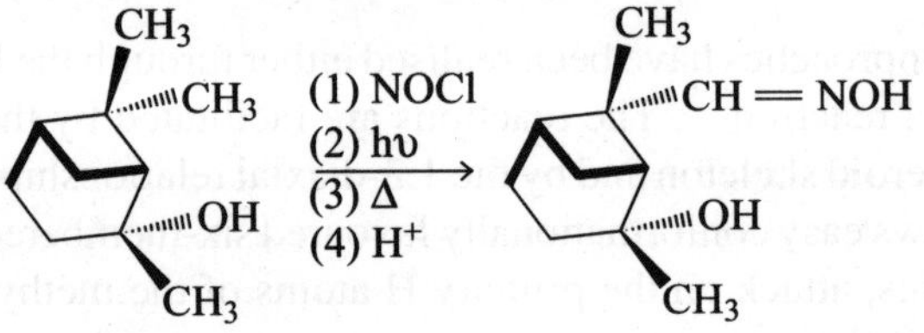

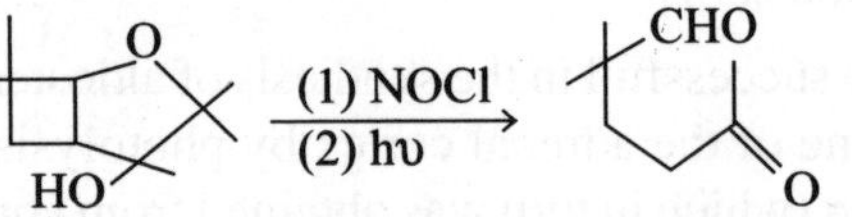

(1) NOCl
(2) hυ
(3) Δ
(4) H⁺

(1) NOCl
(2) hυ

100% yield

REFERENCES

1. D.H.R. Barton, J.M. Reaton, L.E. Geller, M.M. Pechet, J. Am. Chem. Soc., 1960, *82*, 2640; 1961, *83*, 4076; For review, see Barton, Pure Appl. Chem., 1968, *16*, 1-15; Akhtar, Adv. Phytochem., 1964, *2*, 263–304.
2. D.H.R. Barton *et al.* J. Chem. Soc. Perkin. Trans I, 1979, 1159.
3. A.L. Nussbaum and C.H. Robinson, Tetrahedron, 1962, *17*, 35.
4. K. Heusler and J. Kalvoda, Angew. Chem. Internat. edn., 1964, *3*, 525.
5. Kalvoda and K. Heusler, Synthesis, 1971, 501.
6. E.J. Corey, J.F. Arnett, and G.N. Widiger, J. Am. Chem., Soc., 1975, *97*, 430.
7. P.D. Hobbs and P.D. Magnus, J. Am. Chem. Soc., 1976, *98*, 4594.
8. S.W. Baldwin and H.R. Blomquist, J. Am. Chem., Soc., 1982, *104*, 4990.

2.7 BENZOIN CONDENSATION

Aromatic aldehydes (having no α-hydrogen) on treatment with sodium or potassium cyanide undergo selt condensation to give α-hydroxy ketone (benzoin). This is known as benzoin condensation.

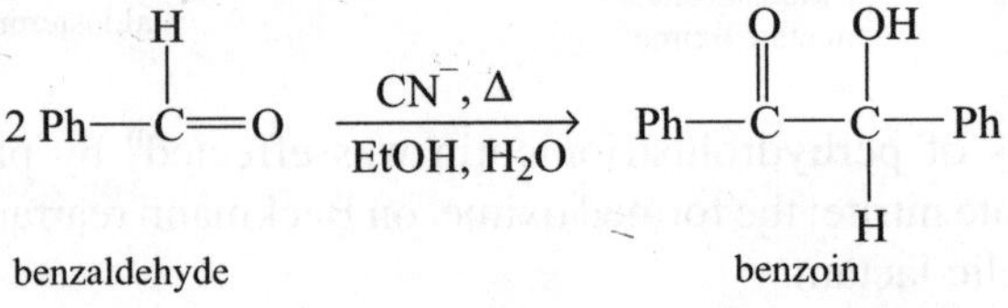

benzaldehyde

benzoin

Benzoin condensation does not take place with aliphatic aldehydes under these conditions.

MECHANISM

Benzoin condensation is believed to occur through a **knoevenagel type addition**. The cyanide ion attacks the carbonyl group of the aldehyde to give carbanion(1), which reacts with a second molecule of the aldehyde.

$$ArCHO \xrightarrow{CN^-} Ar-\overset{\overset{\displaystyle :\ddot{O}:^{\ominus}}{|}}{\underset{\underset{\displaystyle CN}{|}}{C}}-H \rightleftharpoons Ar-\overset{\overset{\displaystyle OH}{|}}{\underset{\underset{\displaystyle CN}{|}}{C}}:^{\ominus} \underset{\xleftarrow{\hspace{1cm}}}{\xrightarrow{ArCHO}}$$

$$Ar-\overset{\overset{\displaystyle OH}{|}}{\underset{\underset{\displaystyle CN}{|}}{C}}-\overset{\overset{\displaystyle :\ddot{O}:^{\ominus}}{|}}{\underset{\underset{\displaystyle H}{|}}{C}}-Ar \rightleftharpoons Ar-\overset{O}{\overset{\|}{C}}-\overset{\overset{\displaystyle OH}{|}}{\underset{\underset{\displaystyle H}{|}}{C}}-Ar$$

2.7.1 Benzoin condensation under catalytic conditions

It has been found[2] that even aliphatic aldehydes like acetaldehyde undergo benzoin condensation with solid potassium hydroxide and using 3-benzyl-4-methylthiazolium chloride as a catalyst. Thus acetaldehyde yields acetoin in quantitative yield.

$$CH_3CHO \xrightarrow[\text{KOH}]{\text{3-benzyl-4-methyl thiazolium chloride}} CH_3-\overset{\overset{\displaystyle}{|}}{\underset{\underset{\displaystyle OH}{|}}{CH}}-\overset{O}{\overset{\|}{C}}-CH_3$$

acetoin
(100%)

It is found that aromatic aldehydes reacted in a few minutes under the above conditions, whereas aliphatic aldehydes required 5–10 hr.

Benzoin condensation of aromatic aldehydes with aqueous sodium cyanide are catalysed[3] by quaternary ammonium salts.

$$ArCHO \xrightarrow[\text{NaCN}]{Q^+X^-} Ar-\overset{\overset{\displaystyle}{|}}{\underset{\underset{\displaystyle OH}{|}}{CH}}-\overset{O}{\overset{\|}{C}}-Ar$$

$$C_6H_5CHO \xrightarrow{Bu_4N^+CN^-} C_6H_5-\overset{\overset{\displaystyle}{|}}{\underset{\underset{\displaystyle OH}{|}}{CH}}-\overset{O}{\overset{\|}{C}}-C_6H_5$$

(63%)

In a similar way, acyloin condensations with aromatic or aliphatic aldehydes proceed remarkably well using N-laurylthiazolium bromide as catalyst with an aqueous phosphate solution[4].

$$R\ CHO \xrightarrow[\text{aqueous phosphate}]{\text{N-laurylthiazolium bromide}} \underset{\underset{16\text{--}95\%}{\overset{OH\ \ \ \ O}{|\ \ \ \ \ |}}}{R\ CH\text{---}C\text{---}R}$$

It was found[5] by Breslow that the benzoin condensation in aqueous media using inorganic salts (e.g. LiCl) is about 200 times faster than in ethanol without any salt. The benzoin condensation was also accelerated by the addition of γ-cyclodextrin, whereas addition of β-cyclodextrin inhibited the condensation.

Unsymmetrical or mixed benzoins may often be obtained in good yield from two different aldehydes

$$R\ CHO + R'\ CHO \longrightarrow R\ CO\ \overset{OH}{\underset{|}{C}}H\ R' + R\ \overset{OH}{\underset{|}{C}}H\ CO\ R'$$

A typical example is

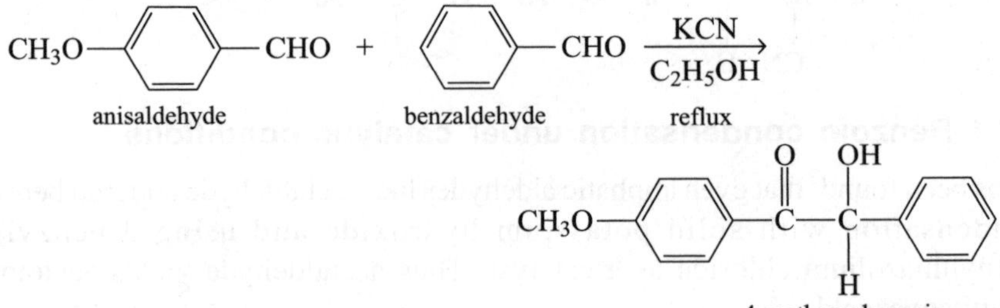

anisaldehyde   benzaldehyde   reflux

4-methoxy benzoin

The above mixed benzoin condensation is catalysed by thiazolium salts[6].

Benzoin condensation can also be brought about by coenzyme 'Thiamine' (R. Breslow, J. Am. Chem. Soc., 1958, *80*, 3719).

For mechanism see Acyloin condensation (Section 2.1.1)

Benzoin condensations of aldehydes are strongly catalysed by quaternary ammonium cyanide in a two phase system (J. Solodav, Tetrahedron, Lett., 1971, 287).

2.7.2 Applications

Anisaldehyde and p-tolualdehyde give the corresponding α-hydroxyketones[7,8]. Similarly furtural gives furoin.

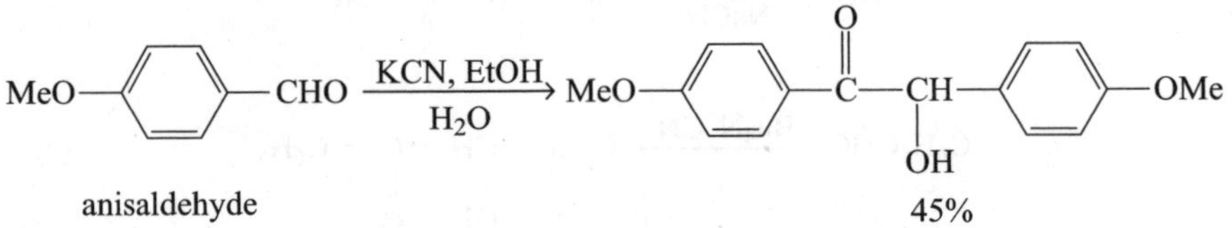

anisaldehyde   45%

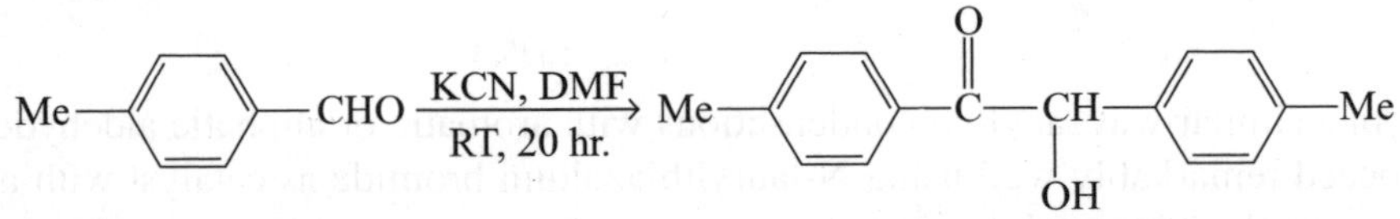

p-tolualdehyde   90%

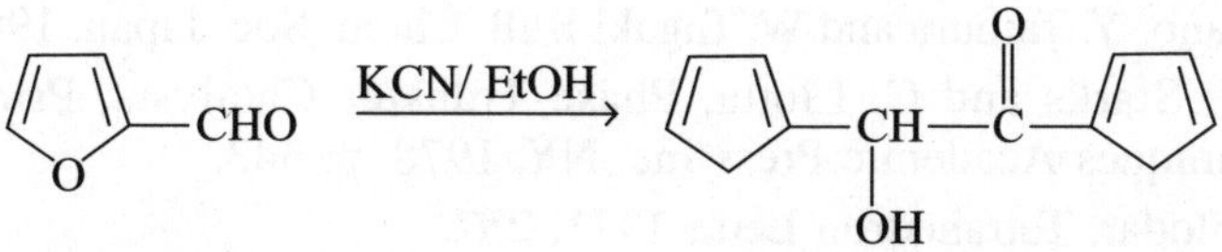

furfural furoin

Another application is benzoin condensation of p-formyl styrene gives[9] the corresponding product.

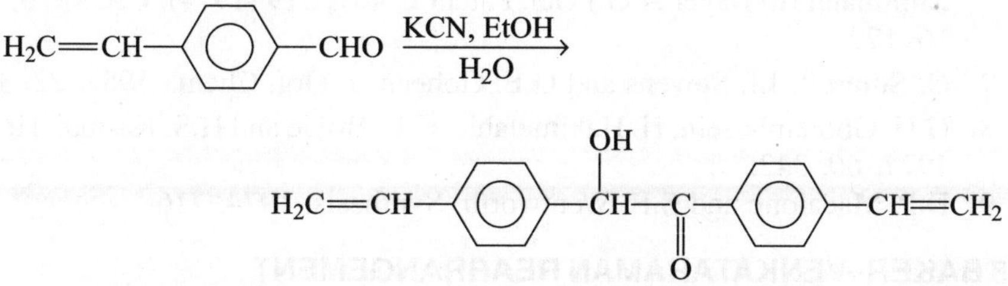

(63 %)

Benzoins are useful intermediates for the synthesis of other compounds, since they can be oxidized to α-diketones and reduced under different conditions to give various products. These reactions are summarised below:

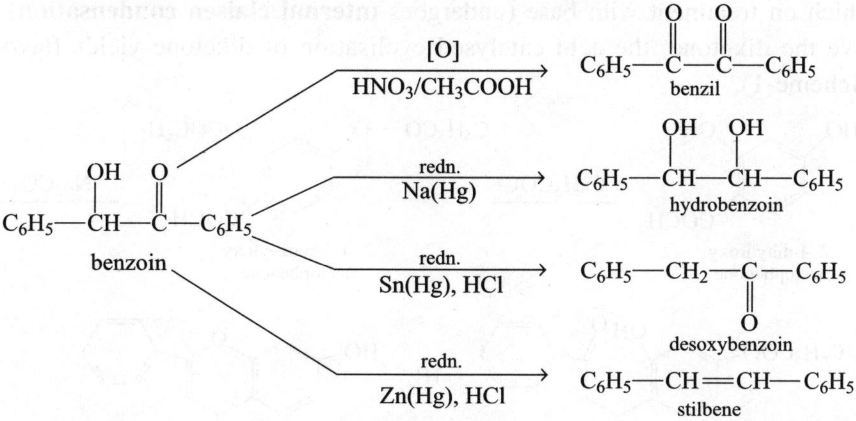

The oxidation product, viz. benzil (a α-diketone) obtained by nitric acid oxidation of benzoin, undergoes an interesting base catalysed rearrangement to the α-hydroxy acid, benzilic acid. (See sect. 2.10.2)

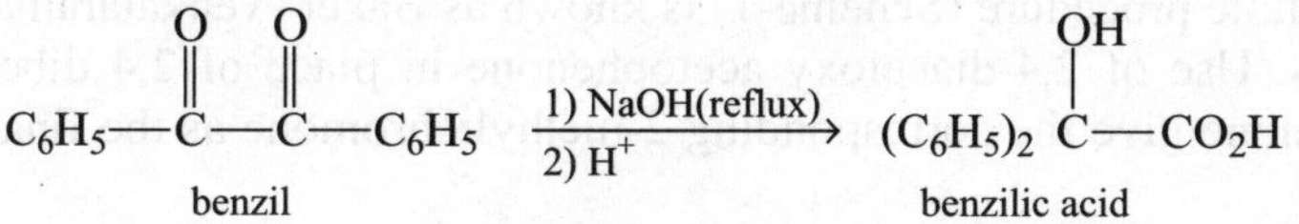

benzil benzilic acid

REFERENCES

1. A.J. Lapworth, J. Chem. Soc., 1903, *83*, 955; 1904, *85*, 1206; Bulk, Organic Reaction Vol. (IV) 269 (1948).

1a. Y. Yano, Y. Tamura and W. Tagaki Bull. Chem. Soc. Japan, 1980, *53*, 740.
2. C.M. Starks and C. Liotta, Phase Transfer Catalysts, Principles and Techniques Academic Press Inc. NY, 1978, p. 343.
3. J. Solodar, Tetrahedron Lett., 1971, 287.
4. W. Tagaki and H. Hara, J. Chem. Soc., Chem. Commun, 1973, 891.
5. E.T. Kool and R. Breslow, J. Am. Chem. Soc., 1988, *110*, 1596.
6. H. Stetter and G. Dambkes, Synthesis, 1977, 403; H. Sletter and H. Kuhlmann (to Bayer A.G.) Ger. Patent 2, 437, 319 (1974); CA, 1976, *84*, 164172.
7. G. Sumrell, J.I. Stevens and G.E. Goheen, J. Org. Chem., 1957, *22*, 39.
8. G.H. Gholamhosein, H. Hakimelahic, C.B. Boyle and H.S. Kasmai, Helv., 1977, *60*, 342.
9. D.P. Macaione and S.E. Wentworth, Synthesis, 1974, 716.

2.8 BAKER–VENKATARAMAN REARRANGEMENT

The base catalysed rearrangement of o-acyloxy (preferably o-benzoyloxy) ketones to β-diketones (which are important intermediates in the synthesis of flavones or chromones) is known as Baker–Venkataraman rearrangement. Thus, 2,4-dihydroxyacetophenone on benzoylation give 2,4-dibenzoyloxyacetophenone, which on treatment with base (undergoes **internal claisen condensation**) to give the diketone; the acid catalysed cyclisation of diketone yields flavone (Scheme-1).

2, 4-dihydroxy
acetophenone

2, 4-dibenzoyloxy
acetophenone

7-hydroxyflavone

(Scheme-1)

The whole procedure (Scheme-1) is known as Baker–Venkataraman Flavone Synthesis. Use of 2,4-diacetoxy acetophenone in place of 2,4 dibenzoyloxy-acetophenone give the corresponding 2-methyl chromone as the final product.

2.8.1 PTC catalysed synthesis of Flavones

A convenient one step process has been developed[2] for the synthesis of β-diketones. In this procedure benzoylchloride is added to a stirred mixture of 2-hydroxyacetophenone, tetrabutylammonium hydrogen sulphate and aqueous potassium carbonate or potassium hydroxide. The mixture is stirred for 2 hr until the starting ketone and the initially formed o-benzoyloxyacetophenone disappears (TLC). Working up gives the required diketone. The diketone can be cyclised with p-toluene sulphonic acid.

MECHANISM

It is believed that the base abstracts a proton from $COCH_3$ group to give a carbanion, which undergoes cyclisation followed by ring opening to give β-diketones (Scheme-2)

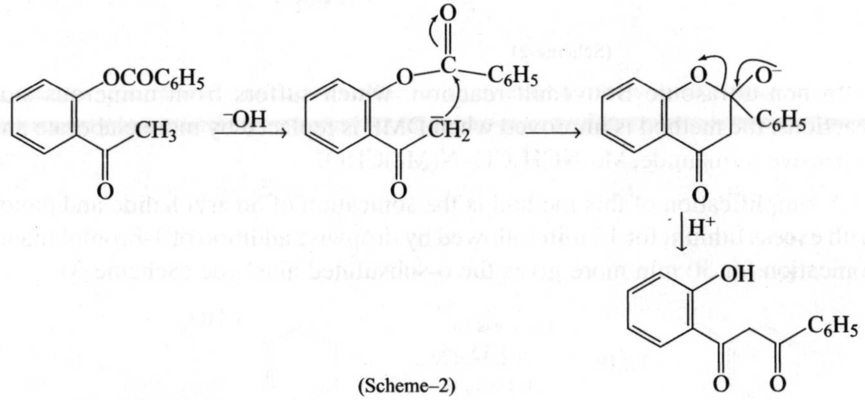

(Scheme–2)

2.8.2 Application

Baker–Venkataraman procedure is used for the synthesis of flavones and chromones.

REFERENCES

1. W. Baker, J. Chem. Soc., 1933, 1381; H.S. Mahal and K. Venkataraman, J. Chem. Soc., 1934, 1767; K. Venkataraman in Zechmeister Progress in the Chemistry of Natural Products, 1959, *17*, 2; E. Levmi, Chem., Rev., 1954, *54*, 493.
2. V.K. Ahluwalia *et al.* unpublished results.

2.9 BOUVEAULT REACTION

The action of Grignard reagents on N,N-disubstituted formamides yield aldehyde. This is known as Bouveault reaction or **Bouveault aldehyde synthesis** (Scheme-1).

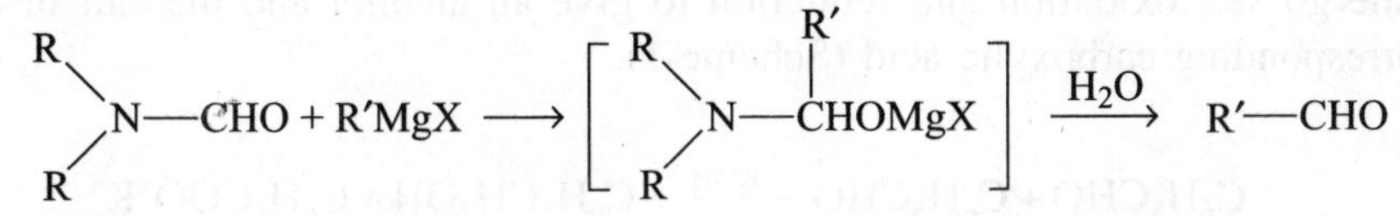

| N,N-disubstituted formamides | grignard reagent | | aldehydeshift |

(Scheme-1)

2.9.1 Bouveault Reactions under Sonication

In place of grignard reagents, organolithium reagents give better yields[2]. The organolithum reagents are obtained by the sonication of aryl halides with lithium using low intensity ultrasonic. These reagents are used in Bouveault reaction and give higher yields than the traditional methods[3] (Scheme-2).

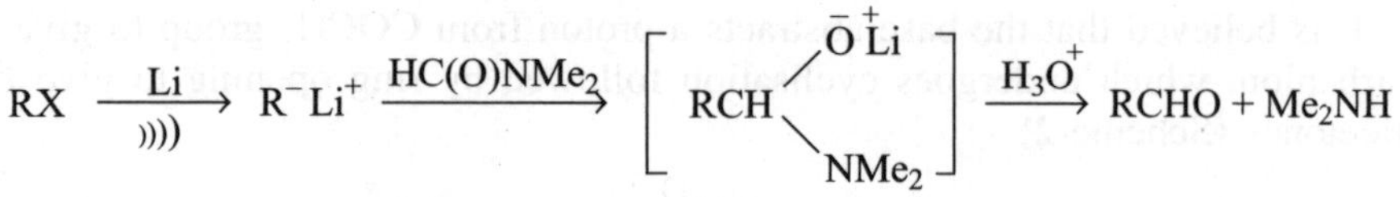

$$RX \xrightarrow[))))]{Li} R^- Li^+ \xrightarrow{HC(O)NMe_2} \left[RCH \begin{smallmatrix} \bar{O} \overset{+}{Li} \\ \diagup \\ \diagdown \\ NMe_2 \end{smallmatrix} \right] \xrightarrow{H_3O^+} RCHO + Me_2NH$$

(Scheme-2)

In non-ultrasonic Bouveault reaction, which suffers from numerous side reactions, the method is improved when DMF is replaced by more elaborate and expensive formamide, $Me_2NCH_2CH_2N(Me)CHO$.

A simplification of this method is the sonication of an aryl halide and amide with excess lithium for 15 min followed by dropwise addition of 1-bromobutane, sonication for 30 min more gives the o-substituted aldehyde (Scheme-3).

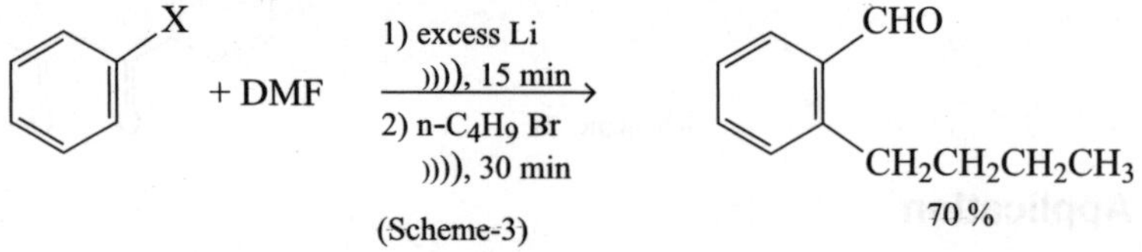

(Scheme-3)

Use of iodomethane in place of n-butyl bromide in the above reaction gives o-tolualdehyde[4].

REFERENCES

1. L. Bouveault, Bull. Soc., Chem. France, 1904, *31*, 1306, 1322.
   N. Smith, J. Org. Chem., 1941, 6, 489; Sice, J. Am. Chem. Soc., 1953, *75*, 3697; Jones *et al*, J. Chem. Soc., 1958, 1054.
2. Evans, Chem. and Ind. (London), 1957, 1956.
3. C. Petrier, A.L. Gemal and J.L. Luche, Tetrahedron Lett., 1982, *23*, 3361.
4. J.L. Luche, ultrasonics, 1987, *25*, 40.

2.10 CANNIZZARO REACTION

Aldehydes without α-hydrogen(s) on treatment with concentrated aqueous alkali undergo self oxidation and reduction to give an alcohol and the salt of the corresponding carboxylic acid (Scheme-1).

$$C_6H_5CHO + C_6H_5CHO \xrightarrow{KOH} C_6H_5CH_2OH + C_6H_5COO^-K^+$$

benzaldehyde   benzyl alcohol   pot. benzoate

(Scheme-1)

This disproportionation or self oxidation and reduction of aromatic aldehydes, devoid of α-hydrogen is known as cannizzaro reaction[1].

Cannizzaro reaction best proceeds with aromatic aldehydes without α-hydrogen, some aliphatic aldehydes like formaldehyde and dimethyl acetaldehyde (which do not have α-hydrogen also undergo cannizzaro reaction (Scheme-2).

$$HCHO + NaOH \xrightarrow{\Delta} CH_3OH + HCOONa$$

formaldehyde   methyl alcohol   sod. formate

$$2(CH_3)_2CHCHO + NaOH \xrightarrow{\Delta} (CH_3)_2CHCH_2OH + (CH_3)_2CHCOONa$$

dimethyl acetaldehyde   2-methyl-1-   sod. 2-methyl-1-
                                     propanol   propionate

(Scheme-2)

MECHANISM

Cannizzaro reaction involves transfer of a hydrogen atom from one molecule of the aldehyde to another. This has been established by using deuterated benzaldehyde instead of benzaldehyde.

$$2C_6H_5CDO + OH^- \xrightarrow{H_2O} C_6H_5CD_2OH + C_6H_5\overset{O}{\overset{\|}{C}}-O^-$$

It is found that the benzyl alcohol formed is exclusively deuterated. So the possibility of exchange of hydrogen with hydrogen atoms in the solvent is ruled out. Thus, a transfer of hydrogen (or deuterium) takes place between the carbonyl atoms of the two aldehyde molecules.

Cannizzaro reaction proceeds by formation of an anion (by reaction with base) which may transfer a hydride ion intermolecularly to the carbonyl of another aldehyde molecule forming the carboxylic acid and the alkoxide ion. Final step is the shifting of a proton from the acid to the alcohol (Scheme-3).

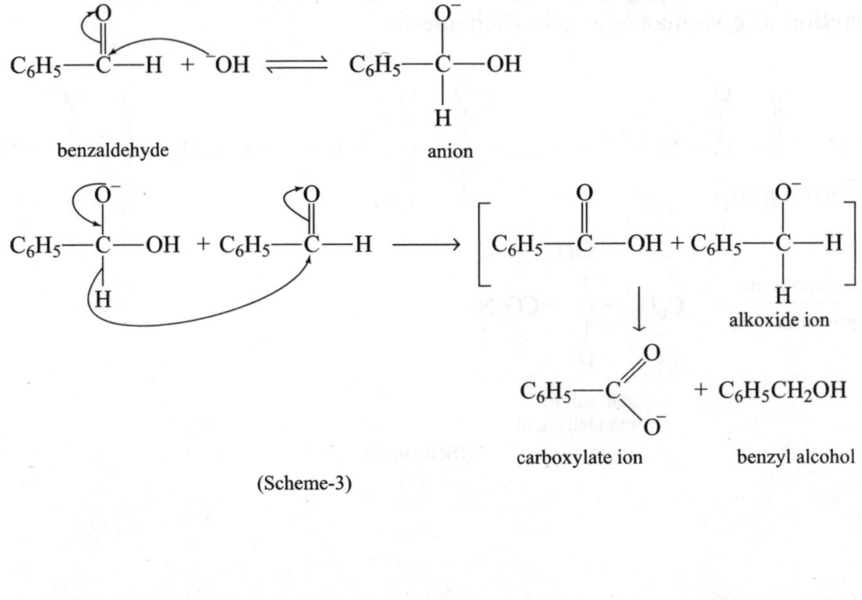

$$C_6H_5-\overset{O}{\overset{\|}{C}}-H + {}^-OH \rightleftharpoons C_6H_5-\overset{H}{\underset{|}{\overset{|}{C}}}-OH$$

benzaldehyde   anion

carboxylate ion   benzyl alcohol

(Scheme-3)

If the cannizzaro reaction is performed in D_2O, the alcohol formed has no deuterium establishing that the mechanism involves a direct transfer of hydrogen from one molecule of aldehyde to another as already depicted in Scheme-3.

2.10.1 Crossed Cannizzaro Reaction

The cannizzaro reaction between two different aldehydes may yield four different products (two carboxylic acids and two alcohols). Such a reaction is called crossed cannizzaro reaction and has no synthetic value. However, if one of the aldehyde is formaldehyde, the formate ion and the alcohol corresponding to the other aldehyde are exclusively formed (Scheme-4).

$$R\text{–}CHO + CH_2O \xrightarrow{\ ^-OH\ } R\,CH_2OH + HCOO^-$$

(Scheme-4)

For other examples of crossed cannizzaro reaction see applications.

2.10.2 Intramolecular Cannizzaro Reaction

Certain compounds, which contain two carbonyl groups undergo **internal cannizzaro reaction**. For example, glyoxal on treatment with base gives glycolic acid (Scheme-5).

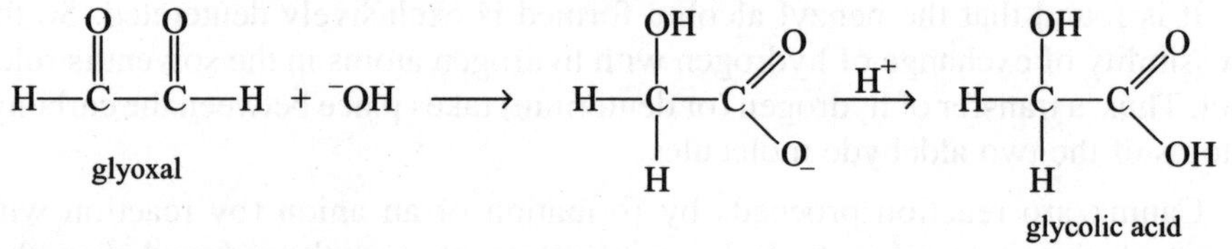

(Scheme-5)

In a similar way, phenyl glyoxal undergoes **intramolecular cannizzaro reaction** to give mandelic acid (Scheme-6).

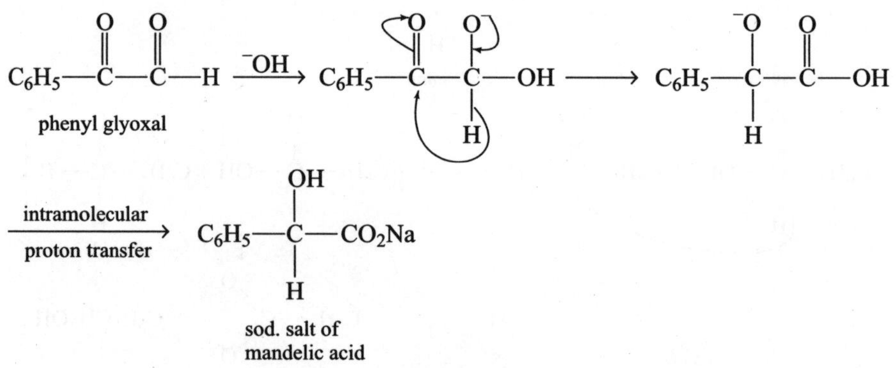

(Scheme-6)

An analogous reaction occurs with benzil, which results in carbon skeleton rearrangement and is known as **benzil-benzilic acid rearrangement** (Scheme-7).

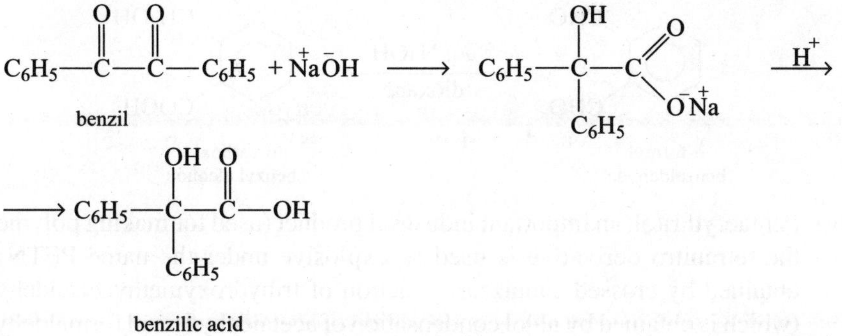

benzil

benzilic acid

(Scheme-7)

2.10.3 Cannizzaro Reactions Under Sonication

The cannizzaro reaction under heterogenous conditions catalysed by barium hydroxide is considerably accelerated (Scheme-8) by sonication. The yields are 100% after 10 min, whereas no reaction is observed during this period without ultrasound[2].

$$C_6H_5CHO \xrightarrow[\text{)))), 10 min}]{\text{Ba(OH)}_2 \text{ EtOH}} C_6H_5CH_2OH + C_6H_5COOH$$

(Scheme-8)

2.10.4 Applications

(i) o-Methoxybenzyl alcohol is prepared in 79% yield by cannizzaro reaction[3] of o-methoxy benzaldehyde. o-Methoxy benzyl alcohol is exclusively obtained by crossed cannizzaro reaction of o-melhoxy benzaldehyde with formaldehyde.

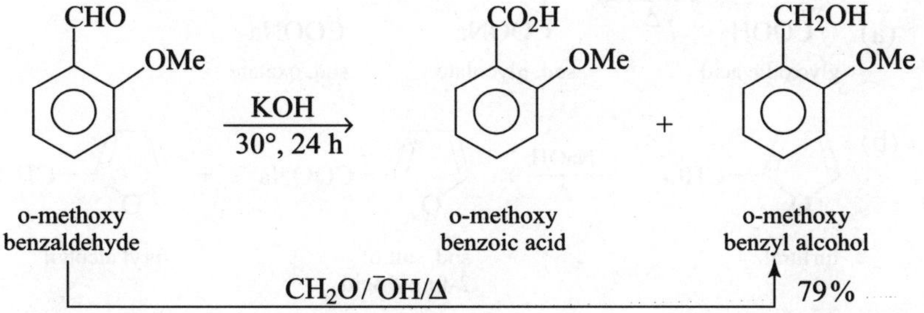

o-methoxy benzaldehyde

o-methoxy benzoic acid

o-methoxy benzyl alcohol

(ii) Benzyl alcohol is obtained by the crossed cannizzaro reaction of benzaldehyde and formaldehyde.

$$C_6H_5CHO + CH_2O \xrightarrow{\text{30\% NaOH}} C_6H_5CH_2OH + HCO_2H$$

benzaldehyde

benzyl alcohol

formic acid

(iii) Internal cannizzaro reaction of o-formyl benzaldehyde gives o-carboxybenzyl alcohol[6].

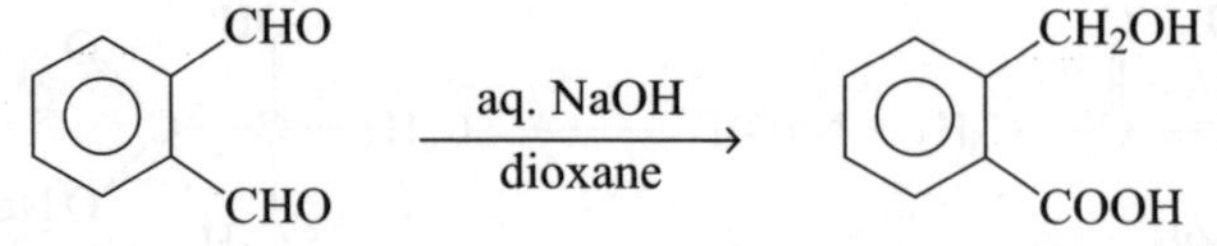

$$\text{o-formyl benzaldehyde (CHO, CHO)} \xrightarrow[\text{dioxane}]{\text{aq. NaOH}} \text{o-carboxy benzyl alcohol (CH}_2\text{OH, COOH)}$$

o-formyl
benzaldehyde

o-carboxy
benzyl alcohol

(iv) Pentaerythritol, an important industrial product (used for making polymers; the tetranitro derivative is used as explosive under the name PETN) is obtained by crossed cannizzaro reaction of trihydroxymethyacetaldehyde (which is obtained by aldol condensation of acetaldehyde and formaldehyde in presence of calcium hydroxide) with formaldehyde.

$$CH_3CHO + 3CH_2O \xrightarrow{\ ^-OH\ } HOCH_2-\underset{\underset{CH_2OH}{|}}{\overset{\overset{CH_2OH}{|}}{C}}-CHO \longrightarrow$$

acetaldehyde

trihydroxymethyl
acetaldehyde

$$\xrightarrow[\text{CH}_2\text{O}/\ ^-\text{OH}]{\text{crossed cannizzaro reaction}} HOCH_2-\underset{\underset{CH_2OH}{|}}{\overset{\overset{CH_2OH}{|}}{C}}-CH_2OH$$

pentaerythritol

(v) Using cannizzaro reaction, a number of carboxylic acids and alcohols have been synthesised, some of these are given below:

(a)
$$2\ \underset{\text{COOH}}{\overset{\text{CHO}}{|}} \xrightarrow[\Delta]{\text{NaOH}} \underset{\text{COONa}}{\overset{\text{CH}_2\text{OH}}{|}} + \underset{\text{COONa}}{\overset{\text{COONa}}{|}}$$

glyoxalic acid

sod. glycolate

sod. oxalate

(b)
furfural $\xrightarrow[\Delta]{\text{NaOH}}$ sod. salt of 2-furoic acid (−COONa) + furyl alcohol (−CH$_2$OH)

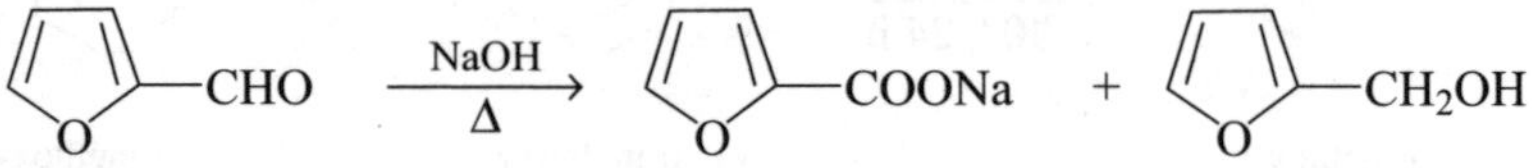

furfural

sod. salt of
2-furoic acid

furyl alcohol

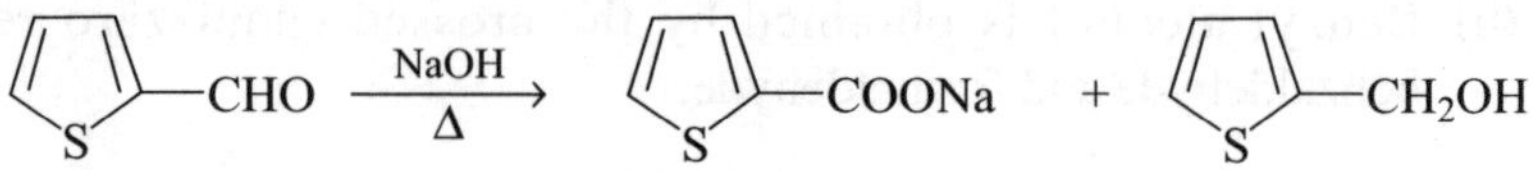

(c)
thiophen-2-aldehyde $\xrightarrow[\Delta]{\text{NaOH}}$ sod. salt of thiophen-2-carboxylic acid (−COONa) + thiophene-2-carbinol (−CH$_2$OH)

thiophen-2-
aldehyde

sod. salt of
thiophen-2-carboxylic
acid

thiophene-2-carbinol

REFERENCES

1. S. Cannizzaro, Ann., 1853, *88*, 129; K. List and H. Limpricht, Ann., 1854, *90*, 180.
2. A. Fuentes and V.S. Sinistera, Tetrahedron Lett., 1986, *27*, 2967.
3. R.A. Bruce, Org. Prep. Proced Int., 1987, 19.
4. T.A. Geissmann, Org. Reactions Vol. II, 1944, 94.
5. T.A. Geissmann, Org. React, 1944, *2*, 95.
6. R.S. McDonald and C.E. Sibley, Can. Jour., Chem., 1981, *59*, 1061.

2.11 CLAISEN REARRANGEMENT

Allyl phenyl ether on heating to 200°C undergoes an intramolecular reaction called claisen rearrangement. Claisen rearrangement is the earliest record of an organic reaction in solid state. The product is o-allylphenol (the allyl group migrates to the ortho position).

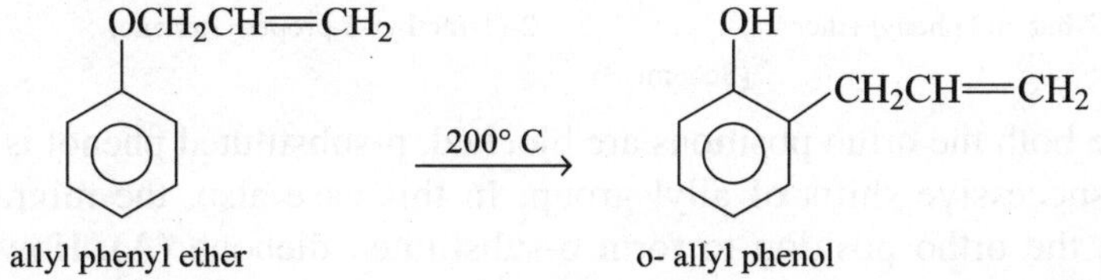

allyl phenyl ether   o- allyl phenol

The reaction does not need any catalyst and is an example of **pericyclic reaction**. Other examples of pericyclic reactions are **cope rearrangement** and **Diels-Alder reaction** (Section 2.17).

MECHANISM[1c,2]

Claisen rearrangement is a [3, 3]-sigmatropic rearrangement and proceeds in a concerted manner, in which the bond between C_3 of the allyl group and the ortho position of the benzene ring form and at the same time the carbon-oxygen bond of the allyl phenyl ether breaks (Scheme-1).

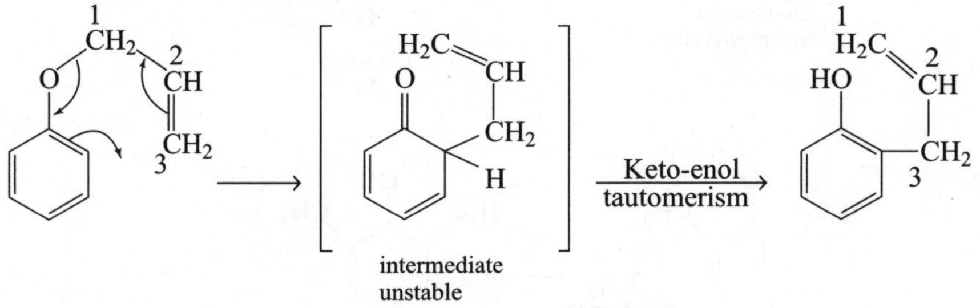

intermediate
unstable

(Scheme-1)

That only C-3 of the allyl group becomes bonded to the benzene ring has been demonstrated by carrying out the rearrangement with allyl phenyl ether containing ^{14}C at C3. Whole of the product obtained in this reaction has the labelled carbon atom bonded to the ring (Scheme-2).

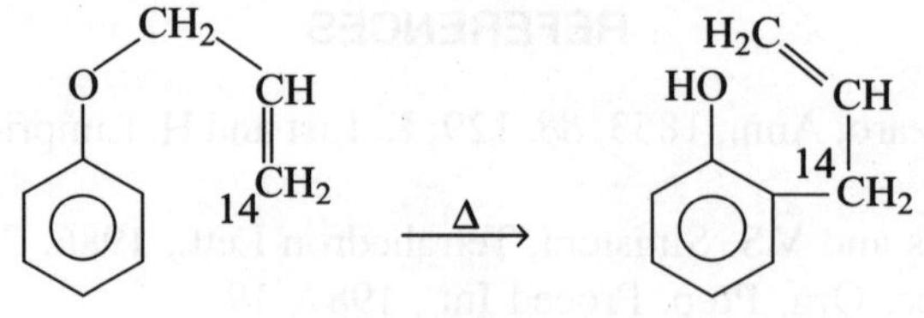

only product

(Scheme-2)

Further evidence for the above mechanism (Scheme-2) is that if 2-butenyl group is present in a ether, the phenol obtained has a methyl branch on the side chain (Scheme-3).

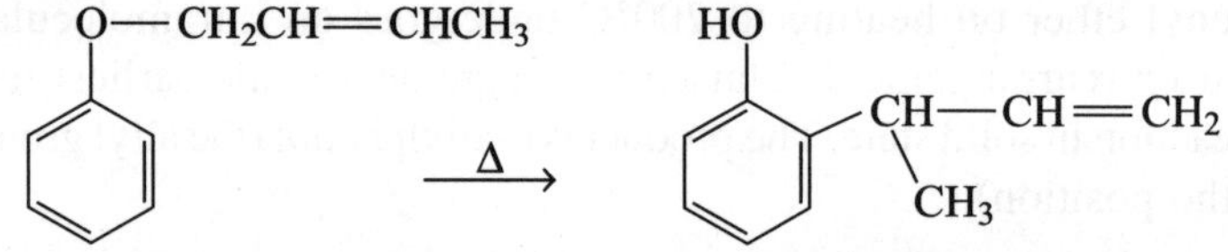

O—CH₂CH=CHCH₃

$\xrightarrow{\Delta}$

2-butenyl phenyl ether

2-(1-methyl-2-propenyl)phenol

(Scheme-3)

In case both the ortho positions are blocked, p-substituted phenol is obtained via two successive shifts of allyl group. In this case also, the migration still occurs at the ortho position to form o-substituted dienone (A). However, the absence of hydrogen at the ortho position prevents enolization (or aromatisation). So the allyl group undergoes a second migration through a similar cyclic transition state to form a dienone (B), which aromatizes to a phenol with allyl group at position 4 (Scheme-4).

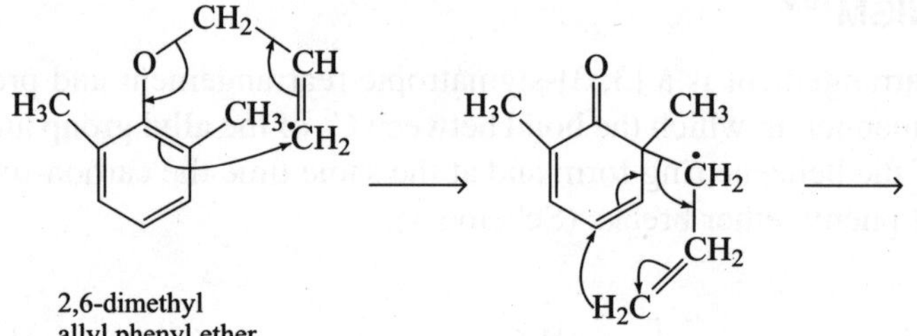

2,6-dimethyl
allyl phenyl ether

Dienone
(A)

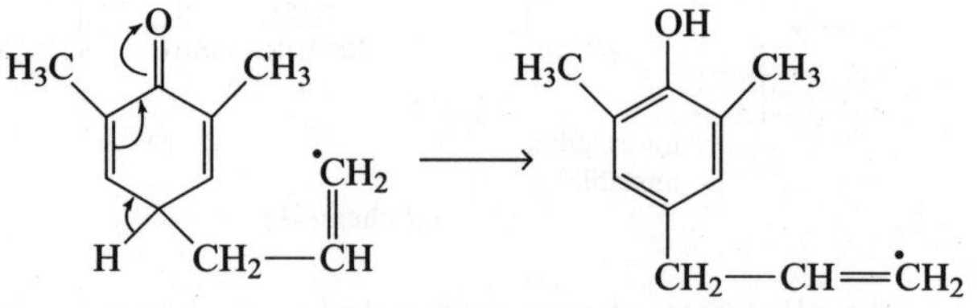

Dienone
(B)

4-allyl-2,6-dimethyl
phenol

(Scheme-4)   [C˙ = ¹⁴C lablled carbon)

The driving force for the para migration (when both the ortho positions are occupied) is regaining of the aromatic character after allyl migration. Migration to meta position has not been observed.

The classic claisen rearrangement involves aromatic allyl ether. The thermal claisen rearrangement of allyl vinyl ethers to γ, δ-unsaturated carbonyl compounds was first described[1a] by Claisen in 1912 (Scheme-5). This is referred to as **aliphatic claisen rearrangement**.

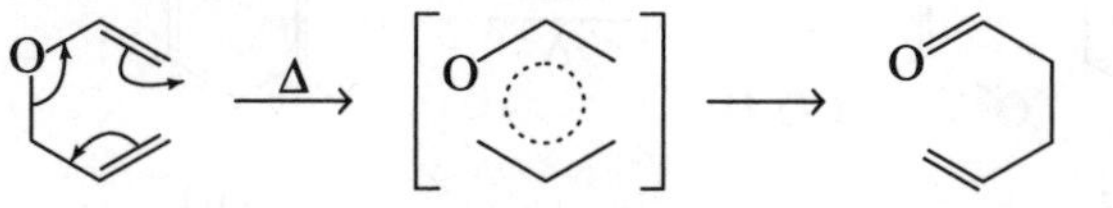

| allyl vinyl ether | transition state (aromatic) | γ, δ-unsaturated carbonyl compounds (4-pentenal) |

(Scheme-5)

The transition state for the above claisen rearrangement (Scheme-5) involves a cycle of six orbitals and six electrons, suggesting that the transition state has aromatic character.

In the case of aliphatic substrate (substituated allyl vinyl ethers), asymmetry at the terminal methylene carbons is transformed into asymmetry at the two new saturated centers in a sense that suggests[3] a chair-like transition state (Scheme-6). In the case of the aromatic claisen rearrangement, the initially formed dienone undergoes tautomerisation to the phenol, so the stereochemistry observable is the transfer of chirality from an optically active starting material with a specific double bond geometry. This indicates that the aromatic claisen rearrangement also proceeds with stereochemistry which is in accordance with a chair-like transition state[4] (Scheme 7).

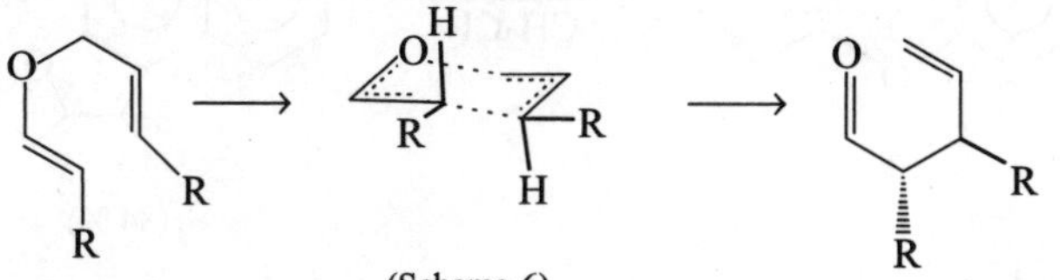

(Scheme-6)

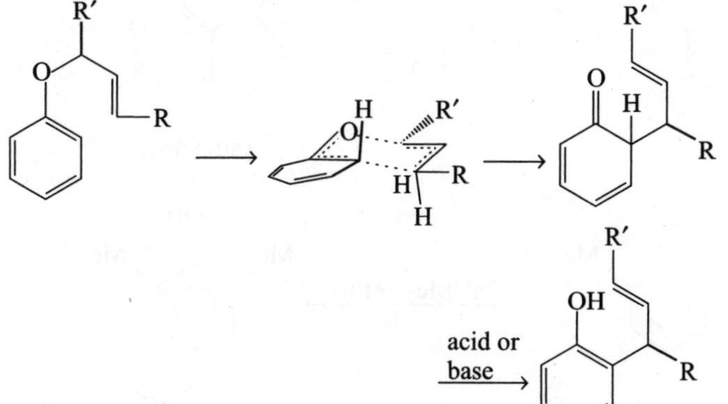

acid or base

(Scheme-7)

2.11.1 Claisen Rearrangement in Water

The use of water in promoting claisen rearrangement was first recorded[5] in 1970. The first example using pure water for claisen rearrangement of chorismic acid[8] is given in Scheme-8.

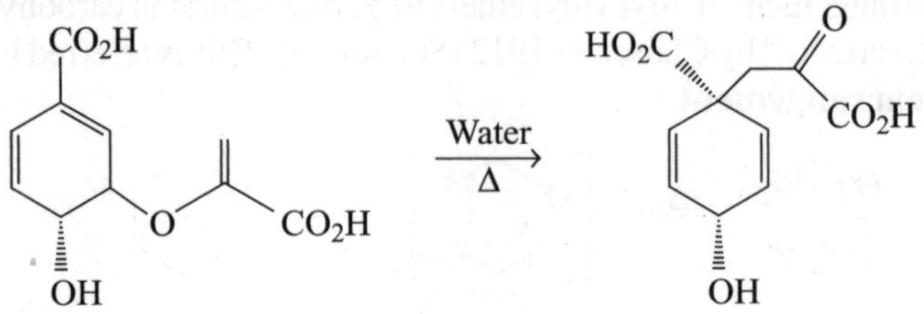

chorismic acid

(Scheme-8)

Claisen rearrangement of allyl-vinyl ether in water gave[6] the aldehyde (4-pentenal, see Scheme-5).

2.11.2 Applications (Classical Claisen Condensation)

(i)

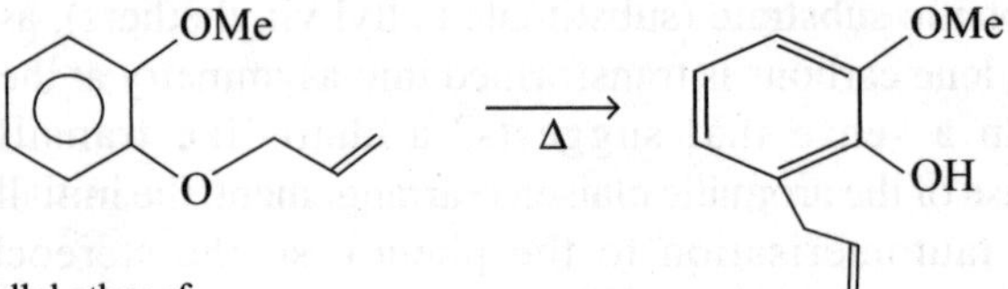

O-allyl ether of
catechol monomethyl ether

(84%)

(Ref 7)

(ii)

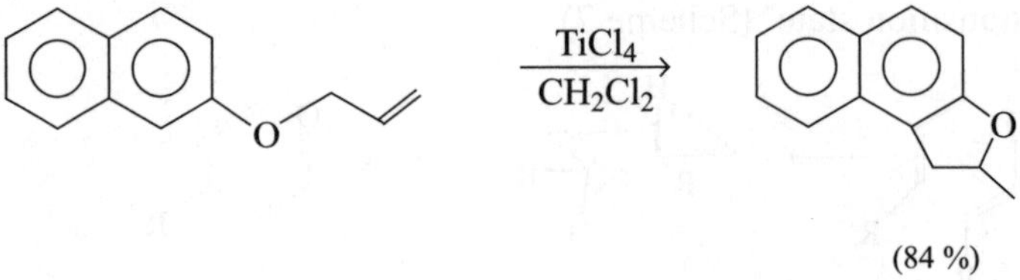

$\xrightarrow{\text{TiCl}_4}{\text{CH}_2\text{Cl}_2}$

(84 %)

(Ref 2)

(iii)

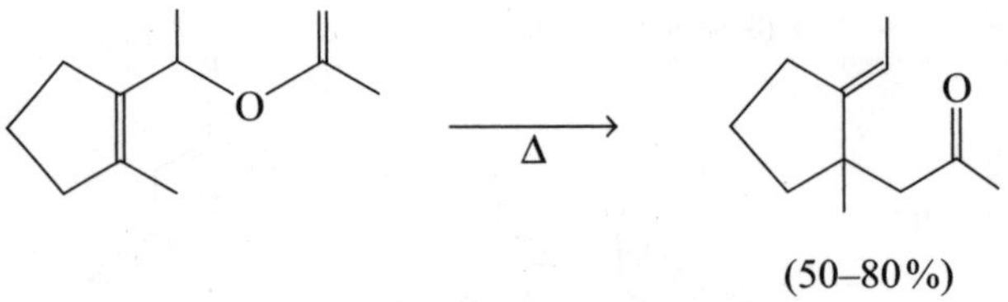

(50–80%)

(Ref 8)

(iv) Me

$\xrightarrow[\text{3h}]{\text{PhNMe}_2\text{/reflux}}$

(67 %)

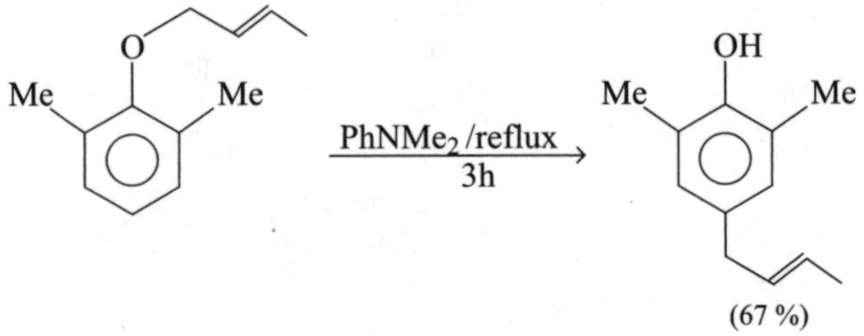

(Ref 8a)

2.11.3 Applications (aqueous phase claisen rearrangement)

(i) A simple aliphatic claisen rearrangement of an allyl vinyl ether gave the aldehyde

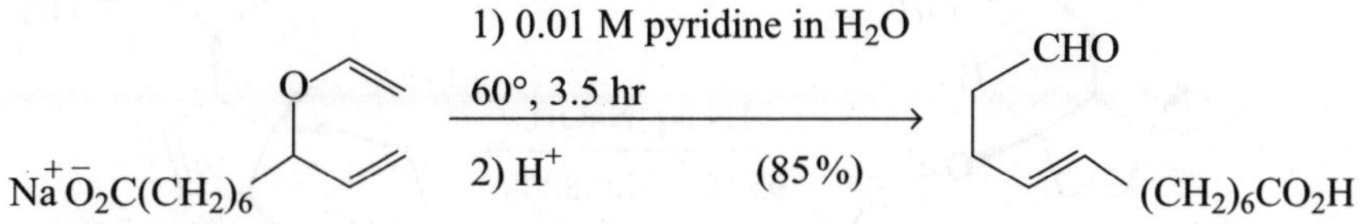

$Na^+ \bar{O}_2C(CH_2)_6$
O
1) 0.01 M pyridine in H_2O
60°, 3.5 hr
2) H^+
(85%)
CHO
$(CH_2)_6CO_2H$

The corresponding ester also undergoes similar rearrangement. Carrying out the above reaction in methanol-water (2.1) increased the yield by about 40 times than those in acetone solvent[10].

(ii) Claisen rearrangement of a vinyl ether substrate results in the formation of an aldehyde; the reaction was performed by heating 2.5 : 1 water-methanol solution at 80° for 24 hr[11]

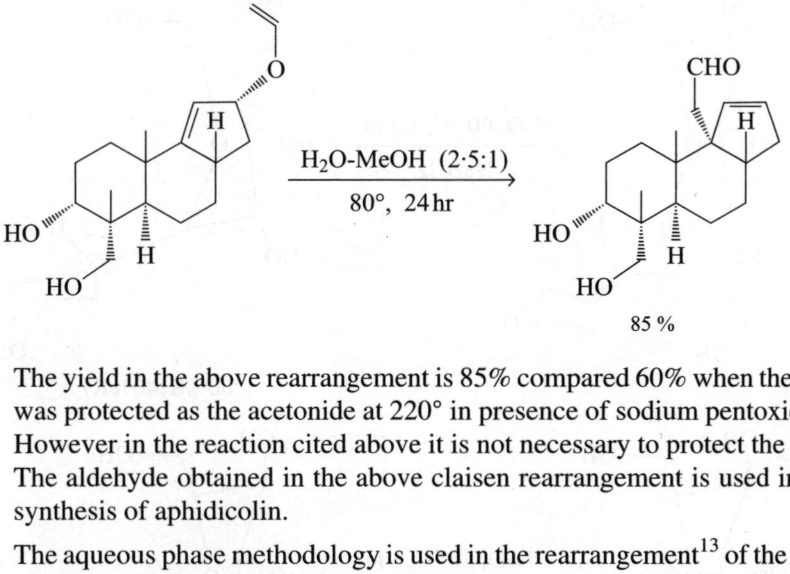

O
H
HO
H
HO
H_2O-MeOH (2·5:1)
80°, 24hr
CHO
H
HO
H
HO
85 %

The yield in the above rearrangement is 85% compared 60% when the diol was protected as the acetonide at 220° in presence of sodium pentoxide[12]. However in the reaction cited above it is not necessary to protect the diol. The aldehyde obtained in the above claisen rearrangement is used in the synthesis of aphidicolin.

(iii) The aqueous phase methodology is used in the rearrangement[13] of the allyl vinyl ether substrate to fenestrene aldehyde having trans ring fusion between the five membered ring.

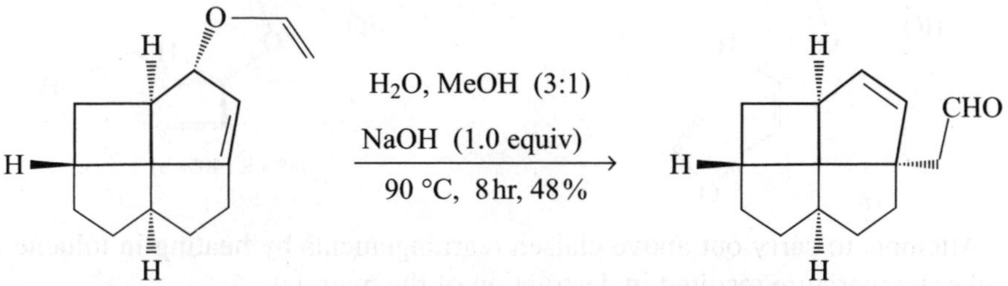

O
H
H
H
H_2O, MeOH (3:1)
NaOH (1.0 equiv)
90 °C, 8hr, 48%
H
CHO
H
H

(iv) Substrates which are particularly reluctant to undergo claisen rearrangement can be induced to undergo rearrangement in aqueous base[13] as shown below.

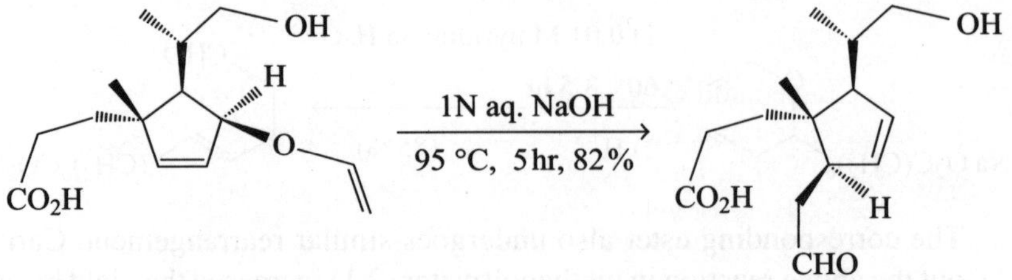

$$\xrightarrow[\text{95 °C, 5 hr, 82\%}]{\text{1N aq. NaOH}}$$

(v) Claisen rearrangement[14] of 6-β-glycosyallyl vinyl ether and 6-α-glucosylallyl vinyl ether in water (80°, 1 hr) was successful. In both the reactions NaBH$_4$ was added so that the formed aldehyde is converted into the corresponding diol.

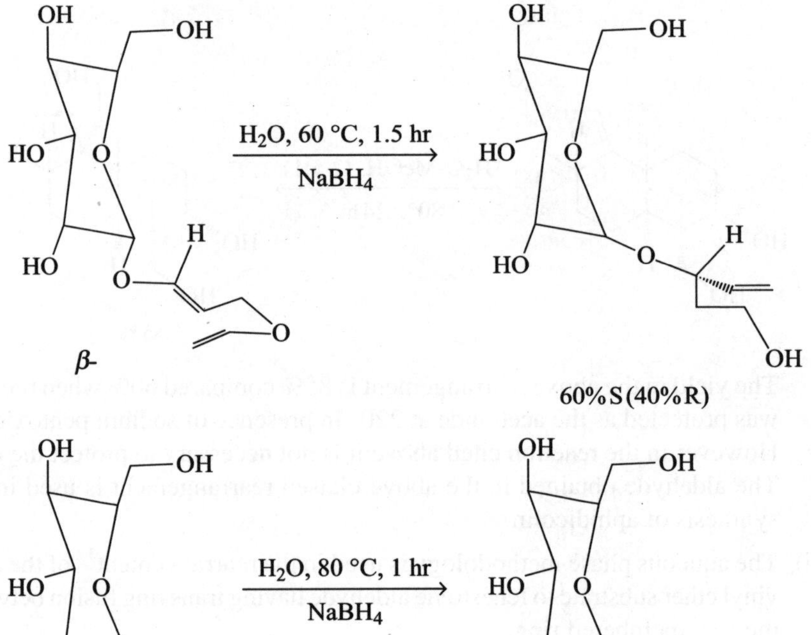

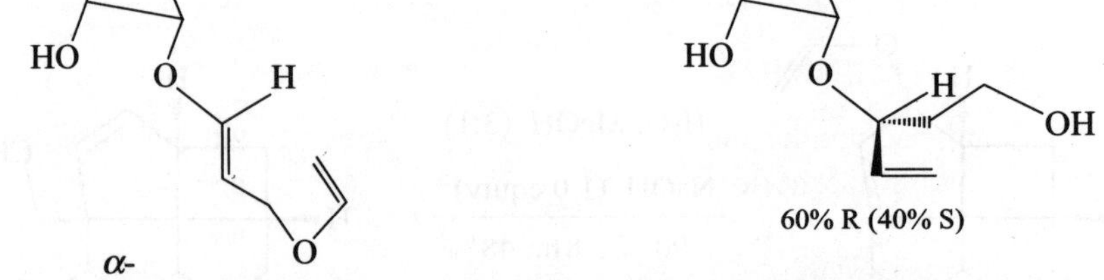

$$\xrightarrow[\text{NaBH}_4]{\text{H}_2\text{O, 60 °C, 1.5 hr}}$$

β-

60% S (40% R)

$$\xrightarrow[\text{NaBH}_4]{\text{H}_2\text{O, 80 °C, 1 hr}}$$

α-

60% R (40% S)

Attempts to carry out above claisen rearrangements by heating in toluene to higher temperature resulted in destruction of the material.

Claisen rearrangement is an important tool available to synthetic organic chemist and extensive reviews are available[15].

REFERENCES

1. (a) L. Claisen, Ber., 1912, *45*, 3157;
    (b) L. Claisen and O. Eisleb, Annalen 1913, *401*, 21;
    (c) L. Claisen and E. Tietze, Ber., 1925, *58*, 275;
    (d) D.S. Tarbell, Org. React., 1944, *2*, 1.
2. M.R. Saidi, Heterocycles, 1982, *19*, 1473.
3. P. Vitorelli, T. Winkler, H.J. Hansen and H. Schmidt, Helv. Chim. Acta, 1968, *51*, 1457;
    R.K. Hill, A.G. Edwards, Tetrahedron Lett., 1964, 3239.
4. E.N. Marvel and J.L. Stephenson, J. Org., Chem., 1960, *25*, 676; H. Hart, J. Am. Chem. Soc., 1954, *76*, 4033.
5. W.N. White and E.F. Wolfartt, J. Org. Chem., 1970, *35*, 2196.
6. E. Brandes, P.A. Grieco and J.J. Gajewski, J. Org. Chem., 1989, *54*, 515.
7. E.J. Corey, R.L. Danheiser, S. Chandrasekaran, P. Siret, G.E. Keck and J.-L., Gras, J. Am. Chem. Soc., 1978, *100*, 8031.
8. J.W.S. Stevenson and T.A. Bryson, Tetrahedron Lett., 1982, 3143.
8a. S.J. Rhoades, J. Am. Chem. Soc., 1955, *73*, 5060.
9. E. Brandes, P.A. Grieco and J.J. Gajewski, J. Org., Chem., 1989, *54*, 515.
10. J.J. Gajewski, J. Jurayj, D.K. Kimbrough, M.E. Gande, B. Ganem and B.K. Carpenter, J. Am. Chem. Soc., 1987, *109*, 1170.
11. P.A. Grieco, E.B. Brandes, S. McCann and J.D. Clark, J. Org. Chem., 1989, *54*, 5849.
12. Mc Murry, A. Andrus, G.M. Ksander, J.J. Musser and M.A. Johnson, Tetrahedron, 1981, *37* (Suppl. No. 1), 319.
13. P.A. Grieco, E.B. Brandes, S. McCann and J.D. Clark, J. Org. Chem., 1989, *54*, 5849.
14. A. Lubineau, J. Auge, N. Bellanger and S. Caillebourdin, J. Chem. Soc. Perkin Trans. I, 1992, *13*, 1631.
15. P. Wipf, comprehensive organic synthesis, B.M. Trost, I. Fleming, L.A. Paquette Eds., Pergamon Press, New York, 1991, Vol. 5, p. 827; F.E. Zieglar Chem. Rev., 1988, *88*, 1423; S.J. Rhoads and N.R. Raulins, Org. React., 1975, *22*, 1. Claisen Rearrangements in aqueous medium in organic synthesis in water. Edited by Paul A. Grieco, Blackie Academic and Professional, London, 1998, pp. 47–101 and the reference cited therein.

2.12 CLAISEN–SCHMIDT REACTION

The base catalysed condensation of an aromatic aldehyde (without α-hydrogen) with an aliphatic aldehyde or a ketone (having α-hydrogen) to form α, β-unsaturated aldehydes or ketones is known as Claisen–Schmidt reaction or Claisen–Schmidt condensation. It is a type of **crossed aldol condensation** (see aldol condensation). For example, benzaldehyde reacts with acetaldehyde to give cinnamaldehyde (Scheme-1).

$$C_6H_5CHO + CH_3CHO \xrightarrow{\text{NaOH}} C_6H_5CH=CHCHO$$

benzaldehyde   cinnamaldehyde

(Scheme-1)

MECHANISM

The mechanism is similar to **aldol condensation**. Thus, as a first step the base ($^-$OH) removes a proton from the α-carbon of one molecule of ketone to give a resonance stabilized enolate ion, which acts as a nucleophile (as a carbanion) and attacks the carbonyl carbon of the molecule of the aldehyde producing an alkoxide anion, which removes a proton from a molecule of water. Finally dehydration occurs readily because the double bond that forms is conjugated both with the carbonyl group and with the benzene ring; the conjugated system is thereby extended. Various steps of the mechanism are shown below (Scheme-2).

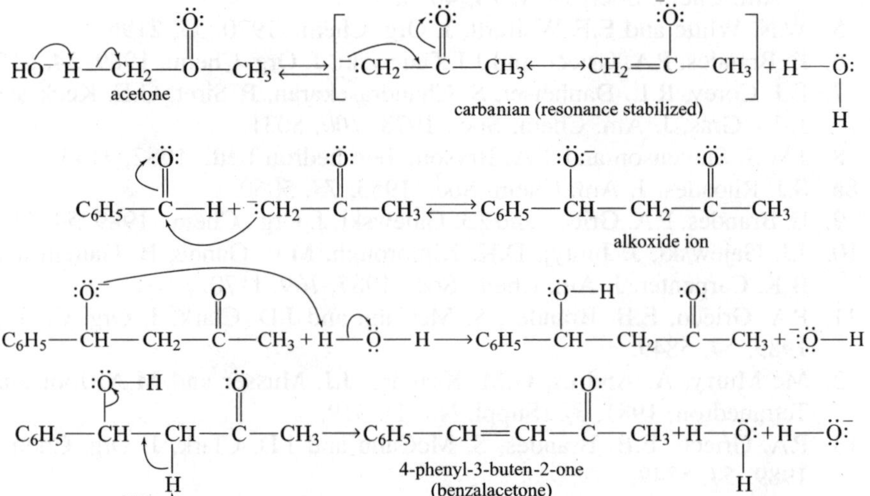

(Scheme-2)

2.12.1 CLAISEN SCHMIDT REACTION IN AQUEOUS PHASE

The claisen-schmidt reaction between a ketone and an aldehyde can be conveniently carried out[2] by using silyl enol ether of the ketone in an organic solvent in presence of TiCl$_4$ (see **Mukaiyama reaction,** sec 2.22). These acidic conditions, are not suitable for acid-sensitive substrates. In such cases, the reaction is carried out under high pressure (in place of catalyst, TiCl$_4$) but the reaction takes long times[3]. It has now been shown[4] that the claisen condensation of trimethylsilyl enol ether of cyclohexanone with benzaldehyde can be carried out in water in heterogenous phase at room temperature and atmospheric pressure; no catalyst is required. No reaction takes place in organic solvents (toluene, tetrahydrofuran, dichloromethane, acetonitrile etc.) at room temperature and atmospheric pressure in the absence of catalyst. Table below gives the result of the claisen condensation of silyl enol ether of cyclohexanone with benzaldehyde under different conditions.

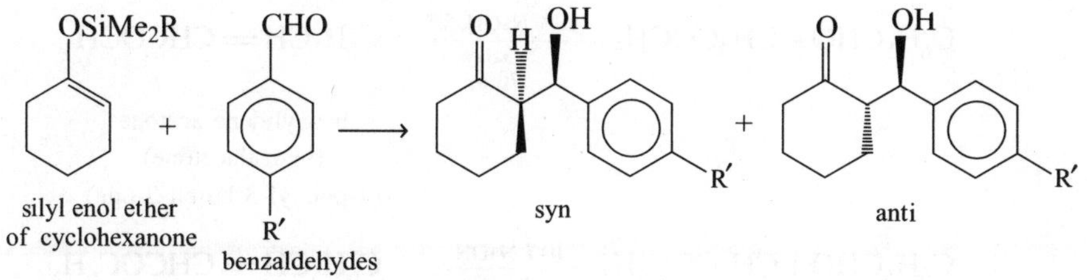
OSiMe2R
CHO
silyl enol ether of cyclohexanone
R′
benzaldehydes
O
H
OH
R′
syn
O
OH
R′
anti

Table

R	*R′*	*Solvent*	*Temp. (°C)*	*Time (b)*	*Reaction conditions*	*Yield %*	*Syn/Ant*
Me	H	CH_2Cl_2	rt.	360	stirring	0	—
Me	H	CH_2Cl_2	20	2	$TiCl_4$	82	1:3
Me	H	CH_2Cl	60	216	10 Kbar	90	3:1
Me	H	H_2O	20	120	stirring	23	7:1
Me	H	H_2O–THF	55	24	stirring	76	2.8:1
t-Bu	H	H_2O–THF	100	16	stirring	84	1.3:1
Me	NO_2	H_2O–THF	55	36	sonication	82	2.3:1
Me	OMe	H_2O–THF	55	36	sonication	29	2.3:1

Phase transfer catalysts such cetyltrimethylammonium compounds (like CTACl, CTABr, $(CTA)_2\ SO_4$ and CTAOH) have been successfully used for claisen-smith reaction of acetophenones with benzaldehydes (Scheme-3) under weakly alkaline conditions. This permits the synthesis of biologically interesting compounds such as chalcones and flavanols in water only[5, 6].

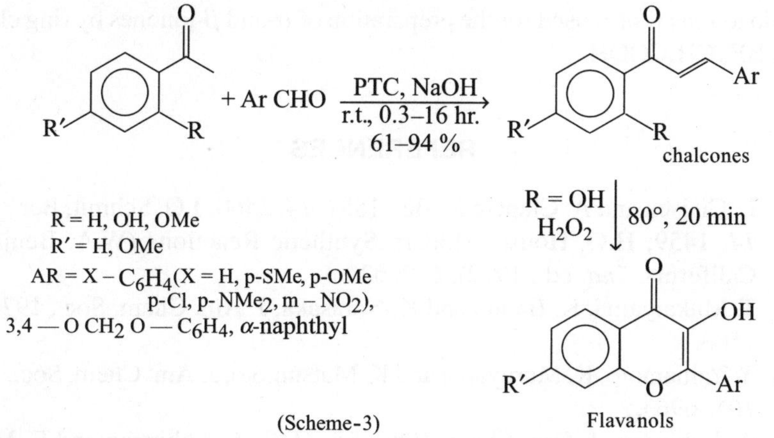
O
R′
R
+ Ar CHO
PTC, NaOH
r.t., 0.3–16 hr.
61–94 %
O
Ar
R′
R
chalcones
R = OH
H2O2
80°, 20 min
R = H, OH, OMe
R′ = H, OMe
AR = X – C6H4(X = H, p-SMe, p-OMe
p-Cl, p- NMe2, m – NO2),
3,4 — O CH2 O — C6H4, α-naphthyl
O
OH
R′
O
Ar
Flavanols

(Scheme-3)

2.12.2 Applications

(i) The α, β-unsaturated carbonyl compounds, (e.g., cinnamaldehyde and benzylidene acetone are useful in perfumary) are commercial products.

$$C_6H_5CHO + CH_3CHO \xrightarrow[8\text{–}10\ \text{days}]{10\%\ \text{NaOH, rt}} C_6H_5CH{=}CHCHO$$

cinnamaldehyde

$$C_6H_5CHO + CH_3COCH_3 \xrightarrow[\substack{2-3\,days \\ (70\%)}]{10\%\ NaOH,\ 30°} C_6H_5CH = CHCOCH_3$$

benzylidene acetone
(benzalacetone)
(4-phenyl-3-buten-2-one)

$$C_6H_5CHO + CH_3COC_6H_5 \xrightarrow[< 30°,85\%]{10\%\ NaOH} C_6H_5CH = CHCOC_6H_5$$

benzylidene acetophenone
(benzalacetophenone)

$$C_6H_5CHO + CH_3COOC_2H_5 \xrightarrow[30°]{10\%\ NaOH} C_6H_5CH = CHCOOC_2H_5$$

ethyl cinnamate

In the above synthesis, the concentration of alkali is crucial, otherwise **cannizzaro reaction** may give the major product(s).

(ii) β-ionone, a key intermediate in the synthesis of vitamin A is obtained by clainsen-schmidt reaction of citral or geranial (a naturally occurring aldehyde obtained from lemon grass oil) and acetone

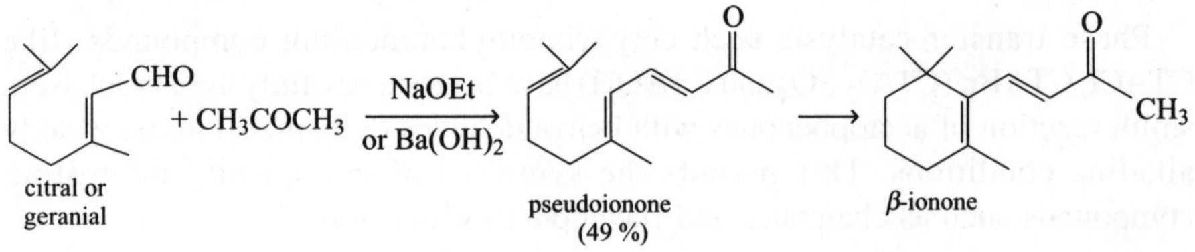

citral or geranial

+ CH₃COCH₃   NaOEt or Ba(OH)₂

pseudoionone (49 %)

β-ionone

pseudo ionone is also used for the preparation of α-and β-ionones by ring closure using BF_3/CH_3COOH.

REFERENCES

1. L. Claisen and A. Claperede, Ber. 1881, *14*, 2460; J.G. Schmit, Ber., 1881, *14*, 1459; H.O. House, Modern Synthetic Reactions (W.A. Benjamin, California, *2nd*. ed., 1972), 632–639.

2. T. Mukaiyama, K. Banno and K. Narasaka, J. Am. Chem. Soc., 1974, *96*, 7503.

3. Y. Yamamoto, K. Maruyama and K. Matsumoto, J. Am. Chem. Soc., 1983, *105*, 6963.

4. A. Lubinean, J. Org. Chem. 1986, 51, 2142; A. Lubineau and E. Meyer. Tetrahedron, 1988, *44*, 6065.

5. K.R. Nivalkar, C.D. Mudaliar and S.H. Mashraqui, J. Chem. Res. (s), 1992, 98.

6. F. Fringulli, G. Pani, O. Piermatti, and F. Pizzo, Tetrahedron, 1994, *50*, 11499.   F. Fringulli, G. Pani, O. Piermatti and F. Pizzo, Life Chem. Rep., 1955, *13*, 133.

2.13 CLEMMENSEN REDUCTION

The carbonyl group of aldehydes and ketones on reduction with zinc amalgam and hydrochloric acid give the corresponding hydrocarbons (i.e., the carbonyl group is converted into CH_2 group) (Scheme-1) is known as clemmensen reduction[1].

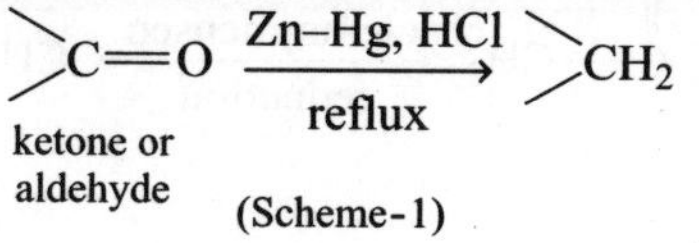

$$\underset{\substack{\text{ketone or}\\ \text{aldehyde}}}{>\!\!C=O} \xrightarrow[\text{reflux}]{\text{Zn–Hg, HCl}} >\!\!CH_2$$

(Scheme-1)

MECHANISM

The mechanism of clemmensen reduction is not well understood. It is, however, clear that in most cases, the alcohol is not an intermediate, since clemmensen reduction do not reduce most alcohols to hydrocarbons. Following mechanism is suggested. It shows that reduction under acidic conditions normally involves protonated species to which the metal offers electron (Scheme-2).

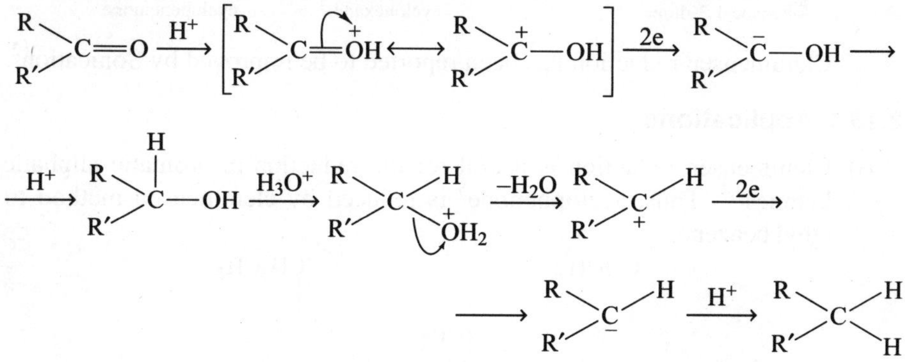

(Scheme-2)

LIMITATIONS

(i) When the substrate is unstable under acidic condition (but stable under alkaline condition), clemmensen reduction cannot be used. In such cases the **Huang-Minlon modification** of the **Wolff-Kishner reduction** can be very conveniently used. Certain compounds like thioacetals can be reduced only in neutral solution; the later procedure is used for compounds that are sensitive to both acids and bases.

(ii) Clemmensens reduction is subject to steric effects. For example, hindered ketone (one example shown below) does not undergo reduction.

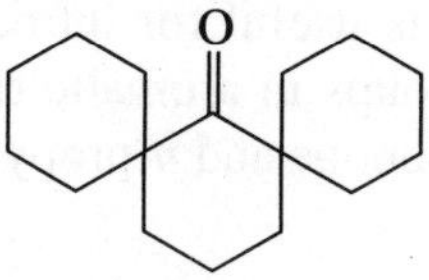

(iii) Certain aldehydes and ketones do not give the normal reduction products only. Thus, α-hydroxy ketones give either ketones through hydrogenolysis of OH group or olefins, and 1,3-diketones give exclusively monoketones accompanied by rearrangement.

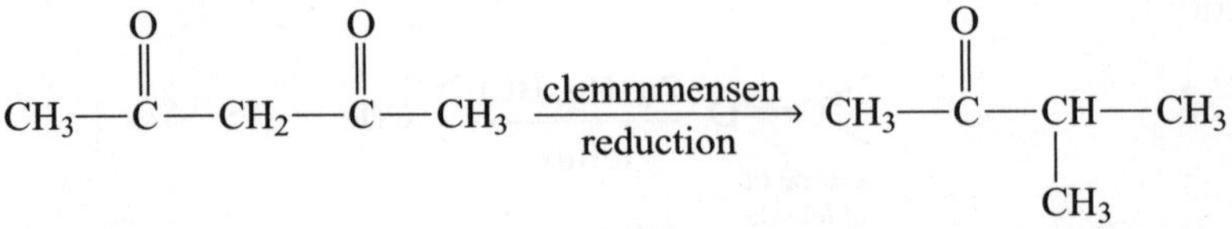

$$CH_3-\overset{\overset{\displaystyle O}{||}}{C}-CH_2-\overset{\overset{\displaystyle O}{||}}{C}-CH_3 \xrightarrow[\text{reduction}]{\text{clemmmensen}} CH_3-\overset{\overset{\displaystyle O}{||}}{C}-\underset{\underset{\displaystyle CH_3}{|}}{CH}-CH_3$$

Clemmensen reduction of cyclic 1,3-diketones give a fully reduced product along with a monoketone with ring contraction.

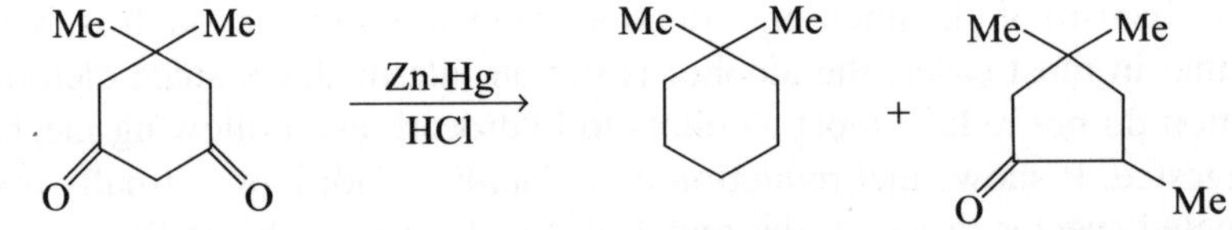

| 5,5-dimethyl cyclohexane-1,3-dione | | 1,1-dimethyl cyclohexane | 2,4,4-trimethyl cyclopentanone |

Clemmensen reduction has been reported to be improved by Sonication[2].

2.13.1 Applications

(i) Clemmensen reduction is useful for the reduction of aromatic aliphatic ketones[3,4]. Thus, acetophenone[5] is reduced by clemmensen method to ethyl benzene.

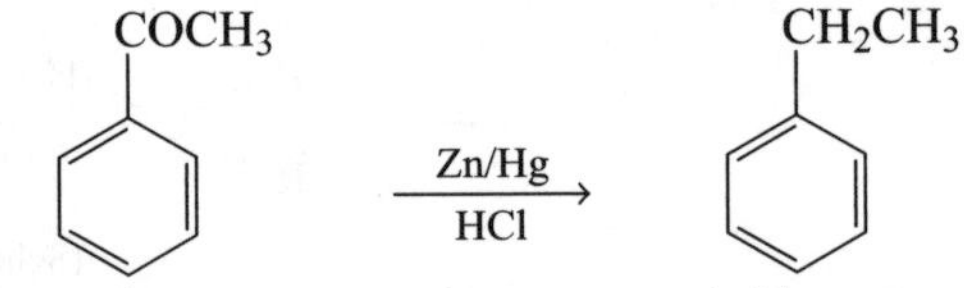

acetophenone   ethyl benzene

Similarly,

$$CH_3(CH_2)_5CHO \xrightarrow[\text{HCl}]{\text{Zn-Hg}} CH_3(CH_2)_5CH_3$$

n-heptaldehyde   *n*-heptane

$$C_6H_5COC_3H_7 \xrightarrow[\text{HCl}]{\text{Zn-Hg}} C_6H_5CH_2C_3H_7$$

n-propylphenyl
ketone

n-butylbenzene

Clemmensen reduction is useful for introducing straight chain (without rearrangement) alkyl groups in aromatic rings and subsequent reduction as in the case of acetophenone and *n*-propylphenyl ketone (examples give above).

(ii) Reduction of cyclic ketones

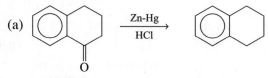

(a) α-tetralone → Zn-Hg / HCl →

(b) α-hydridone → Zn-Hg / HCl → indane

(c)

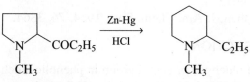

Ph Ph → HCl, Et₂O / Zn-dust → Ph Ph 60 %   (Ref 6)

(d) → Zn–Hg / HCl → 80 %   (Ref 7)

(e) cis-10-methyl-2-decalone → Zn(Hg) / HCl reflux → cis-9-methyldecalin   (Ref 7a)

(iii) Reduction with ring expansion

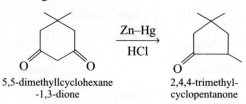

1-methyl-2-propionyl pyrrolidine → Zn-Hg / HCl → 2-ethyl-1-methyl--piperidine

(iv) Reduction with ring contraction

5,5-dimethyllcyclohexane-1,3-dione → Zn–Hg / HCl → 2,4,4-trimethyl-cyclopentanone

(v) Synthesis ot cycloparafins

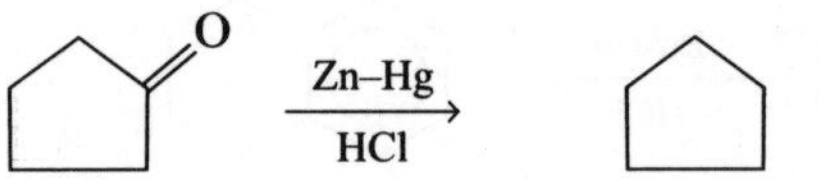

cyclopentanone → cyclopentane (Ref 8, 9)

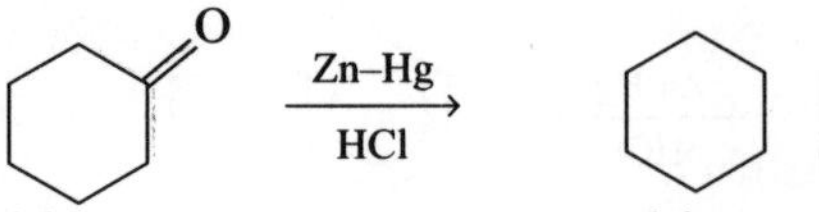

cyclohexanone → cyclohexane (Ref 8, 9)

(vi) Reduction of γ-keto acids

$$C_6H_5COCH_2CH_2CO_2H \xrightarrow[HCl]{Zn-Hg} C_6H_5CH_2CH_2CH_2CO_2H$$

β-benzoyl propionic acid γ-phenylbutanoic acid
(γ keto acid)

α- and β-ketone are normally not reduced.

REFERENCES

1. E. Clemmensen, Ber., 1913, 46, 1838; 1914, 51, 681; E.L. Martin in Organic Reactions I (New York, 1942), 155; J. G. St. C. Buchamam, P.G Woodgate, Quart. Rev., 1961, 23, 522; V. Vedejs, Org. Reactions, 1975, 22, 401.
2. W.P. Reeves, J.A. Murry, D.W. Willoughby, and W.I. Friedrich, Synthetic Communication, 1988, 18, 1961.
3. E. Clemmensen, Ber., 1913, 46, 1838.
4. B. Bannister and B.B. Elsner, J. Chem. Soc., 1951, 1055.
5. E.L. Martin, J. Am. Chem., Soc., 1936, 58, 143.
6. Reported in review by Vedejs (reference 1).
7. A.K. Banerjee, J. Alvarez, M. Santan and M.C. Carrasco, Tetrahedron, 1986, 42, 6615.
7a. W. G. Dauben, J. Am. Chem. Soc., 1954, 76, 3864.

2.14 DAKIN REACTION

The oxidation of aldehyde or acetyl group in phenolic aldehydes or ketones by reaction with alkaline hydrogen peroxide results in replacement of —CHO or —COCH$_3$ group with OH. This is known as Dakin reaction or Dakin oxidation[1]. For example, salicyladehyde is converted[1a] into catechol in 70% yield (Scheme-1).

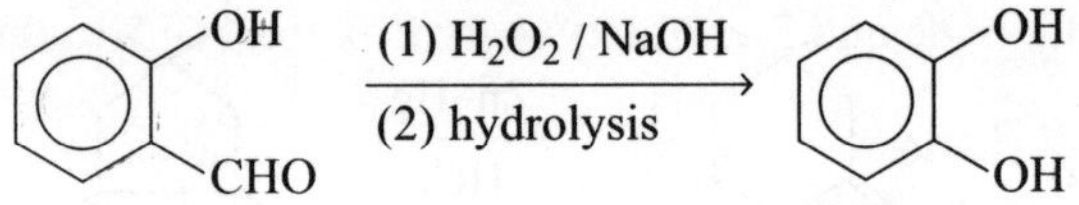

salicylaldehyde   (1) H$_2$O$_2$ / NaOH   catechol
(2) hydrolysis   70 %

(Scheme=1)

Dakins oxidation is applicable in aromatic aldehydes having hydroxyl group in ortho or para positions.

MECHANISM

The mechanisms of Dakin reaction is uncertain. However, a mechanism similar to **Baeyer–Villiger oxidation** is suggested. The carbonyl carbon is attacked by the hydroperoxide anion to give a tetrahectral intermediate (1), which undergoes migration of the aryl group with subsequent removal of hydroxide ion to give formate ester (2). An electron-releasing group (such as OH or NH_2) is necessary for efficient migration of the aryl group. The formed intermediate formate ester (2) can be isolated and converted into catechol on hydrolysis under aqueous alkaline conditions of the reaction (Scheme-2).

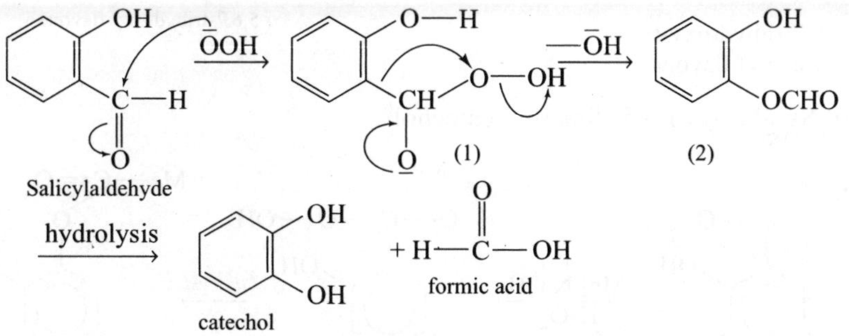

Scheme-2

The above mechanism has been confirmed[2] (Scheme-3) by taking o-hydroxybenzaldehyde in which the oxygen of CHO group is labelled by ^{18}O. The product, formic acid had all the labelled oxygen.

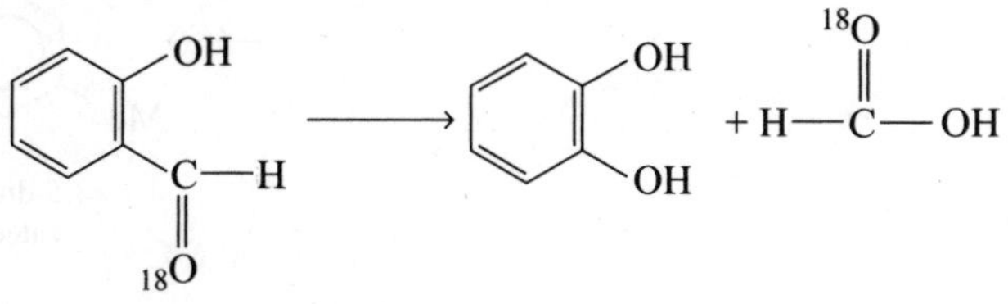

(Scheme=3)

Generally, the yields in Dakin reaction are low. This reaction has now been carried out in high yields using sodium percarbonate (SPC, Na_2CO_3, $1.5H_2O$) in H_2O–THF under **ultrasonic irradiation**[3]. Using this procedure the following aldehydes have been oxidised in 85–90% yields: o-hydroxybenzaldehyde, p-hydroxybenzaldehyde, 2-hydroxy-4-methoxybenzaldehyde, 2-hydroxy-3-methoxybenzaldehyde and 3-methoxy-4-hydroxybenzaldehyde.

2.14.1 Applications

Dakins oxidation has a number of synthetic applications.

   (i) Synthesis of quinol[4]

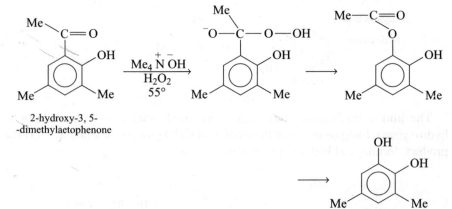

(ii) Dakin reaction can also be used in case of flavones having acetyl group[5].

(iii) Synthesis of 3,5-dimethyl catechol[6].

REFERENCES

1. H.D. Dakin, OS, 1941, *1*, 149; J.E. Lettler, Chem. Rev., 1949, *45*, 385; H.D. Dakin, J. Am. Chem. Soc., 1909, *42*, 477.
1a. H.D. Dakin, Org., Synth., Coll. Vol. 1, 1941.
2. C.A. Bunton in Peroxide Reaction Mechanism, edited by J.O. Edwards, Interscience.
3. G.W. Kabalka, N.K. Reddy and C. Narayana, Tetrahedron Lett., 1992, *33*, 865.
4. M.B. Hocking and J.H. Ong., Can. J. Chem., 1977, *55*, 102. M.B. Hocking, M.K.O. and T.A. Smyth, Can. J. Chem., 1978, *56*, 2646.
5. Y. Agasimundin and S. Siddappa, J. Chem. Soc., Perkin I, 1973, 503.
6. J. Baker, J. Chem. Soc., 1953, 1615.

2.15 DARZEN REACTION

The condensation of aldehydes or ketones with α-haloester in presence of a base (like potassium tert. butoxide, a hindered base to avoid the S_N2 displacement of the chloride) gives α, β-epoxy esters, called glycidic esters is known as Darzen Reaction[1] or Darzen glycidic ester condensation. As an example, the condensation of acetophenone with ethyl chloroacetate in presence of base give the α, β-epoxy ester (Scheme-1).

$$
\underset{\text{acetophenone}}{C_6H_5 \overset{\overset{\displaystyle O}{\|}}{C} CH_3} + \underset{\text{ethyl chloroacetate}}{ClCH_2COOC_2H_5} \xrightarrow{\text{base}} \underset{H_3C}{\overset{H_5C_6}{>}}\!\!\!\underset{O}{\triangle}\!\!- COOC_2H_5
$$

(Scheme-1)

In place of α-haloester (ethyl chloroacetate), α-halonitriles (e.g., chloro acetonitrile, $ClCH_2CN$) can also be used.

2.15.1 Darzen reaction in presence of phase transfer catalyst

Darzen reaction has been found the occur in alkaline solution in presence of a phase transfer catalyst (benzyl triethylammonium chloride)[2] (Scheme-1).

$$
\underset{R'}{\overset{R}{>}}C=O + ClCH_2CN + \underset{aq}{NaOH} \xrightarrow{C_6H_5CH_2N^+Et_3Cl^-} \underset{R'}{\overset{R}{>}}\!\!\!\underset{O}{\overset{}{C}}\!\!-CH-CN
$$

(Scheme-2)

R	R′	% Yield
Ph	H	75
CH$_3$	CH$_3$	60
Ph	CH$_3$	80
(CH$_2$)$_3$	H	79
(CH$_2$)$_4$	H	65
(CH$_2$)$_5$	H	78
Ph	Ph	55

With aldehydes and unsymmetrical ketones both possible steroisomers are formed. However with more acidic ketones, such as phenylacetone, the ketone carbanion is formed rather than from the nitrile (which is normally the case), leading to alkylation of the ketone (Scheme-3).

$$
\underset{\text{phenyl acetone}}{C_6H_5CH_2COCH_3} + \underset{\substack{\text{chloroaceto} \\ \text{nitrile}}}{ClCH_2CN} + \underset{aq}{NaOH} \xrightarrow{Q^+X^-} \underset{\underset{CH_2CN}{|}}{C_6H_5CHCOCH_3}
$$

(Scheme-3)

In the above darzen reaction using quaternary ammonium salts, it has been shown[3] that the structure of quarternary ammonium salt has virtually no effect on the yield of the glycidonitrile. It is best to perform the reaction at ~ 20°C and using sufficient sodium hydroxide (about 3 moles per mole of nitrile).

In place of quarternary ammonium salts (PTC), crown ethers (e.g., dibenzo-18-crown-6) (1 mole %) can also be used. Thus the reaction of benzaldehyde with chloroacetonitrile gives 78% of the corresponding glycidonitrile (Scheme-4).

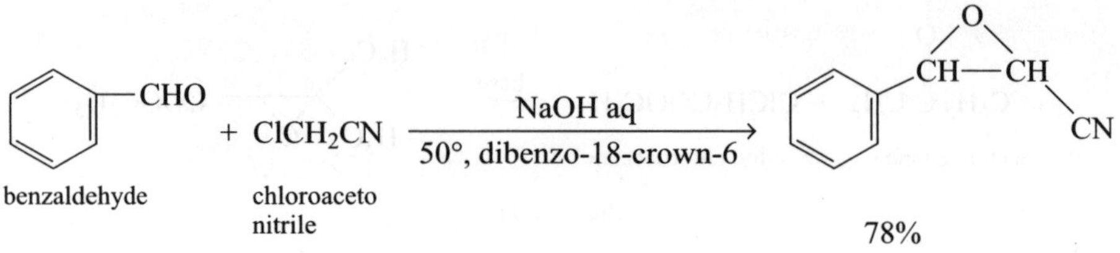

benzaldehyde   chloroaceto
                      nitrile

NaOH aq
50°, dibenzo-18-crown-6

78%

(Scheme-4)

The PTC catalysed condensation of α-thiocarbonyl compounds with 2-chloroacrylonitrile yielded[5] 2-cyano-2,3-epoxytetrahydrothiophenes. The reaction probably involved first the thiol addition followed by an **intramolecular Darzen reaction** (Scheme-5).

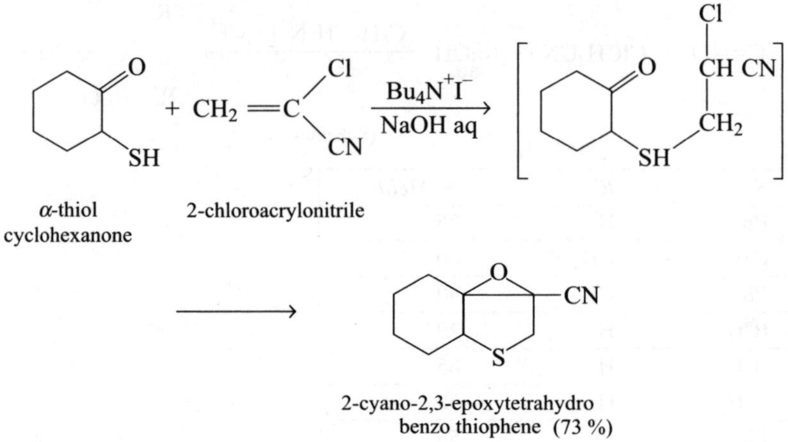

α-thiol   2-chloroacrylonitrile
cyclohexanone

Bu$_4$N$^+$I$^-$
NaOH aq

2-cyano-2,3-epoxytetrahydro
benzo thiophene (73 %)

(Scheme-5)

MECHANISM

The reaction of ethyl chloroacetate with a ketone in presence of a base (potassium tert. butoxide) is believed to proceed as follows (Scheme-6).

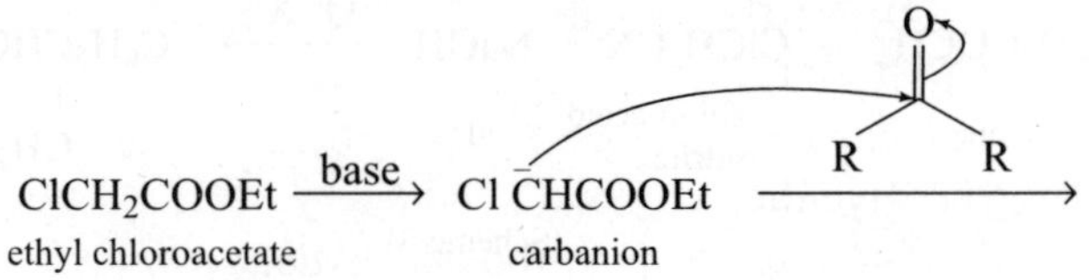

ClCH$_2$COOEt   base   Cl $\overset{-}{C}$HCOOEt
ethyl chloroacetate   carbanion

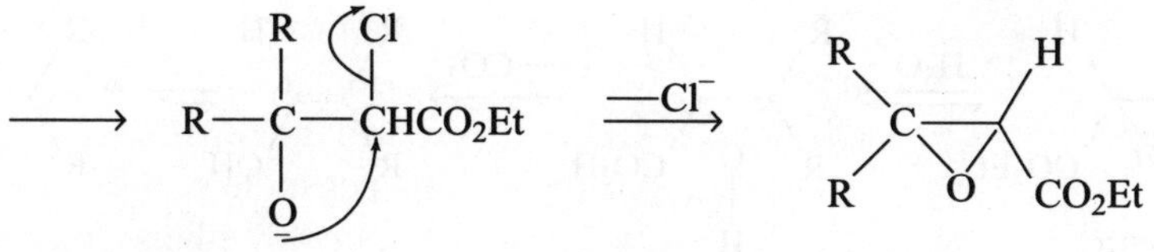
R
Cl
R—C—CHCO2Et
O
—Cl⁻
R
H
C
R
O
CO2Et

(Scheme-6)

2.15.2 Applications

Some important applications are given below

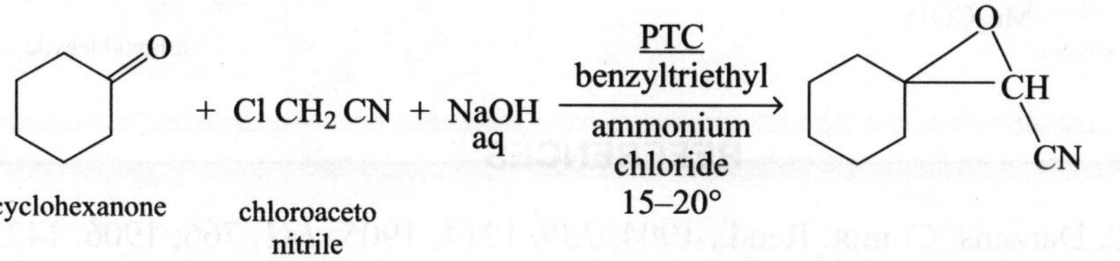
O
+ Cl CH2 CN + NaOH
aq
PTC
benzyltriethyl
ammonium
chloride
15–20°
O
CH
CN
cyclohexanone
chloroaceto
nitrile

(Ref 2)

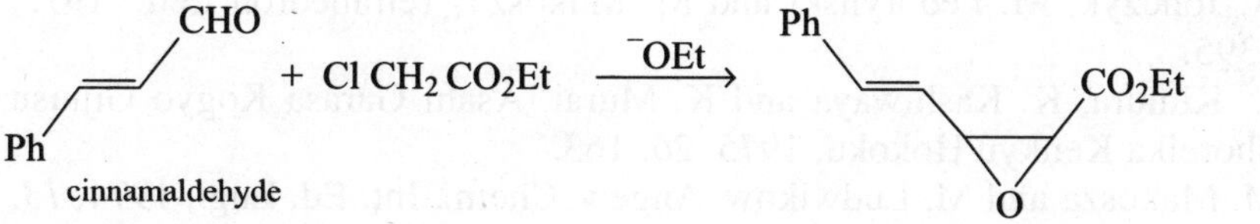
CHO
+ Cl CH2 CO2Et
⁻OEt
Ph
Ph
CO2Et
O
cinnamaldehyde

(Ref 6)

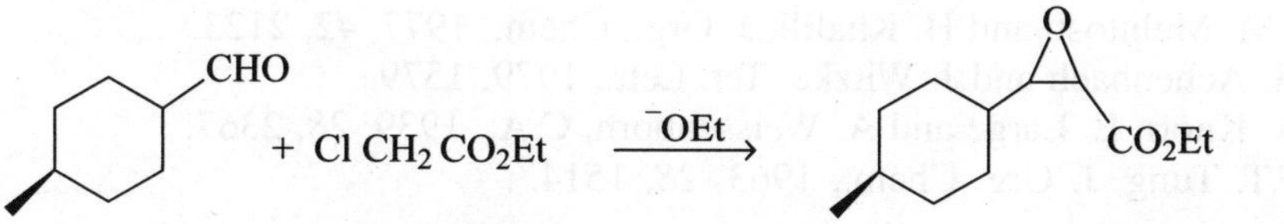
CHO
+ Cl CH2 CO2Et
⁻OEt
O
CO2Et

(Ref 7)

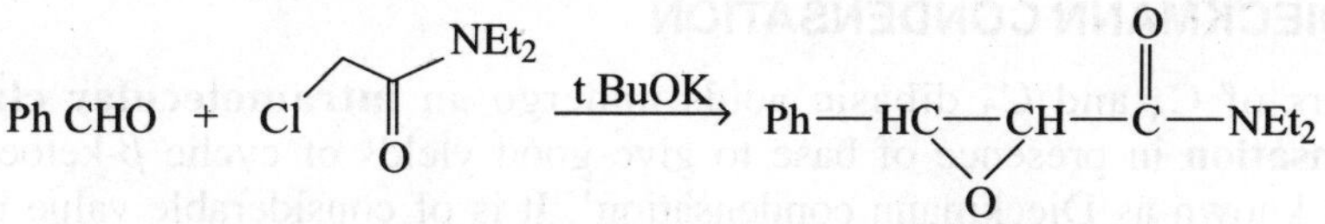
NEt2
Ph CHO + Cl
O
t BuOK
O
Ph—HC—CH—C—NEt2
O

(Ref 8)

cis and trans-
epoxy-N,N-diethylpropanamide
(88 %)

The glycidic esters (or the glycidonitriles) obtained by the darzen reaction are important intermediates for organic synthesis. For example, the glycidic ester or glycidonitrile on alkaline hydrolysis gives glycidic acid, which undergoes decarboxylative rearrangement in presence of acid to give an aldehyde or a ketone. The overall process involves the addition of one more carbon atom, as aldehyde, to a carbonyl group. Thus an aldehyde, RCHO is converted to a longer chain aldehyde homologue (RCH_2CHO) and a ketone (R_2CO) is converted to a longer chain aldehyde (R_2CHCHO). This is known as chain extension procedure. For example[8], cyclohexanone on Darzen reaction with ethyl chloroacetate followed by hydrolysis of the glycidic ester and subsequent hydrolysis give the epoxy acid; it on decarboxylation produces carboxaldehyde.

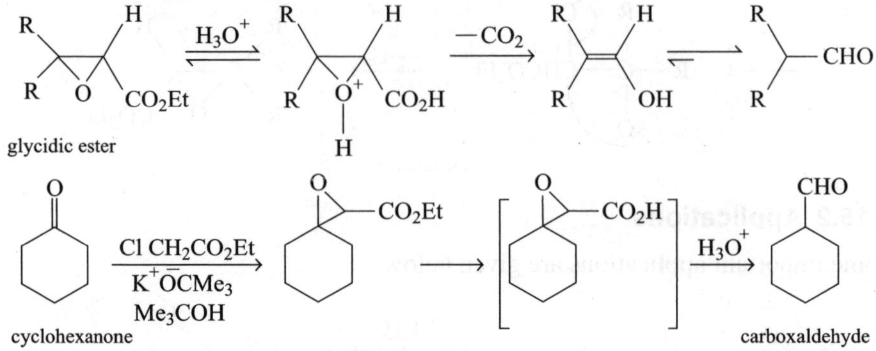

glycidic ester

cyclohexanone

carboxaldehyde

REFERENCES

1. C. Darzens, Compt. Rend., 1904, *139*, 1214; 1905, *141*, 766; 1906, 142, 214; M. Ballester, Chem., Rev., 1955, *55*, 283.
2. A. Jonczyk, M. Fedorynski and M. Makosza, Tetrahedron Lett., 1972, 2395.
3. C. Kimura, K. Kashiwaya and K. Murai, Asahi Garasa Kogyo Gijutsu Shoreika Kenkyu Hokoku, 1975, *26*, 163.
4. M. Makosza and M. Ludwikow, Angew. Chem., Int. Ed. Engl, 1974, *13*, 655.
5. J.M. Melntosh and H. Khalili, J. Org., Chem., 1977, *42*, 2123.
6. H. Achenbach and J. Witzke, Tet. Lett., 1979, 1579.
7. A. Knorr, E. Large and A. Weissenborn, C.A., 1939, *28*, 2367.
8. T.T. Tung, J. Org. Chem., 1963, *28*, 1514.

2.16 DIECKMANN CONDENSATION

Diesters of C_6 and C_7 dibasic acids undergo an **intramolecular claisen condensation** in presence of base to give good yields of cyclic β-ketoesters. This is known as Dieckmann condensation[1]. It is of considerable value in the synthesis of cyclic compounds. For examples, ethyl esters of adipic acid and pimelic acids give 2-carbethoxycyclopentanone and 2 carbethoxycyclohexanone respectively (Scheme-1).

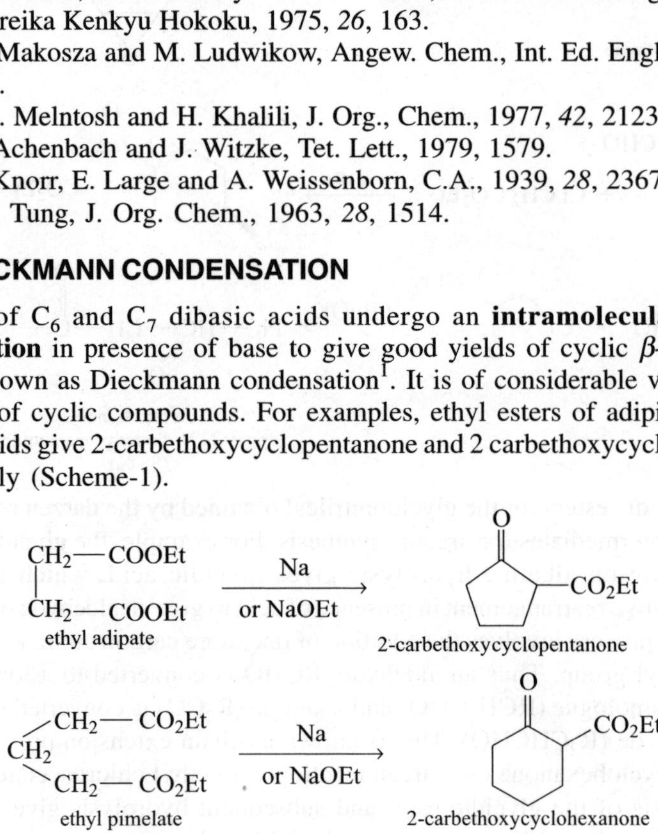

ethyl adipate

2-carbethoxycyclopentanone

ethyl pimelate

2-carbethoxycyclohexanone

(Scheme-1)

Dieckmann condensation best proceeds with esters having 6, 7 or 8 carbon atoms and gives stable rings with 5, 6, or 7 carbons.

MECHANISM

Abstraction of a proton from one of the α-carbons gives the carbanion, which attacks the carbonyl carbon of the other ester group. Finally, the β-ketoester is formed by expulsion of ethoxide anion ($\bar{O}$ Et) (Scheme-2).

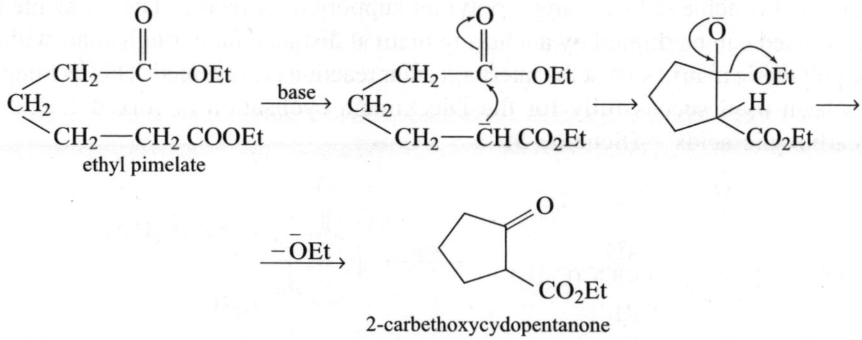
ethyl pimelate
base
2-carbethoxycydopentanone

(Scheme-2)

2.16.1 Dieckmann Condensation in Solid State

Dieckmann condensation of diesters has been carried out in solid state in the absence of any solvent in presence of a base (Na or NaOEt). The reaction products are obtained by direct distillation of the powdered reaction mixture, which was neutralised with p-TSOH · H_2O (Scheme-3).

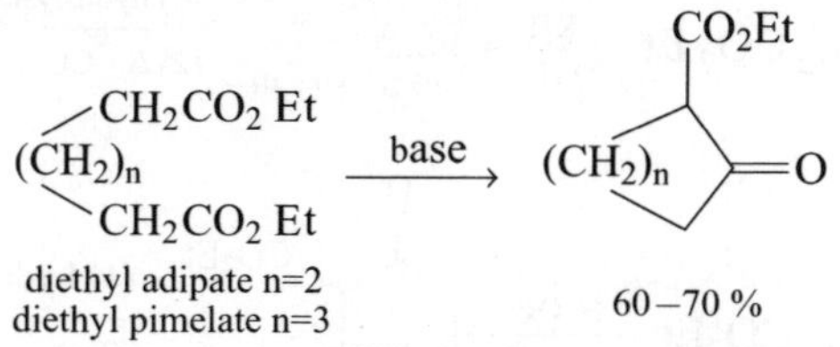
base
diethyl adipate n=2
diethyl pimelate n=3
60–70 %

(Scheme-3)

2.16.2 Dieckmann Condensation Under Sonication

It is found[3] that Dieckmann condensation proceeded very well on sonication in a short time (Scheme-4). On soniation[3a] potassium is easily transformed to a silver blue suspension in toluene. The ultrasonically dispersed potassium is extremely helpful in Dieckmann condensation (cyclisation) (Scheme-4).

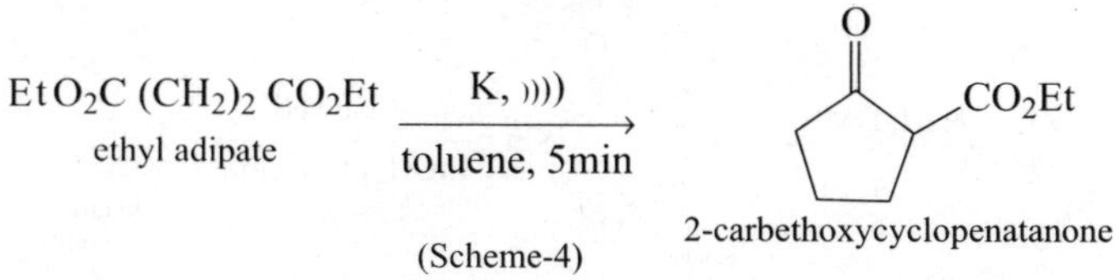
EtO2C (CH2)2 CO2Et
ethyl adipate
K,))))
toluene, 5min
CO2Et
2-carbethoxycyclopenatanone

(Scheme-4)

In the above condensation bases like Bu^{t}OK, Bu^{t}ONa, EtOK or EtONa could be used.

2.16.3 Dieckmann Condensation Using Polymer Support Technique

In Dieckmann cyclisation, it is necessary to control the relative rates of two competing reactions, viz. the interamolecular cyclisation and the intermolecular reaction; the former must be faster. Normally, this is achieved by having the compound in large volume of the solvent (dilution technique). Similar dilution technique is achieved by using a polymer supported molecule. The molecule to be cyclised can be diluted by anchoring them at distance far enough apart within the polymeric matrix so that the intermolecular reaction is prevented. This technique has been used successfully for the Dieckmann cyclisation of mixed esters of dicarboxylic acids[4-7] (Scheme-5).

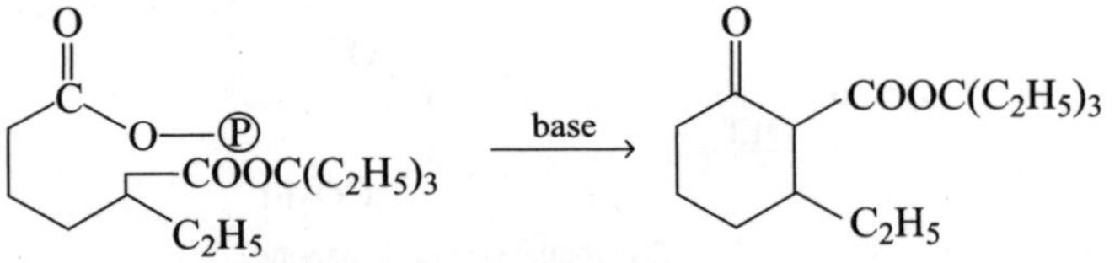

(Scheme-5)

2.16.4 Applications

Dieckmann condensation has been used for the synthesis of various cyclopentane and cyclohexane derivatives. Some of the important applications of Dieckmann condensation are given below.

(i) 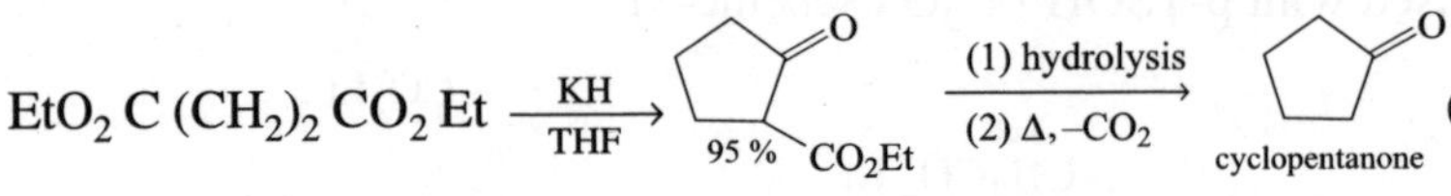 EtO$_2$ C (CH$_2$)$_2$ CO$_2$ Et $\xrightarrow[\text{THF}]{\text{KH}}$ ... 95 % ... $\xrightarrow[\text{(2) Δ,--CO$_2$}]{\text{(1) hydrolysis}}$ ... cyclopentanone (Ref. 8)

(ii) 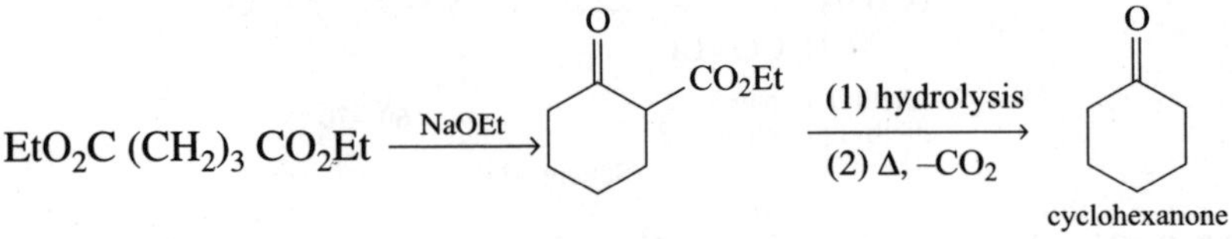 EtO$_2$C (CH$_2$)$_3$ CO$_2$Et $\xrightarrow{\text{NaOEt}}$ ... $\xrightarrow[\text{(2) Δ, --CO$_2$}]{\text{(1) hydrolysis}}$ ... cyclohexanone

(iii) 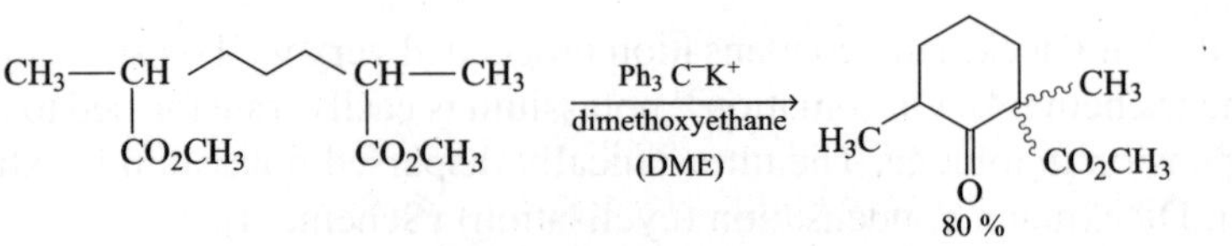 CH$_3$—CH ... CH—CH$_3$ ($\overset{|}{CO_2CH_3}$) ($\overset{|}{CO_2CH_3}$) $\xrightarrow[\text{dimethoxyethane}]{\text{Ph}_3 \text{C}^-\text{K}^+}$ ... (DME) ... 80 % (Ref. 9)

(iv) 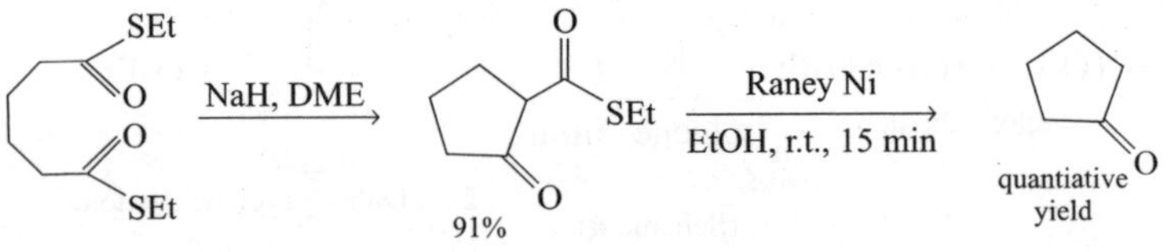 $\xrightarrow{\text{NaH, DME}}$ ... 91% $\xrightarrow[\text{EtOH, r.t., 15 min}]{\text{Raney Ni}}$ ... quantitative yield (Ref. 10)

(v) 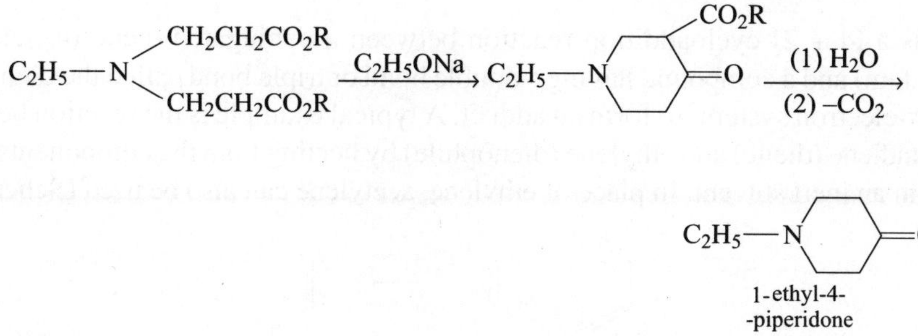 C_2H_5—N $\diagdown$ $CH_2CH_2CO_2R$ / $CH_2CH_2CO_2R$ $\xrightarrow{C_2H_5ONa}$ C_2H_5—N ... CO_2R =O $\xrightarrow[(2) -CO_2]{(1) H_2O}$

C_2H_5—N ... =O

1-ethyl-4-
-piperidone

(vi) Synthesis of thiophene derivatives

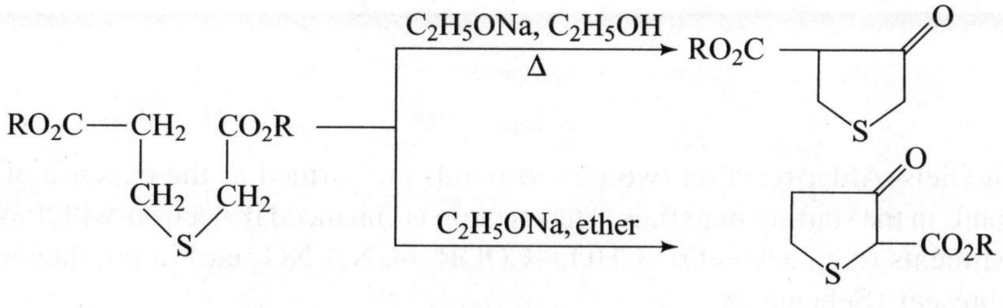

RO_2C—CH_2 CO_2R / CH_2 CH_2 \ S

$\xrightarrow[\Delta]{C_2H_5ONa, C_2H_5OH}$ RO_2C ... O / S

$\xrightarrow{C_2H_5ONa, ether}$ ... O / S—CO_2R

(vii) In the synthesis of steroids, five or six membered ring is built up.

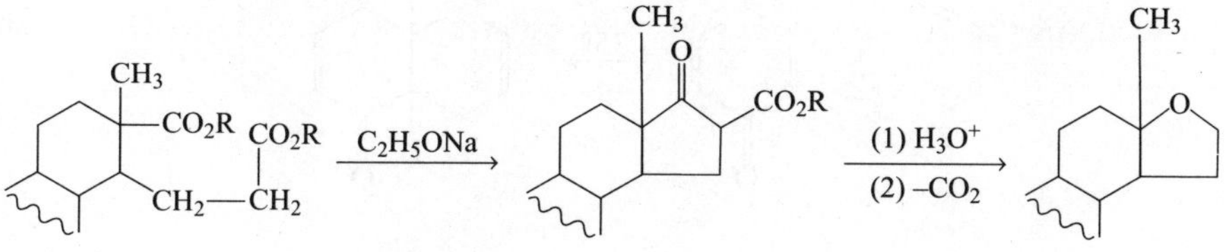

CH_3 ... —CO_2R CO_2R / CH_2—CH_2 $\xrightarrow{C_2H_5ONa}$ CH_3 O ... CO_2R $\xrightarrow[(2) -CO_2]{(1) H_3O^+}$ CH_3 ... O

REFERENCES

1. W. Dieckmann, Ber., 1894, *27*, 102, 965; 1900, *33*, 2670; Ann., 1901, 317, 53, 93; C.R. Housen and B.E. Hudson, Organic Reactions, 1942, I, 274.
2. F. Toda, T. Suzuki and S. Higa, J. Chem. Soc., Perkin Trans. I. 1998, 3521.
3. Citied in a review by J.M. Khurana, Chemistry Education, 1990 (Oct, Dec.) p. 27.
3a. J.L. Luche, C. Petrier and C. Duputy, Tetrahedron Lett., 1985, *26*, 753.
4. J.L. Crowley and H. Rapoport, J. Chem., Soc., 1970, *92*, 6363.
5. J. Schalffer, J.J. Bloomfield, Org., React., 1967, *15*.
6. H.L. Lochte and A.G. Pinman, J. Org. Chem., 1960, *25*, 1462.
7. O.S. Bhanot, Indian J. Chem., 1967, *5*, 127.
8. C.A. Broun, Synthesis, 1975, *5*, 326.
9. G. Nee and B. Tehboubar, Tetrahedron Lett., 1979, 3717.
10. H.-J. Liu and H.K. Lai, Tetrahectron Lett., 1979, 1193.

2.17 DIELS-ALDER REACTION

It is a [4 + 2] cycloaddition reaction between a conjugated diene (4π-electron system) and a compound having a double bond or triple bond called the dienophile (2π-electron system) to form an adduct. A typical example is the reaction between butadiene (diene) and ethylene (dienophile) by heating both the components alone or in an inert solvent. In place of ethylene, acetylene can also be used (Scheme-1).

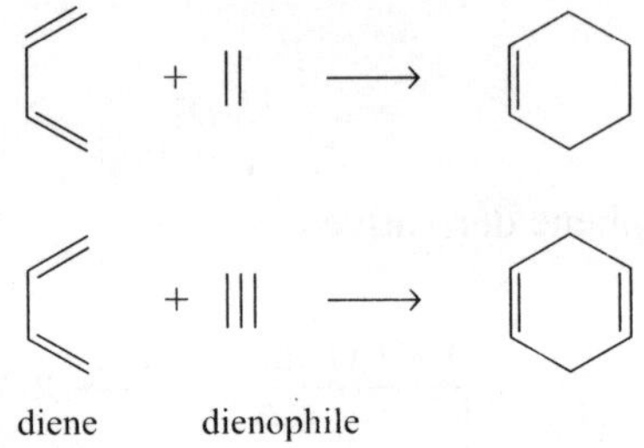

diene   dienophile

(Scheme-1)

In Diels–Alder reaction two new σ-bonds are formed as the expense of two π-bonds in the starting materials. The reaction is enhanced if electron-withdrawing substituents (e.g., $> C=O$, $-CHO$, $-COOR$, $-CN$, $-NO_2$ etc) in the dienophile are present (Scheme-2).

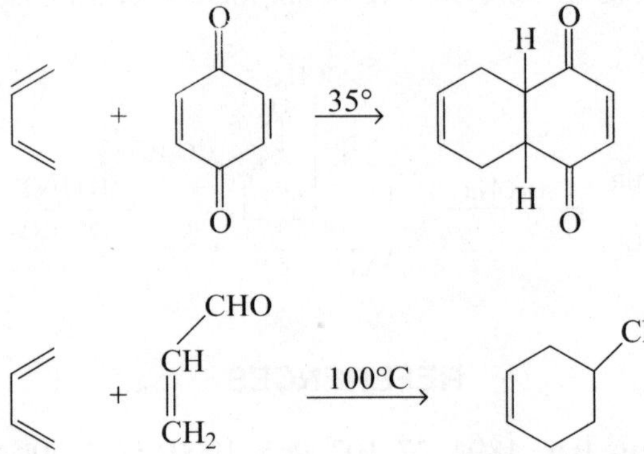

(Scheme-2)

It is not essential that the dienes should be acyclic hydrocarbons; these may be acyclic hydrocarbons in which two conjugated double bonds may be present partially inside the ring, some heterocyclic and some aromatic compounds. Some of the examples are given below (Scheme-3).

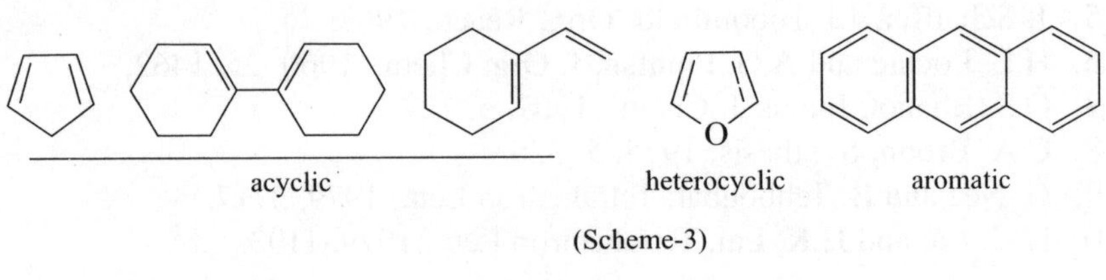

acyclic   heterocyclic   aromatic

(Scheme-3)

Amongst the aromatic compounds, benzene and naphthalene and phenanthrene are unreactive. However anthracene reacts readily. Some examples of anthracene are given below (Scheme-4).

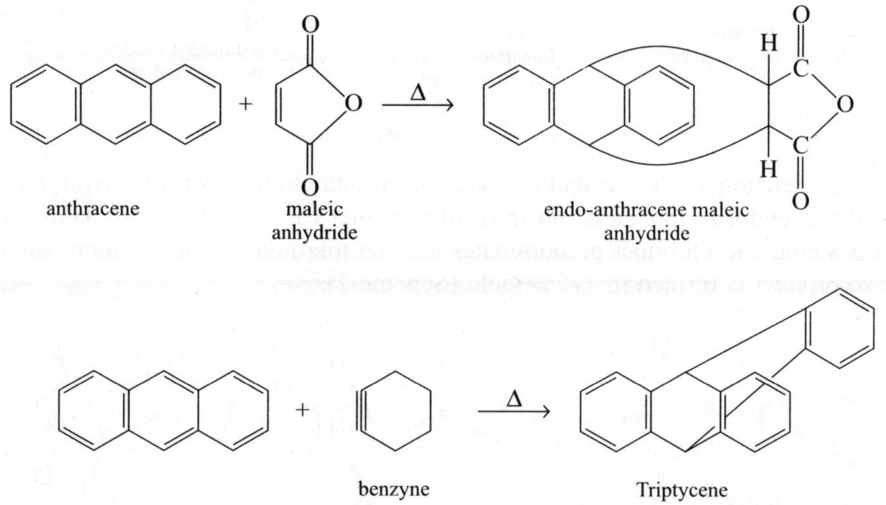

anthracene   maleic anhydride   endo-anthracene maleic anhydride

benzyne   Triptycene

(Scheme-4)

The Diels–Alder reaction occurs with only the cisoid form (S-cis conformation) of the diene. Cyclic dienes in cisoid form also react but cyclic dienes in transoid conformation do not react (Scheme-5).

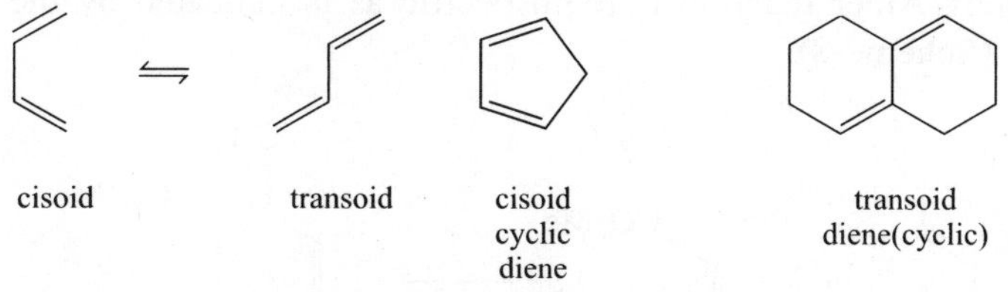

cisoid   transoid   cisoid cyclic diene   transoid diene(cyclic)

(Scheme-5)

Diels–Alder reaction is highly stereospecific as shown by the following examples (Scheme-6).

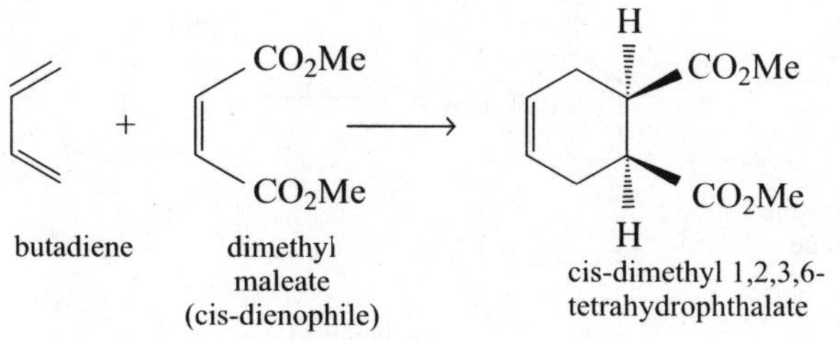

butadiene   dimethyl maleate (cis-dienophile)   cis-dimethyl 1,2,3,6-tetrahydrophthalate butadiene   dimethyl fumarate (trans dienophile)   trans-dimethyl 1,2,3,6-tetrahydrophthalate

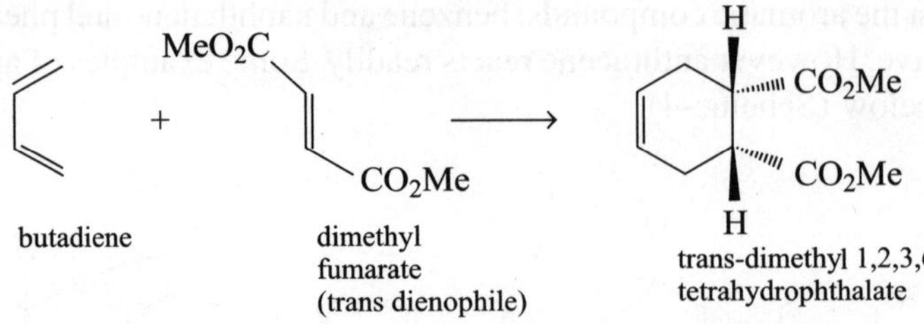

(Scheme-6)

The reaction of cyclopentadiene with maleic anhydride may give two products, exo and endo. In this case and most of the other cases the thermodynamically less stable endo product predominates and the thermodynamically more stable exo product is formed in 1–2% yield (Scheme-7).

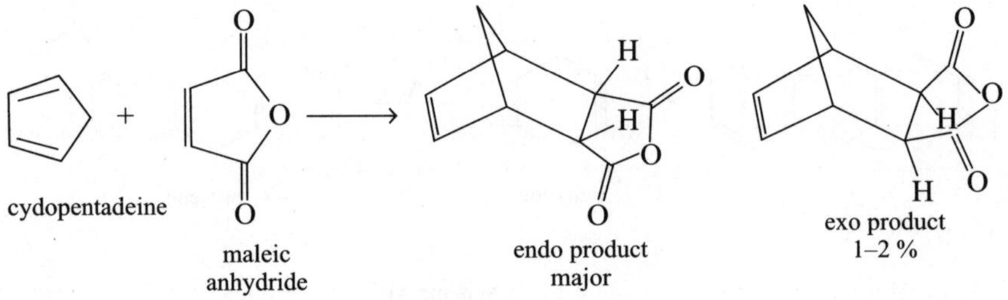

cydopentadeine   maleic anhydride   endo product major   exo product 1–2 %

(Scheme-7)

However reaction of furan with malimide gives thermodynamically more stable exo product. In this case the initially formed endo product gives the more stable exo-product.

The Diels–Alder reaction is regiospecific as is indicated by the following examples (Scheme-8).

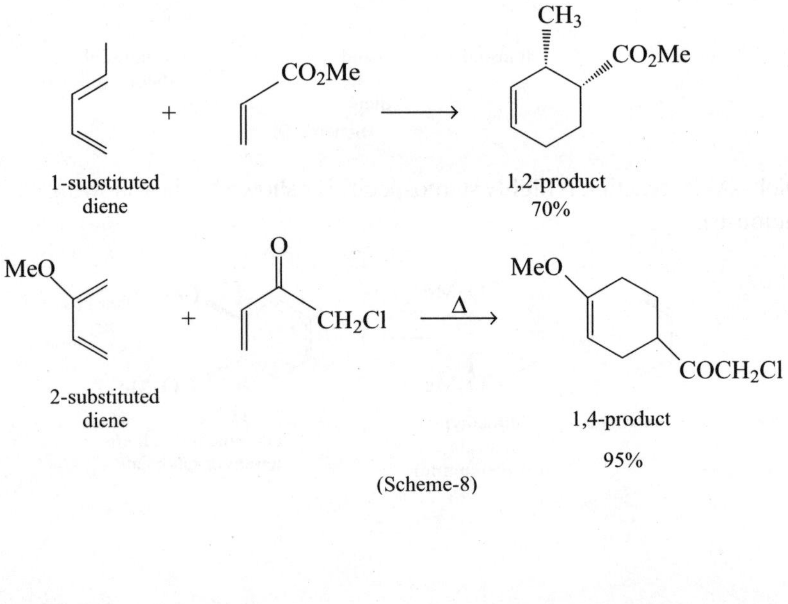

1-substituted diene   1,2-product 70%

2-substituted diene   1,4-product 95%

(Scheme-8)

MECHANISM

It has a concerted mechanism, in which there is simultaneous breaking and making of bonds through a six-centered transition state with no intermediate. It is a $4\pi + 2\pi$ cycloaddition and sterospecifically cis with respect to both the reactants (Scheme-9).

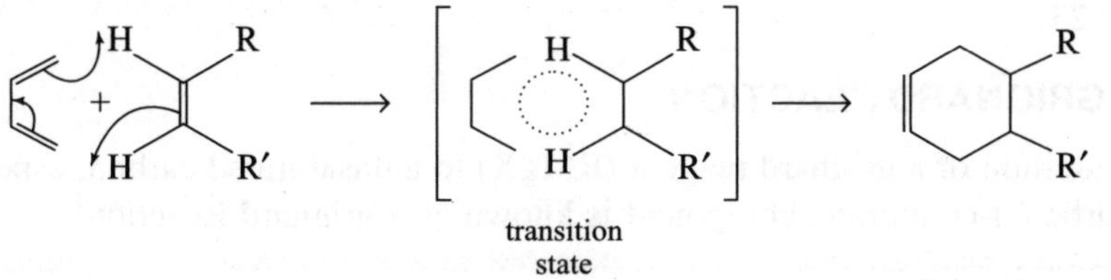

transition
state

(Scheme-9)

2.17.1 Diels–Alder Reactions Under Microwave Irradiation

The Diels–Alder reaction proceeds[2] very satisfactory under microwave irradiation; this procedure reduces the time for completion of the reaction to 90 seconds.

2.17.2 Diels–Alder Reactions in Aqueous Phase

Diels–Alder reaction also proceeds satisfactory in aqueous phase[3]. Thus the reaction of furan with maleic acid in hot water gives the adduct (Scheme-10).

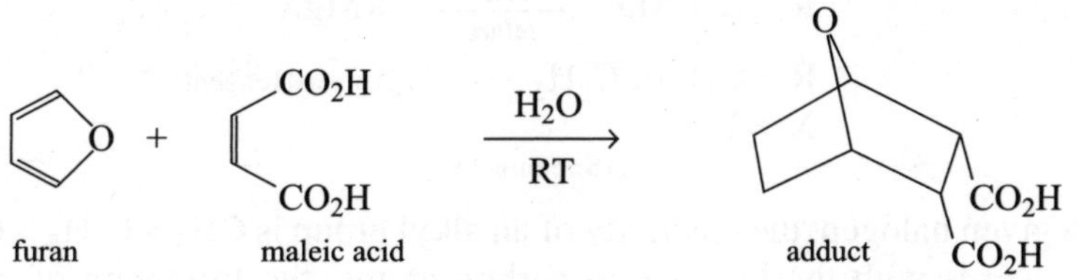

furan   maleic acid   adduct CO_2H

(Scheme-10)

2.17.3 Diels–Alder Reaction in Ionic Liquids

The great usefulness of Diels–Alder reaction lies in its high yield and high stereospecificity. When ionic liquids such as [bmin] [BF_4], [bmin] [ClO_4], [emim] [CF_3SO_3] etc. are used for Diels–Alder reaction between cyclopentadiene and methyl acrylate, rate enhancement, high yields and strong endo selectivities have been observed[4].

REFERENCES

1. O. Diels, K. Alder, Ann., 1928, *460*, 98; 1929, *62*, 470; Ber., 1929, *62*, 2081, 2087; J.A. Norton, Chem. Res., 1943, *31*, 319; D.A. Oppolzer, Angew Chem. Int. Ed., 1977. *16*, 10.

2. O.C. Dermer, J. King, J. Amer. Chem. Soc., 1941, *63*, 3232; R.J. Geguere, T.L. Bray, S.M. Duncan, Tetrahedron Lett., 1986, *27(41)*, 1945.

3. D.C. Rideout and R. Breslow, J. Am. Chem. Soc., 1980, *102*, 7816, R.B. Woodward and Harold Baer. J. Am. Chem. Soc., 1948, *70*, 1161.

4. T. Fischer, T. Sethi, T. Welton, J. Woolf, Tetrahedron Lett., 1999, *40*, 793; M.J. Earle, P.B. McMormal, K.R. Sedden, Green Chem., 1999, *1*, 23.

2.18 GRIGNARD REACTION

Any addition of a grignard reagent (RMgX) to a unsaturated carbon, especially the carbonyl-containing compound is known as a grignard Reaction[1].

GRIGNARD REAGENT

The organomagnesium halides are known as grignard reagents. These were discovered by Victor grignard and so named grignard reagents. In view of the tremendous synthetic potentials of grignard reagents, Victor grignard was awarded Nobel Prize in 1912. The grignard reagents are represented as R – Mg – X, where R = alkyl, alkenyl, alkynyl or aryl group and X = Cl, Br, I. Grignard reagents act as nucleophiles and attack unsaturated carbon—especially the carbon of a carbonyl group.

The grignard reagents are prepared by the action of magnesium on alkyl halide in anhydrous ether (Scheme-1).

$$R - X + Mg \xrightarrow[\text{reflux}]{\text{ether}} RMgX$$

$R = CH_3$ or C_2H_5 grignard reagent

$X = I$

(Scheme-1)

For a given halogen, the reactivity of an alkyl group is $CH_3 > C_2H_5 > C_3H_7$ or in other words with the increase in carbon atoms, the formation of grignard reagent becomes difficult. Also for an alkyl halide, the ease of formation of grignard reagent is of the order RI > RBr > RCl.

2.18.1 Grignard Reaction Under Sonication

The reaction is performed under anhydrous conditions. In case the reaction is slow or sluggish, small amount of iodine is added to start the reaction.

It is found that it is best to activate magnesium by sonication. This activated magnesium finds applications[2] in the synthesis of grignard reagents without the use of activators (Scheme-2).

$$R - X + Mg \xrightarrow[))))]{\text{ether}} R\,Mg\,X$$

90%

(Scheme-2)

Use of sonication makes the preparation of grignard reagents very convenient. Using this procedure even halides containing olefinic groups can be made to react with activated magnesium to give[3] the corresponding grignard reagents.

Ary halides react with magnesium to form arylmagnesium halides (e.g. phenyl magnesium bromide, C_6H_5MgBr). In a similar way, vinyl halides ($CH_3CH=CHX$) react with magnesium to form vinyl grignard reagents (e.g., $CH_3CH=CHMgBr$).

In addition to the method described above, the grignard reagents are prepared from substrates that have acidic hydrogens. Some examples are

$$R-C\equiv CH + R\,Mg\,Br \longrightarrow R-C\equiv C\,Mg\,Br + RH$$
alkynyl magnesium bromide

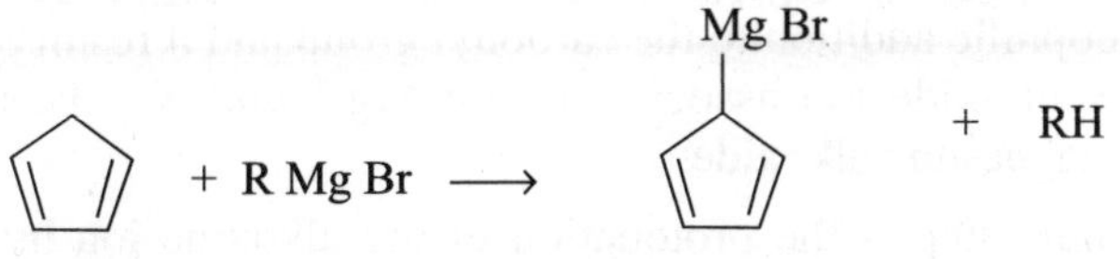

cyclopentadiene   cyclopentadienyl
magnesium bromide

STRUCTURE OF GRIGNARD REAGENT

A grignard reagent is represented as RMgX. Its exact structure has been a matter of discussion for a long time. It is believed that a grignard reagent exists as a co-ordination complex with ether as shown below.

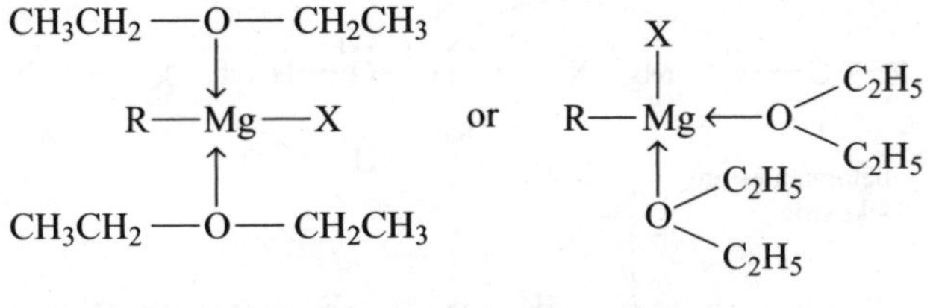

The structure of phenyl magnesium bromide as given below has been established by x-ray diffraction studies.

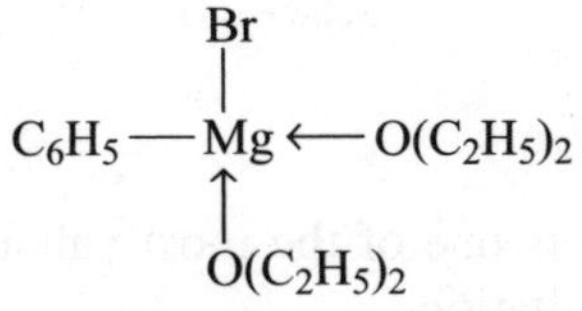

phenylmagnesium bromide

GRIGNARD REACTION

The grignard reagents react with carbonyl compounds (aldehydes and ketones) to give alcohols (Scheme-3). This reaction is known as grignard reaction.

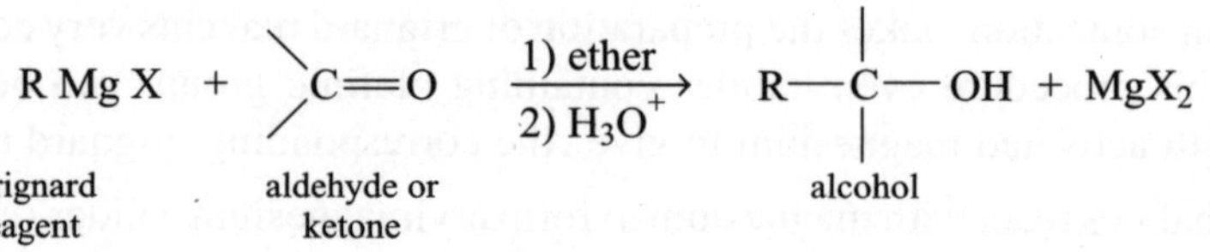

$$\text{R Mg X} + \overset{\diagdown}{\underset{\diagup}{\text{C}}}=\text{O} \xrightarrow[\text{2) H}_3\text{O}^+]{\text{1) ether}} \text{R}-\overset{|}{\underset{|}{\text{C}}}-\text{OH} + \text{MgX}_2$$

grignard aldehyde or   alcohol
reagent   ketone

(Scheme-3)

REACTION MECHANISM

It is believed that the grignard reaction occurs in the following way.

 (i) The grignard reagent bring strongly nucleophilic uses its electron pair to form a bond to the carbon atom of the carbonyl group. One electron pair of the carbonyl group shifts out to the oxygen. This reaction is a nucleophilic addition to the carbonyl group and it results in the formation of an alkoxide ion associated with Mg^{2+} and X^-; the adduct is called halomagnesium alkoxide.

 (ii) The next step is the protonation of the alkoxide ion by the addition of aqueous hydrogen halide (HX). This results in the formation of the alcohol and MgX_2.

  The various steps are represented as shown below (Scheme-4).

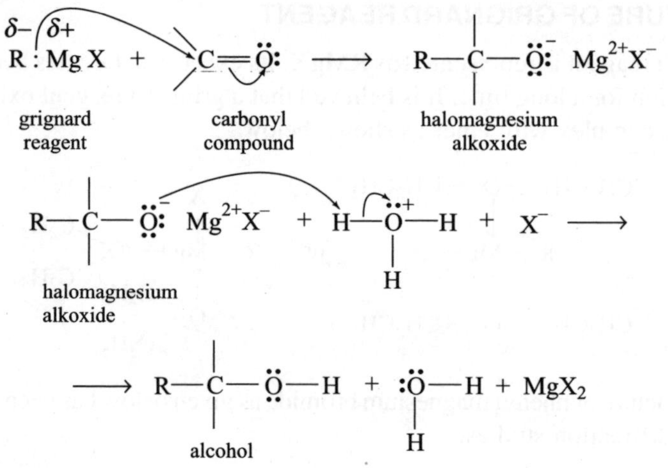

$$\overset{\delta-}{\text{R}} : \overset{\delta+}{\text{Mg}} \text{ X} + \overset{\diagdown}{\underset{\diagup}{\text{C}}}=\ddot{\text{O}}: \longrightarrow \text{R}-\overset{|}{\underset{|}{\text{C}}}-\ddot{\text{O}}:^- \text{ Mg}^{2+}\text{X}^-$$

grignard   carbonyl   halomagnesium
reagent   compound   alkoxide

$$\text{R}-\overset{|}{\underset{|}{\text{C}}}-\ddot{\text{O}}:^- \text{ Mg}^{2+}\text{X}^- + \text{H}-\overset{+}{\ddot{\text{O}}}-\text{H} + \text{X}^- \longrightarrow$$
                |
                H

halomagnesium
alkoxide

$$\longrightarrow \text{R}-\overset{|}{\underset{|}{\text{C}}}-\ddot{\text{O}}-\text{H} + :\ddot{\text{O}}-\text{H} + \text{MgX}_2$$
                  |
alcohol   H

(Scheme-4)

LIMITATIONS

Though grignard synthesis is one of the most valuable of all general synthetic procedures, it has some limitations.

 (i) Since grignard reagent contains a carbanion, (and so is a very powerful base), it is not possible to prepare a grignard reagent from an organic group that contains an acidic hydrogen, i.e, hydrogen atom which is more acidic than the hydrogen atoms of an alkane or alkene. It is thus not possible to prepare a grignard reagent from a compound containing an –OH group,

—NH$_2$-group, —SH group, —COOH group or an —SO$_3$H group.

(ii) Since grignard reagents are powerful nucleophiles, it is not possible to prepare a grignard reagent from any halide that contains a carboxyl, epoxy, nitro or cyano group.

This implies that the preparation of grignard reagents is limited to alkyl halides or to analogous organic halides containing carbon-carbon double bonds, internal triple bonds, ether linkages and —NR$_2$ group.

(iii) A number of cases are known where a grignard reagent is made to react with a ketone, which does not give the usual product (i.e., tert. alcohol). Instead, 'abnormal' products are obtained. It is found that branching of the carbon chain near the carbonyl group hinders the nucelophilic addition of the reagent due to steric hinderance. Also, in case the grignard reagent has bulky alkyl or aryl group, it does not attack the electrophilic center of the carbonyl compound. Thus, even methylmagnesium bromide does not react with di-tert-butyl ketone. In a similar way, methyl isopropyl ketone reacts with methylmagnesium halides but not with tertiary-butylmagnesium halides. There are certain reactions in which substituted ketones and reagents react to form products which are generally not obtained in grignard reactions. One such reaction is given below:

$$[(CH_3)_2CH]_2CO + (CH_3)_2CHMg\ Br \longrightarrow (CH_3)_2CHCH—CH(CH_3)_2 + CH_3CH = CH_2$$
$$\overset{|}{OH}$$

di-isopropyl ketone	isopropylmagnesium bromide	di-isopropyl carbinol	propene

2.18.2 Grignard Reaction in Solid State

The results obtained by carrying out the usual grignard reaction are different than in the solid state[4]. Thus the reaction of ketone (e.g., benzophenone) with grignard reagent [the reaction is carried out by mixing ketone and powdered grignard reagent, obtained by evaporating the solution of the grignard reagent (prepared as usual) in vacuo (caution)] in solid state gives more of the reduced product of the ketone than the adduct (Scheme-5).

$$Ph_2CO + RMgX \xrightarrow[\text{solid}]{0.5\ hr} Ph_2RCOH + Ph_2CHOH$$

benzophenone	grignard reagent	adduct (A)	reduced product of the ketone (B)

grignard reagent RMgX		% products obtained in solid state	
R	X	(A)	(B)
Me	I	No reaction	
Et	Br	30	31
iPr	Br	2	20
Ph	Br	59	—

(Scheme-5)

2.18.3 Applications

Grignard reaction has tremendous synthetic potentials. Some of the applications are given below.

(i) **Synthesis of alcohols**

The reaction of grignard reagent, to carbonyl compounds are especially useful since they can be used to prepare primary, secondary or tertiary alcohols.

(a) grignard reagent reacts with formaldehyde to give primary alcohol.

$$\overset{\delta-}{R} : \overset{\delta+}{Mg} X \;+\; \overset{H}{\underset{H}{C}}=\overset{\cdot\cdot}{\underset{\cdot\cdot}{O}} \longrightarrow R-\overset{H}{\underset{H}{\overset{|}{C}}}-\overset{\cdot\cdot}{\underset{\cdot\cdot}{O}}:\; Mg\,X \xrightarrow{H_3O^+} R-\overset{H}{\underset{H}{\overset{|}{C}}}-\overset{\cdot\cdot}{\underset{\cdot\cdot}{O}}H$$

1° alcohol

If $R = C_6H_5$, benzyl alcohol is obtained in 90% yield.

Grignard reagent also reacts with an epoxide (ethylene oxide) to give primary alcohol with lengthening of the carbon chain by two carbon atoms.

$$CH_3MgBr \;+\; H_2C\overset{\diagdown}{\underset{O}{\diagup}}CH_2 \longrightarrow CH_3CH_2CH_2\,\overset{+}{\bar{O}}\,MgBr \xrightarrow{H^+} CH_3CH_2CH_2OH$$

methyl magnesium bromide   ethylene oxide   propylalcohol

(b) All aldehydes except formaldehyde react with grignard reagent to give secondary alcohols.

$$\overset{\delta-}{R} : \overset{\delta+}{Mg} X \;+\; \overset{R'}{\underset{H}{C}}=\overset{\cdot\cdot}{\underset{\cdot\cdot}{O}} \longrightarrow R-\overset{R'}{\underset{H}{\overset{|}{C}}}-OMg\,Br \xrightarrow{H_3O^+} R-\overset{R'}{\underset{H}{\overset{|}{C}}}-OH$$

2° alcohol

Thus 2-butanol is obtained in 80% yield by the reaction of ethylmagnesium bromide with acetaldehyde.

$$CH_3CH_2MgBr \;+\; \overset{CH_3}{\underset{H}{C}}=O \xrightarrow{Et_2O} CH_3CH_2-\overset{CH_3}{\underset{H}{\overset{|}{C}}}-OMgBr \xrightarrow{H_3O^+} CH_3CH_2\underset{\underset{OH}{|}}{C}HCH_3$$

ethyl magnesium bromide   acetaldehyde   2-butanol (80%)

The reaction of grignard reagent with ethylformate followed by hydrolysis gives aldehydes, which further react with grignard reagent to give secondary alcohols. If ethyl orthoformate is used in place of ethyl formate, aldehydes can be isolated. Use of ethyl acetate gives ketones, which further react with grignard reagent to give tertiary alcohol.

$$\delta - \overset{\delta +}{CH_3MgI} + HC\overset{O}{\underset{OEt}{\diagup}} \longrightarrow \left[\overset{H}{\underset{H_3C}{\diagdown}}C\overset{\overset{+}{O}MgI}{\underset{OEt}{\diagup}} \right] \xrightarrow{- Mg(OEt)I}$$

methyl magnesium bromide   ethyl formate

$$CH_3-C\overset{O}{\underset{H}{\diagdown}} \xrightarrow[2)\ H^+]{1)\ CH_3MgI} CH_3-\overset{CH_3}{\underset{|}{CH}}-OH$$

acetaldehyde   isopropyl alcohol
2° alcohol

Secondary alcohols can also be obtained by the reaction of grignard reagent with substituted epoxide (e.g., propylene oxide). In this case the attack of the grignard reagent is from the less substituted ring carbon of the epoxide.

$$C_6H_5MgBr + H_2C\overset{\diagup\diagdown}{\underset{O}{}}CHCH_3 \longrightarrow C_6H_5CH_2-\overset{|}{\underset{CH_3}{CH}}-\bar{O}\ \overset{+}{Mg}\ Br \xrightarrow{H^+} C_6H_5CH_2\underset{CH_3}{CHOH}$$

phenyl magnesium bromide   propylene oxide   2° alcohol

(c) Grignard Reagents react with ketones to give tertiary alcohols

$$\overset{\delta-}{R}:\overset{\delta+}{Mg}X + \overset{R'}{\underset{R''}{\diagdown}}C=\ddot{O} \longrightarrow R-\overset{R'}{\underset{R''}{\overset{|}{C}}}-OMg\ Br \xrightarrow[H_2O]{NH_4Cl} R-\overset{R'}{\underset{R''}{\overset{|}{C}}}-OH$$

grignard reagent   ketone   3° alcohol

Thus, butylmagnesium bromide on reaction with acetone gives 2-methyl-2-hexanol in 92% yield.

$$CH_3CH_2CH_2CH_2MgBr + \overset{CH_3}{\underset{CH_3}{\diagdown}}C=O \longrightarrow CH_3CH_2CH_2CH_2-\overset{CH_3}{\underset{CH_3}{\overset{|}{C}}}-\overset{+}{O}\overset{}{Mg}\ Br$$

butyl magnesium bromide   acetone

$$\xrightarrow[H_2O]{NH_4Cl} CH_3CH_2CH_2CH_2-\overset{CH_3}{\underset{OH}{\overset{|}{C}}}-CH_3$$

2-methyl-2-hexanol
92%

Tertiary alcohols are also obtained by the reaction of esters with grignard reagent (2 equivalent); in this case the initial product is an ketone, which further reacts (ketones are more reactive than esters) with grignard reagent to give 3° alcohol.

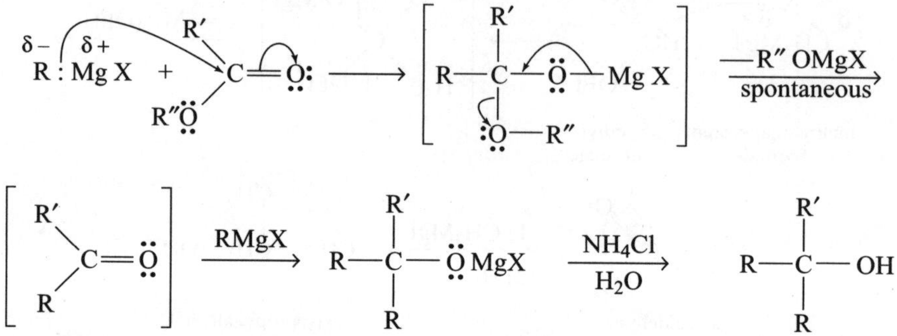

$$\underset{R'\overset{\cdot\cdot}{O}}{\overset{\delta-\text{\Large(}\,\delta+}{R:MgX}} + {}^{R'}_{R''\overset{\cdot\cdot}{O}}C{=}\overset{\cdot\cdot}{O}: \longrightarrow \left[R-\underset{\overset{|}{\overset{\cdot\cdot}{O}-R''}}{\overset{R'}{\overset{|}{C}}}\overset{\cdot\cdot}{O}-MgX \right] \xrightarrow[\text{spontaneous}]{-R''\,OMgX}$$

$$\left[{}^{R'}_{R}C{=}\overset{\cdot\cdot}{O} \right] \xrightarrow{RMgX} R-\underset{R}{\overset{R'}{\overset{|}{\underset{|}{C}}}}-\overset{\cdot\cdot}{O}\,MgX \xrightarrow[\text{H}_2\text{O}]{\text{NH}_4\text{Cl}} R-\underset{R}{\overset{R'}{\overset{|}{\underset{|}{C}}}}-OH$$

Ketone   3° alcohol

Thus, by the reaction of ethylmagnesium bromide with ethyl acetate (2 moles), 3-methyl-3-pentanol is obtained in 67% yield.

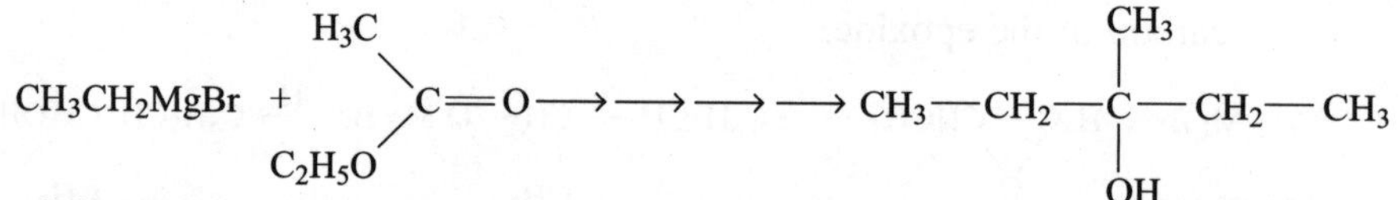

$$\text{CH}_3\text{CH}_2\text{MgBr} + {}^{\text{H}_3\text{C}}_{\text{C}_2\text{H}_5\text{O}}\text{C}{=}\text{O} \longrightarrow \longrightarrow \longrightarrow \longrightarrow \text{CH}_3-\text{CH}_2-\underset{\overset{|}{\text{OH}}}{\overset{\text{CH}_3}{\overset{|}{\text{C}}}}-\text{CH}_2-\text{CH}_3$$

ethyl magnesium bromide   ethyl acetate   3-methyl-3-pentanol
                                                       67%

Tertiary alcohols are also obtained by the reaction of cyclic ketones with grignard reaction.

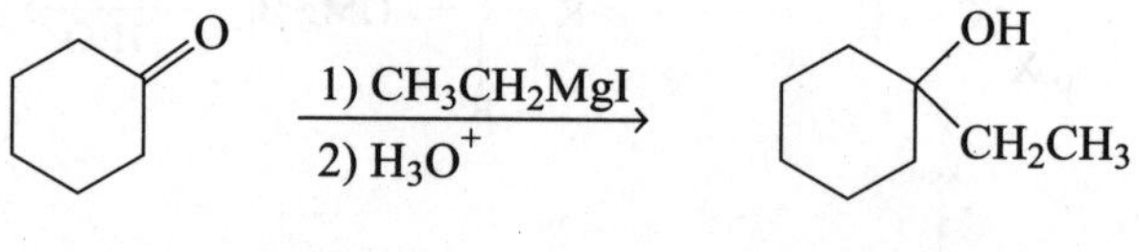

$$\text{cyclohexanone} \xrightarrow[\text{2) H}_3\text{O}^+]{\text{1) CH}_3\text{CH}_2\text{MgI}} \text{3° alcohol (OH, CH}_2\text{CH}_3\text{)}$$

cyclohexanone   3° alcohol

In the above reaction, the formed carbinol looses a molecule of water on treatment with conc. H_2SO_4 to give the corresponding cyclohexene, which on treatment with per acid gives the expoxide.

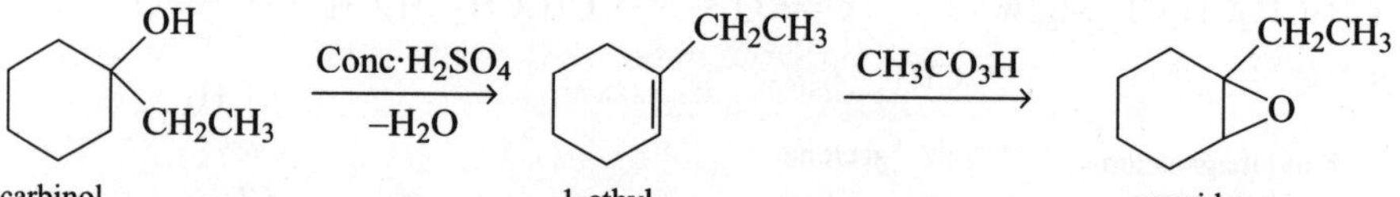

$$\text{carbinol (OH, CH}_2\text{CH}_3\text{)} \xrightarrow[-\text{H}_2\text{O}]{\text{Conc·H}_2\text{SO}_4} \text{1-ethyl cyclohexene (CH}_2\text{CH}_3\text{)} \xrightarrow{\text{CH}_3\text{CO}_3\text{H}} \text{epoxide (CH}_2\text{CH}_3\text{, O)}$$

carbinol   1-ethyl cyclohexene   epoxide

The above sequence of reaction is a convenient route for the synthesis of 1,2-epoxy-1-ethycylclohexane.

Tertiary alcohols are also obtained by the reaction acid chloride, acid anhydride or amides with grignard reagents.

$$CH_3—\overset{\overset{\displaystyle O}{\|}}{C}—Cl + CH_3MgI \longrightarrow \left[\overset{\displaystyle CH_3}{\underset{\displaystyle CH_3}{\overset{|}{\underset{|}{C}}}}\overset{\overset{\displaystyle \bar{O}\,\overset{+}{M}gI}{}}{\underset{\displaystyle Cl}{}} \right] \overset{HOH}{\longrightarrow}$$

acetyl chloride

$$\overset{\displaystyle CH_3}{\underset{\displaystyle CH_3}{C}}{=}O \quad \overset{1)\ CH_3MgI}{\underset{2)\ H_3O^+}{\longrightarrow}} \quad \overset{\displaystyle CH_3}{\underset{\displaystyle CH_3}{\overset{|}{\underset{|}{C}}}}\overset{\displaystyle CH_3}{\underset{\displaystyle OH}{}}$$

tert. butyl alcohol

$$\overset{\delta-\ \ \delta+}{RMg\,X} + \overset{\displaystyle R'—C}{\underset{\displaystyle R'—C}{}} \longrightarrow R'—\overset{\overset{\displaystyle \bar{O}MgX}{}}{\underset{\displaystyle R}{\overset{|}{\underset{|}{C}}}}—O—\overset{\overset{\displaystyle O}{\|}}{C}—R' \longrightarrow R'—\overset{\overset{\displaystyle O}{\|}}{C}—R'$$

anhydride

$$\overset{RMgX}{\longrightarrow} R—\overset{\overset{\displaystyle OH}{}}{\underset{\displaystyle R}{\overset{|}{\underset{|}{C}}}}—R'$$

3° alcohol

$$RMg\,X + \overset{\displaystyle R'}{\underset{\displaystyle H_2N}{}}C{=}O \longrightarrow R—H + \left[R'—\overset{\overset{\displaystyle O}{\|}}{\underset{\displaystyle \underset{-}{N}H\,\overset{+}{M}gX}{C}} \right] \overset{RMgX}{\longrightarrow}$$

$$\left[\overset{\displaystyle R}{\underset{\displaystyle R'}{\overset{|}{\underset{|}{C}}}}\overset{\overset{\displaystyle \bar{O}\overset{+}{M}gX}{}}{\underset{\displaystyle \underset{-}{N}H\,\overset{+}{M}gX}{}} \right] \overset{H_3O^+}{\longrightarrow} \overset{\displaystyle R}{\underset{\displaystyle R'}{}}C{=}O \quad \overset{1)\ RMgX}{\underset{2)\ H_3O}{\longrightarrow}} \quad 3°\ alcohol$$

By using grignard reaction (or synthesis) skilfully, any alcohol can be synthesised by using correct grignard reagent and the correct aldehyde, ketone, ester or epoxide.

(ii) **Synthesis of carboxylic acids**

Grignard reagents on treatment with solid carbon dioxide followed by hydrolysis yields carboxylic acids. Alternatively, a good yield of carboxylic acid can be obtained by passing CO_2 into the grignard solution at low temperature (0°C).

$$\overset{\delta-}{RMgX} + \overset{\overset{\displaystyle O}{\|}}{\underset{\displaystyle O}{C}} \longrightarrow R—\overset{\overset{\displaystyle O}{\|}}{C}\bar{O}\overset{+}{M}gX \overset{HCl}{\underset{H_2O}{\longrightarrow}} RCOOH + Mg(X)Cl$$

The carboxylic acid obtained corresponds to the alkyl group of the grignard reagent.

(iii) Synthesis of alkyl cyanides

Grignard reagents react with cyanogen or cyanogen chloride to yield alkyl cyanide.

$$\underset{\substack{\text{methyl magnesium}\\\text{iodide}}}{\overset{\delta-\quad\delta+}{CH_3MgI}} + \underset{\text{cyanogen}}{NC-CN} \longrightarrow \underset{\text{methyl cyanide}}{CH_3C\equiv N} + Mg(CN)I$$

$$\underset{\substack{\text{cyanogen}\\\text{chloride}}}{\overset{\delta-\quad\delta+}{CH_3MgI}} + NC-Cl \longrightarrow \underset{\text{methyl cyanide}}{CH_3C\equiv N} + Mg(Cl)I$$

(iv) Synthesis of ethers

Grignard reagents react with lower halogenated ethers to produce higher ethers.

$$\underset{\substack{\text{ethyl magnesium}\\\text{bromide}}}{\overset{\delta-\quad\delta+}{CH_3CH_2Mg\,Br}} + \underset{\substack{\text{methoxy methyl}\\\text{chloride}}}{CH_3O\,CH_2-Cl} \longrightarrow \underset{\text{methyl propyl ether}}{CH_3CH_2CH_2OCH_3}$$

(v) Synthesis of primary amines

Grignard reagents react with chloramine to give primary amines.

$$\underset{\substack{\text{methyl magnesium}\\\text{iodide}}}{\overset{\delta-\quad\delta+}{CH_3MgI}} + \underset{\text{chloramine}}{H_2N-Cl} \longrightarrow \underset{\substack{\text{methyl}\\\text{amine}}}{CH_3NH_2} + Mg(Cl)I$$

$$\underset{\substack{\text{isopropyl methyl magnesium}\\\text{iodide}}}{\overset{\delta-\quad\delta+}{(CH_3)_2CH\,CH_2\,MgI}} + \underset{\text{chloramine}}{H_2N-Cl} \longrightarrow \underset{\text{2-methyl propyl amine}}{(CH_3)_2CH\,CH_2NH_2} + Mg(Cl)I$$

(vi) Synthesis of aldehydes

The reaction of grignard reagents with ethyl orthoformate gives aldehyde as the main product. In case of ethyl orthoformate, the formation of secondary alcohol is not possible due to formation of acetal. The acetal can be isolated and treated with acid to give aldehyde.

$$\underset{\substack{\text{methyl magnesium}\\\text{iodide}}}{\overset{\delta-\quad\delta+}{CH_3\,MgI}} + \underset{\text{ethyl orthoformate}}{H-\overset{\overset{\displaystyle OC_2H_5}{|}}{\underset{\underset{\displaystyle OC_2H_5}{|}}{C}}-OC_2H_5} \longrightarrow \underset{\text{acetaldehyde acetal}}{CH_3CH(OC_2H_5)_2} + Mg(OC_2H_5)I$$

$$\downarrow H^+$$

$$\underset{\text{acetaldehyde}}{CH_3CHO} + 2C_2H_5OH$$

This procedure is useful for the synthesis of formyl compounds which are difficult to prepare by usual methods.

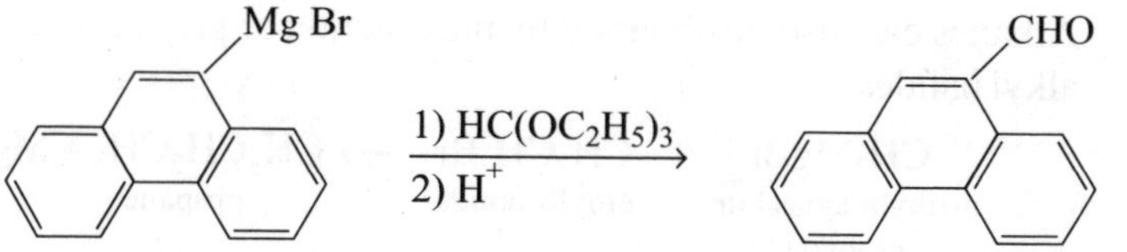

45%

(vii) Synthesis of ketones

Grignard reagents react with nitriles (alkyl cyanides) to give ketones. (However, use of HCN in place of alkyl cyanide give an aldehyde). The reaction proceeds via the formation of a ketimine, which on acid hydrolysis gives ketone. (See also reaction of grignard reagent with acid chlorides and esters, which yield alcohols).

$$R\,MgX + R'C\equiv N \longrightarrow R\,R'C\!\!=\!\!N\,MgX$$

$$\downarrow H_2O$$

$$R\,COR' \xleftarrow{H_3O^+} [R\,R'C\!\!=\!\!NH]$$

ketone   ketimine

The above procedure is used for the preparation of ketones which are difficult to prepare by conventional methods.

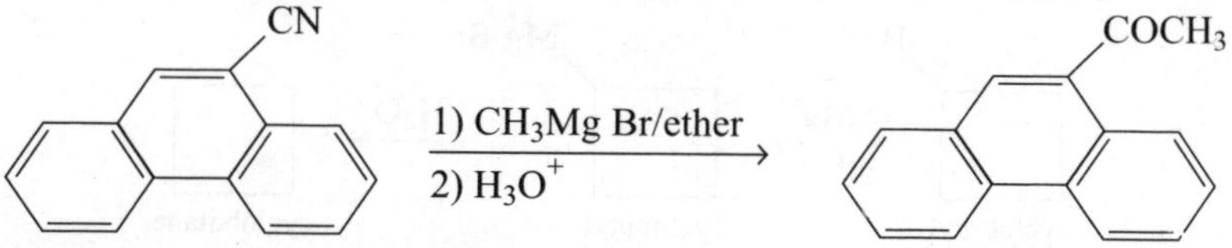

(viii) Synthesis of hydrocarbons

(a) Alkanes are obtained by the reaction of grignard reagent with compounds containing active hydrogen like water, alcohol, ammonia or amines.

$$\underset{\substack{\text{methylmagnesium}\\ \text{iodide}}}{CH_3MgI} + \underset{\text{water}}{HOH} \longrightarrow \underset{\text{methane}}{CH_4} + Mg(OH)I$$

$$\underset{\substack{\text{methylmagnesium}\\ \text{iodide}}}{CH_3MgI} + \underset{\text{alcohol}}{R\text{–}OH} \longrightarrow CH_4 + Mg(OR)I$$

$$\underset{\substack{\text{methylmagnesium}\\ \text{iodide}}}{CH_3MgI} + \underset{\text{ammonia}}{NH_3} \longrightarrow CH_4 + Mg(NH_2)I$$

$$\underset{\substack{\text{methylmagnesium}\\ \text{iodide}}}{CH_3MgI} + \underset{\text{amine}}{RNH_2} \longrightarrow CH_4 + Mg(NHR)I$$

This method forms the basis for the **Zerewitinoff determination of active hydrogens**. Thus, by measuring the amount of methane liberated from a known weight of the compound, containing active hydrogens, the number of active hydrogens present in a molecule can be obtained.

Alkanes can also be obtained by the reaction of grignard reagents with alkyl halides.

$$CH_3MgBr + CH_3CH_2Br \longrightarrow CH_3CH_2CH_3 + Mg(Br)I$$

methylmagnesium   ethylbromide   propane
bromide

(b) Alkenes can be similarly obtained by the reaction of grignard reagents with unsaturated alkyl halides.

$$CH_3MgI + CH_2{=}CHCH_2Br \longrightarrow CH_2{=}CHCH_2CH_3 + Mg(Br)I$$

methyl magnesium   allyl bromide   1-butene
iodide

(c) Higher alkynes are obtained by the treatment of terminal alkynes with grignard reagents followed by treatment of the formed alkynylmagnesium halide with alkyl halides.

$$CH_3C{\equiv}CH + CH_3MgI \longrightarrow CH_3C{\equiv}CH-MgI \xrightarrow[SN_2]{CH_3I} CH_3C{\equiv}CCH_3 + MgI_2$$

propyne   methyl magnesium   propyenyl magnesium   2-butyne
iodide   iodide

(d) Cycloalkanes are obtained by the treatment of cyclobutylmagnesium bromide with water

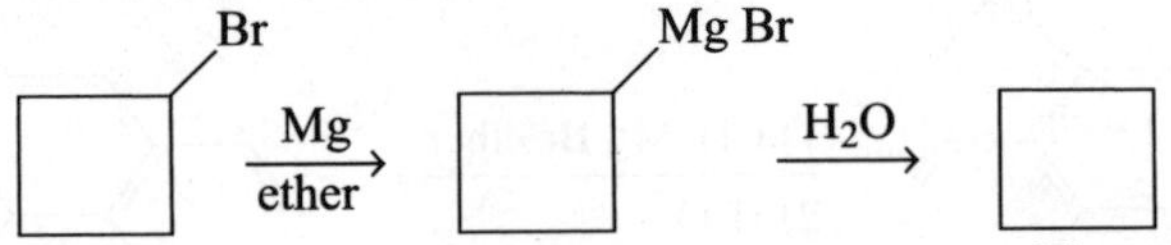

cyclobutyl   cyclobutyl   cyclobutane
bromide   magnesium
bromide

The above procedure is used for the replacement of a halogen with hydrogen in alkyl halides

$$R-X \xrightarrow[ether]{Mg} RMgX \xrightarrow{H_2O} RH + Mg(X)OH$$

(e) Deuterated hydrocarbons
The reaction of grignard reagent with D_2O gives deuterated hydrocarbons.

$$R-I \xrightarrow[ether]{Mg} RMgI \xrightarrow{D_2O} RD + Mg(I)OD$$

alkyl halide   grignard reagent

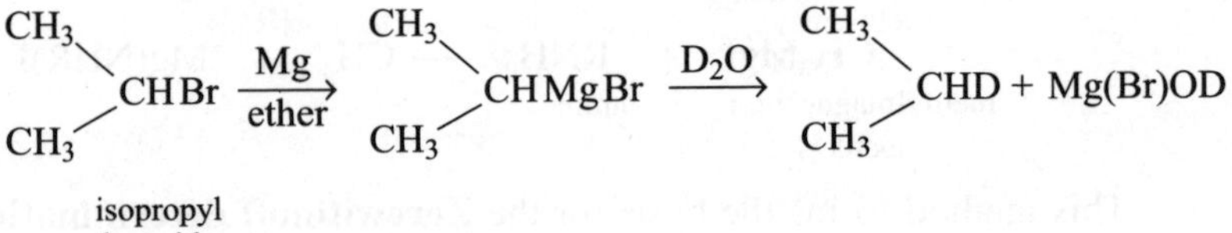

isopropyl
bromide

(ix) **Synthesis of alkyl iodides**
The reaction of grignard reagents (RMgCl or RMgBr) with iodine in acetone give the corresponding alkyl iodides.

$$\overset{\delta-}{R}\overset{\delta+}{MgI} + I{-}I \longrightarrow RI + MgI_2$$

$$(CH_3)_3C\ CH_2Cl \xrightarrow[\text{ether}]{Mg} (CH_3)_3C\ CH_2MgCl$$

neopentyl chloride

$$\Big\downarrow I_2/\text{acetone}$$

$$(CH_3)_3C\ CH_2I$$
neopentyl iodide

(x) Synthesis of thioalcohols, dithioacetic acids and sulfinic acids

Grignard reagents react with sulphur, carbon disulphide and sulphur dioxide to give the corresponding thioalcohols, dithioacetic acids and sulfinic acids respectively.

$$R\ MgX + S \longrightarrow Mg\!\!\begin{array}{c}SR\\\ \\X\end{array} \xrightarrow{H_3O^+} RSH + Mg(OH)X$$

sulfur   thioalcohols

$$\overset{\delta-}{CH_3}\overset{\delta+}{MgX} + S{=}C{=}S \longrightarrow S{=}\overset{\overset{\textstyle CH_3}{|}}{C}{-}\overset{+}{S}\ MgI \xrightarrow{H_3O^+} CH_3CS_2H + Mg(OH)I$$

carbon disulphide   dithioacetic acid

$$R\ MgX + SO_2 \longrightarrow R{-}\overset{\overset{\textstyle \bar{O}\ \overset{+}{M}gX}{|}}{\underset{\underset{\textstyle O}{\|}}{S}} \xrightarrow{H_3O^+} R{-}\overset{\underset{\textstyle O}{\|}}{S}{-}OH + Mg(OH)X$$

sulfur dioxide   sulfinic acid

(xi) Synthesis of organometallic compounds

Grignard reagents react with inorganic halides to give organometallic compounds. Some examples are

$$4C_2H_5MgBr + PbCl_4 \longrightarrow (C_2H_5)_4Pb + 4Mg(Br)Cl$$
ethylmagnesium   lead   lead tertra
bromide   chloride   ethyl

$$4CH_3MgI + SiCl_4 \longrightarrow (CH_3)_4Si + 4Mg(Br)Cl$$
methylmagnesium   silicon   tetramethyl
iodide   tetrachloride   silane

$$3CH_3MgBr + PCl_5 \longrightarrow (CH_3)_3P + 3Mg(Cl)Br$$
methylmagnesium   Phosphorus   Trimethyl
bromide   pentachloride   phosphene

$$3CH_3MgBr + CdCl_2 \longrightarrow (CH_3)_2Cd + 2Mg(Cl)Br$$
methylmagnesium   cadmium   dimethyl
chloride   chloride   cadmium

(xii) Miscellaneous applications of grignard reaction

(a) Synthesis of hydroperoxides

Oxygenation of grignard reagent at low temperature provides an excellent method for the synthesis of hydroperoxides

$$RMgX + O_2 \xrightarrow{-70°} ROOMgX \xrightarrow{H^+} ROOH$$

To prevent the formation of excessive amounts of alcohol, inverse addition is best, i.e., a solution of grignard reagent is added to ether through which is bubbled rather than have the oxygen bubble through a solution of the grignard reagent.

(b) Synthesis of stilbene

The reaction of benzylmagnesium bromide with benzaldehyde followed by dehydration of the formal alcohol gives stilbene

$$C_6H_5CH_2MgBr + C_6H_5CHO \longrightarrow C_6H_5CH_2\underset{\underset{OH}{|}}{CH}\,C_6H_5 \xrightarrow{-H_2O} C_6H_5CH = CH\,C_6H_5$$

benzylmagnesium   benzaldehyde
bromide   stilbene

(c) Synthesis of 1,2-divinyl cyclohexanol[5]

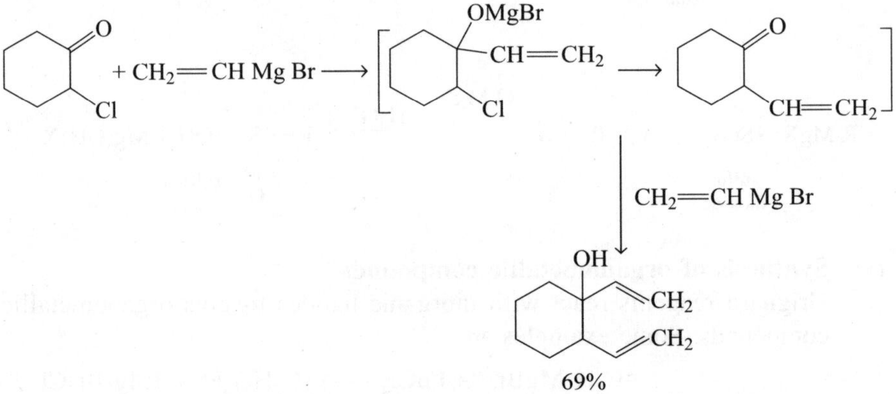

69%
(cis/trans, 85:15)
1,2-divinyl cyclohexanol

REFERENCES

1. V. Grignard, Compt. Rend., 1900, 130, 1322.
2. J.L. Luche and J.C. Damino, J. Am. Chem. Soc., 102, 7964; J.D. Sprinch and G.S. Lewandos, Inorg. Chem. Acta, 1982, 76, 1241; W. Oppolzer and A, Nakao, Tetrahedtron Lett., 1986, 27, 5471.
3. W. Oppolzer and A. Nakao, Tetrahedron Lett., 1986, 27, 5471.
4. F. Todd, H. Takumi and H. Yamaguchi. Chem. Exp., 1980, 4, 507.
5. D. Holt, Tetrahedron Lett., 1981, 2243.

2.19 HECK REACTION

The coupling of an alkene with a halide or triflate in the presence of Pd(O) catalyst to form a new alkene is known as Heck reaction[1].

$$RX + \underset{R'}{\overset{H}{\diagdown}} \xrightarrow[\text{base}]{Pd\,(O)} \underset{R'}{\overset{R}{\diagdown}} + HX$$

R = aryl, vinyl or alkyl group without β-hydrogen on a sp^3 carbon atom
X = halide or triflate (OSO$_2$CF$_3$).

The base used in Heck reaction is a mild base like Et$_3$N or anions like $^-$OH, $^-$OCOCH$_3$, CO$_3$$^{2-}$ etc. and the reaction is carried out in anhydrous polar solvents like DMF, MeCN etc.

Heck reaction is one of the most synthetically useful palladium catalysed reaction. It is used for the alkylation or arylation of alkenes.

MECHANISM

The bases like $^-$OH, $^-$OCOCH$_3$, CO$_3$$^{2-}$ etc. may serve as ancillary ligands for palladium. The mechanism involves the oxidative addition of the halide (RX), insertion of the olefin, and elimination of the product by a β-elimination. The base regenerates the palladium (O) catalyst and the cycle continues (Scheme-1).

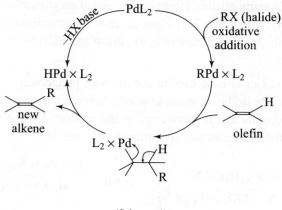

(Scheme-1)

2.19.1 Heck Reaction in Aqueous Phase

Though traditionally, Heck reaction uses anhydrous solvents (e.g., DMF and MeCN), it has been found that the reaction can proceed very well *in water*. The role of water in the Heck reaction, as well as other reactions catalysed by Pd(O) in presence of phosphine ligand is transformation of catalyst precursor into Pd(O) species and the generation of zero-valent palladium species capable of oxidative addition by oxidation of phosphine ligands by the Pd(II) precursor can be effected by the water content of the reaction mixture.

2.19.2 Heck Reaction under PTC conditions

It has been found that the Heck reaction can be successfully carried out under PTC conditions[2] with inorganic carbonates as base under very mild conditions even at room temperature. Using this procedure, even substrates such as methyl vinyl ketone that is unstable under the conventional condition of Heck arylation react (action of base at high temperature). Subsequently it was shown that water and aqueous organic solvents can be used successfully for carrying out Heck reaction in presence[3] of palladium salts and inorganic bases like K_2CO_3, Na_2CO_3, $NaHCO_3$, KOH etc.

An interesting application of Heck reaction is the synthesis of cinnamic acid by the reaction of aryl halides and acrylic acid (Scheme-2).

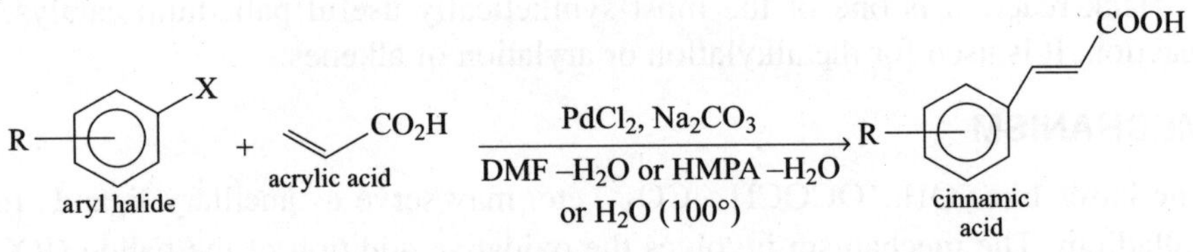

(Scheme-2)

Use of acrylonitrile in place of acrylic acid in the above reaction gives the corresponding cinnamonitriles. However, in case of acrylonitrile a mixture of (E) and (Z) isomers with ratio 3:1, close to that observed under conventional conditions[4] is obtained in comparison to almost exclusively (E) isomers in the Heck reactions.

Heck reaction can also be carried out at room temperature[5] if diaryliodium salts are taken as the arylating agent in water. At room temperature, only are aryl group of the iodinium salt is transferred to the product. However at 100°, both aryl groups of the iodonium salt are utilised (Scheme-3).

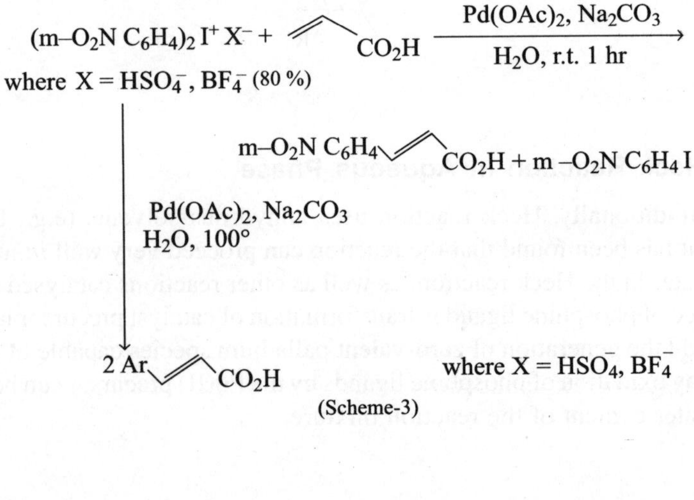

(Scheme-3)

2.19.3 Heck Reaction in Ionic Liquids

The Heck reaction has also been performed in neutral ionic liquids which are excellent solvents. The special advantage of using neutral ionic liquids is that many palladium complexes are soluble in ionic liquids[5] and that the product or products of the reaction can be extracted with water or alkane solvents. A typical Heck reaction in ionic liquid is given below (Scheme-4).

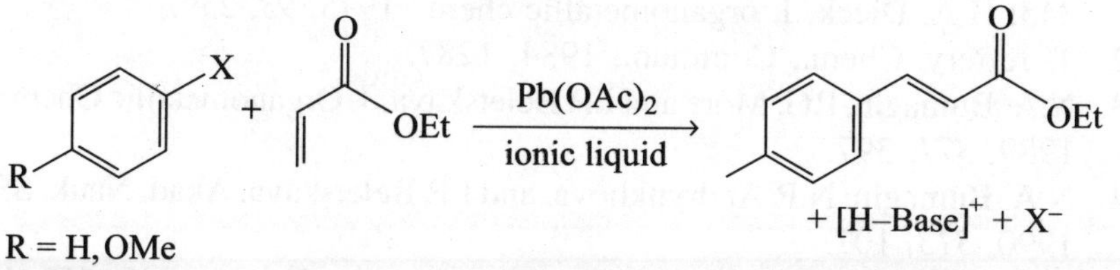

R = H, OMe
X = Br, I

(Scheme-4)

An alternative to Heck reaction is to use aromatic anhydrides as a source of the aryl group (Scheme-5).

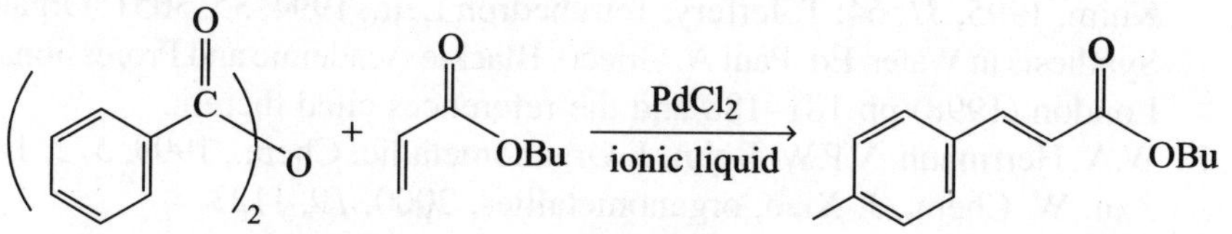

(Scheme-5)

In the above procedure benzoic acid is obtained as a by product, but it can be recycled.

2.19.4 Applications

Some applications of Heck reaction under conventional conditions are given below.

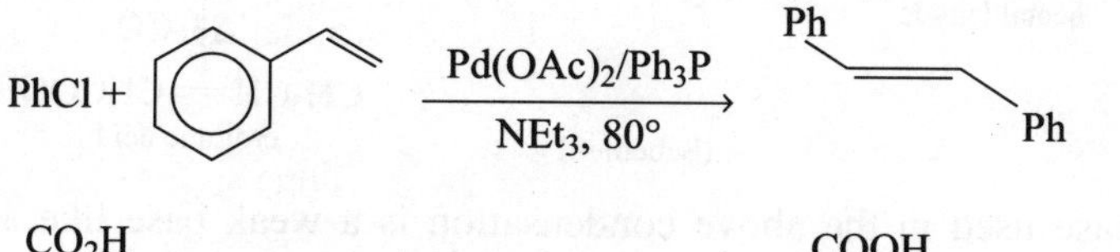

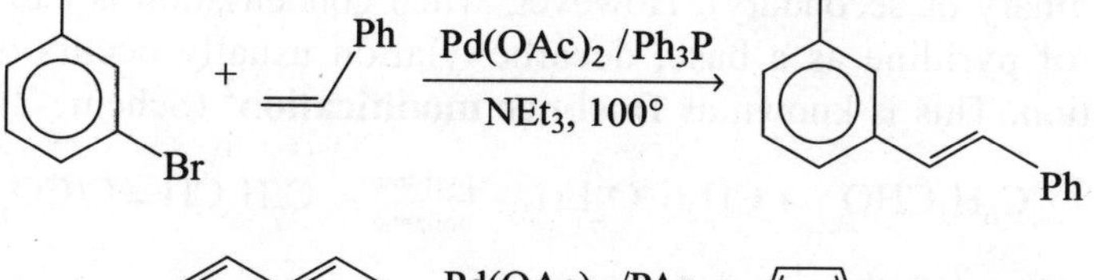

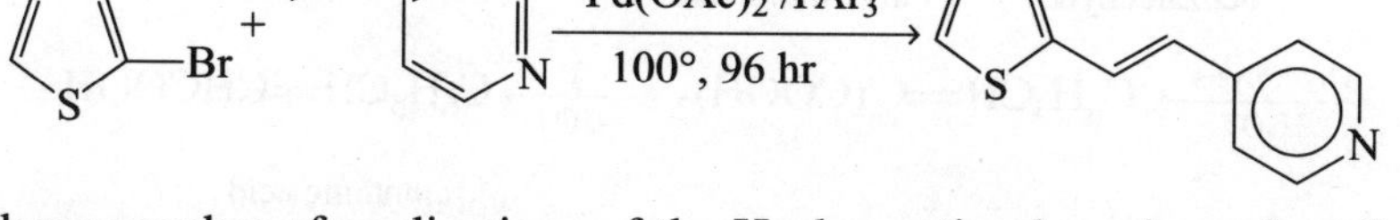

A large number of applications of the Heck reaction have been described in literature[6].

The Heck reaction of chlorobenzene with styrene in presence of ionic liquid has been carried out[7].

REFERENCES

1. R.F Heck, Acc. Chem. Res., 1979, *12*, 146; Organic Reactions, 1982, *27*, 345; H.A. Dieck, J. organometallic chem., 1975, *93*, 259.
2. T. Jeffery, Chem., Commun., 1984, 1287.
3. N.A. Bumagin, P.G. More and I.P. Beletskaya, J. Organometallic Chemistry, 1989, *371*, 397.
4. N.A. Bumagin, N.P. Andryukhova, and I.P. Beletskaya, Akad. Nauk. SSSR, 1990, *313*, 107.
5. W.A. Herrann and V.P.W. Bohn. J. Organomet. Chem., 1999, *572*, 141.
6. N.A. Bumagin, L.I. Sukhomlinova, A.N. Vanchikov, T.P. Tolstaya and I.P. Beletskaya, Bull. Russ. Acad. Sec., Div. Chem. Sec. 1992, *41*, 2130; N.N. Denik, M.M. Rabachnik, M.M. Novikova and I.P. Beletskaya, Zh. Org. Khim, 1995, *31*, 64; T. Jeffery, Tetrahedron Lett., 1994, *35*, 3051; Organic Synthesis in Water. Ed. Paul A. Grieco, Blackie Academic and Professionals, London (1998) pp 181–188 and the references cited therein.
7. W.A. Herrmann, V.P.W. Bohn, J. Organometallic. Chem., 1999, *572*, 141; Zxu, W. Chem., J. Xiao, organometallics, 2000, *19*, 1123.

2.20 KNOEVENAGEL CONDENSATION

Base catalysed condensation between an aldehyde or ketone, with any compound having an active methylene group (especially malonic ester) is called Knoevenagel condensation[1] (Scheme-1).

$$CH_3CHO + CH_2(COOH)_2 \xrightarrow{\text{Base}} CH_3CH = C(COOH)_2$$

acetaldehyde

$$\Delta \downarrow -CO_2$$

$$CH_3CH = CHCOOH$$

crotonic acid

(Scheme-1)

The base used in the above condensation is a weak base like ammonia or amine (primary or secondary). However, when condensation is carried out in presence of pyridine as a base, decarboxylation usually occurs during the condensation. This is known as **Doebner modification**[2] (Scheme-2).

$$C_6H_5CHO + CH_2(CO_2Et)_2 \xrightarrow[\text{benzene}]{\text{pyridine}} C_6H_5CH=C(CO_2Et)_2$$

benzaldehyde   ethyl malonate

$$\xrightarrow[H_3O^+]{\text{hydrolysis}} C_6H_5CH=C(COOH)_2 \xrightarrow[-CO_2]{\Delta} C_6H_5CH=CHCOOH$$

cinnamic acid

(Scheme-2)

The Knoevenagel reaction is more useful with aromatic aldehydes, since with aliphatic aldehydes, the product obtained undergoes **Michael condensation**. As an example, tetraethyl propane-1,1,3,3-tetracarboxylate is obtained by Knoevenagel condensation of formaldehyde with diethyl malonate in the presence of diethylamine, followed by Michael addition reaction to yield the final product (Scheme-3).

$$O{=\!\!=}CH_2 \; + \; \bar{C}H(CO_2Et)_2 \xrightarrow[]{Et_2NH} HOCH_2{-\!\!\!-}CH(CO_2Et)_2$$

formaldehyde

$$\text{(EtO}_2C)_2\,CH\,CH_2\,CH(CO_2Et)_2 \xleftarrow[\text{Michael addn.}]{\bar{C}H(CO_2Et)_2} CH_2{=\!\!=}C(COOEt)_2$$

adduct
tetraethyl propane-
-1,1,3,3-tetracarboxylate

$\Big\downarrow -H_2O$

(Scheme-3)

The effectiveness of various activating groups in the active methylene compounds is found to be in the order

$$NO_2 > CN > COCH_3 > COC_6H_5 > COOC_2H_5$$

Ketones do not undergo Knoevenagel condensation with malonic ester but can react with more active cyanoacetic acid or its ester. For example, acetone forms[2a] isopropylidene cyanoacetic ester when condensed with ethyl cyanoacetate (Scheme-4).

$$(CH_3)_2\,C{=\!\!=}O + CH_2 \overset{CN}{\underset{CO_2Et}{\big<}} \longrightarrow (CH_3)_2\,C{=\!\!=}C \overset{CN}{\underset{CO_2Et}{\big<}}$$

acetone ethyl cyanoacetate isopropylidene cyanoacetic ester

(Scheme-4)

The Knoevenagel reaction is reversible and the equilibrium normally lies towards left giving low yields. The yield can be improved by carrying out the reaction in benzene and removal of the formed water by azeotropic distillation using a Dean stark apparatus. This is referred to as **Cope-Knoevenagel reaction**.

MECHANISM

The base removes a proton from the active methylene compound to give a carbanion (which is resonance stabilized). The carbonion then attacks the carbonyl carbon of the aldehyde or ketone. Subsequent protonation of the anion followed by dehydration yields the product (Scheme-5).

$$CH_2\,(CO_2Et)_2 \xrightarrow{:B} \bar{C}H\,(CO_2Et)_2$$

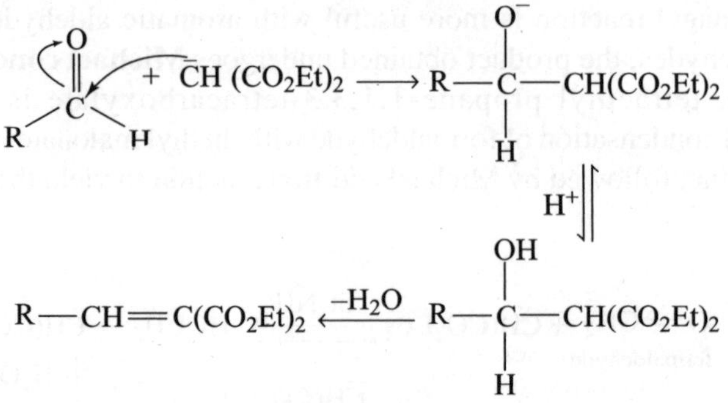

(Scheme-5)

Ketones or aldehydes also react with a succinic ester in presence of sodium hydride to give the corresponding condensation product (Scheme-6).

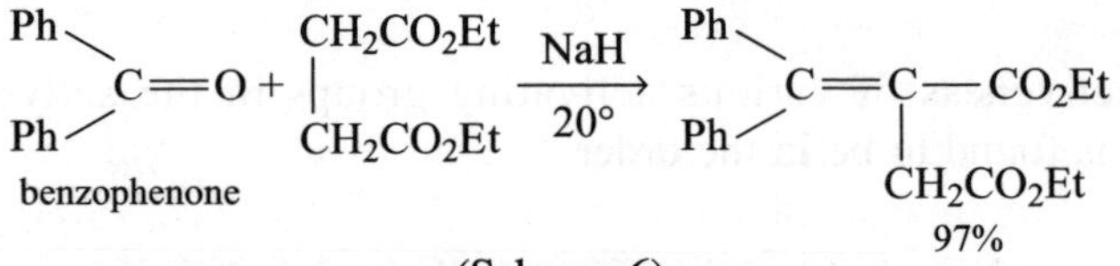

benzophenone

(Scheme-6)

This condensation is known as **Knoevenagel-stobbe condensation**[3].

2.20.1 Knoevenagel Reaction in Water

The Knoevenagel reaction has been carried out between aldehydes and acetonitrile in water. Thus salicyaldehydes react with malononitrile in heterogeneous aqueous alkaline medium at room temperature to give o-hydroxy benzylidenemalononitrile[4], which are converted by acidification and heating to give 3-cyanocumarins in good yield (Scheme-7).

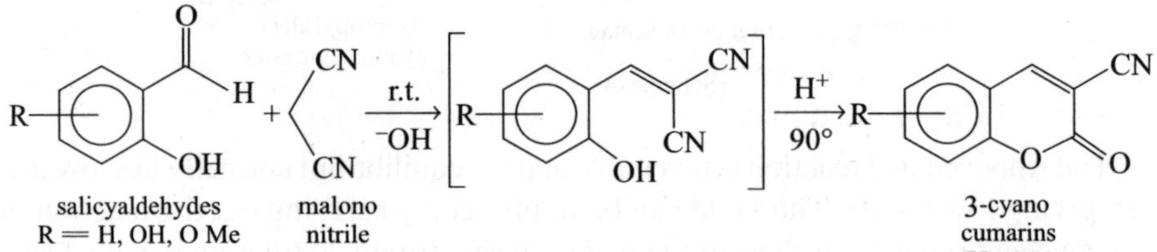

salicyaldehydes   malono   3-cyano
R = H, OH, O Me   nitrile   cumarins
                                          75−95 %

(Scheme-7)

The condensation of substituted acetonitriles with salicyaldehydes requires the presence of catalytic amount of CTABr. It is found that the aqueous phase reaction[4] gives better yield (Scheme-8).

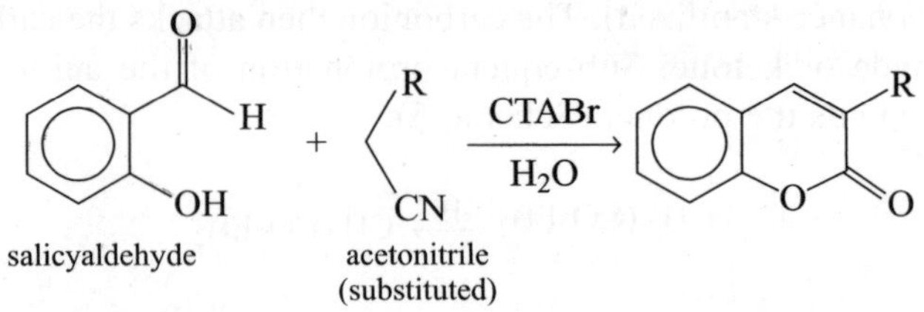

salicyaldehyde   acetonitrile
                        (substituted)

R	Yield of Curmarin	
	in water	in ethanol
CN	90	70
CO$_2$Et	66	35
NO$_2$	87	80
2-Py	98	55

(Scheme-8)

The condensation of benzaldehyde with aryl acetonitrile does not take place in water but requires the presence of CTACl or TBACl to give high yields of aryl cinnamonitriles[5] (Scheme-9).

Ar CH$_2$CN + Ph CHO $\xrightarrow[\text{r.t. 0.5-9 h}]{\text{CTACl / NaOH}}$

Aryl acetonitrile   benzaldehyde
Ar = Ph, p-or NO$_2$C$_6$H$_4$,
PhSO$_2$

Ph $\diagup$ CN
H $\diagdown$ Ar
85–90 %
aryl cinnamonitriles

(Scheme-10)

Knoevenagel-type addition products can be obtained[6] by the reaction of acrylic derivatives in presence of 1,4-diazabicyclo[2·2·2] octane (DABCO) (Scheme-11).

CN (acrylo nitrile) + Ph CHO $\xrightarrow[\text{DABCO}]{\text{RT / H}_2\text{O}}$

benzaldehyde

Ph $\diagdown$ OH
$\diagup$ CN
90–98 %

(Scheme-11)

(See also Weiss-cook reaction, section 2.27)

2.20.2 Knoevenagel Reaction in Solid State

Knoevenagel reaction has been carried out in dry media[6a]. The method consist in adding solid inorganic support to a solution of aromatic aldehyde and diethylmalonate in acetone. The adsorbed material was mixed properly, dried in air (beaker) and placed in an alumina bath[6b] inside the microwave oven for 2–3 min at medium power level (600 W) intermitently at 0.5 min. intervals, at 102°. The product was isolated by extraction of the reaction mixture with alcohol.

2.20.3 Knoevenagel Reaction in Ionic Liquids

Knoevenagel condensation has been carried out by using chloroaluminate based ionic liquids[9], which have variable Lewis acidity such as 1-butyl-3-methylimidazolium chloroaluminate, [bmin]Cl · AlCl$_3$ · X (AlCl$_3$) = 0.67, where X is the mole fraction. The extent of product obtained varies with the Lewis acidity and the molar proportion of the ionic liquids.

2.20.4 Applications

Knoevenagel reactions (or condensation) is of great importance. Some applications are given below.

(i) Synthesis of α, β-unsaturated carboxylic acids[7]

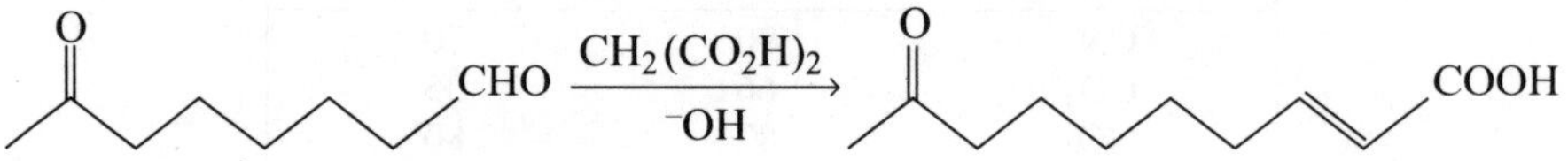

$$\text{(keto-aldehyde)} \xrightarrow[\text{$^-$OH}]{CH_2(CO_2H)_2} \text{(keto-unsaturated acid)}$$

(ii) Synthesis of conjugated carboxylic acids[8]

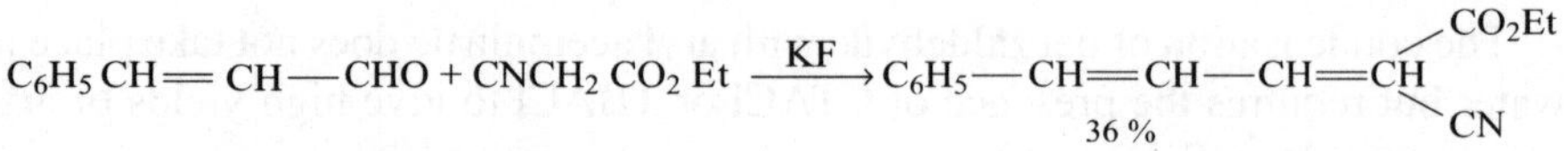

$$C_6H_5\,CH = CH\text{—}CHO + CNCH_2\,CO_2\,Et \xrightarrow{KF} C_6H_5\text{—}CH = CH\text{—}CH = CH\underset{CN}{\overset{CO_2Et}{<}}$$

36 %

(iii) Synthesis of malic acid

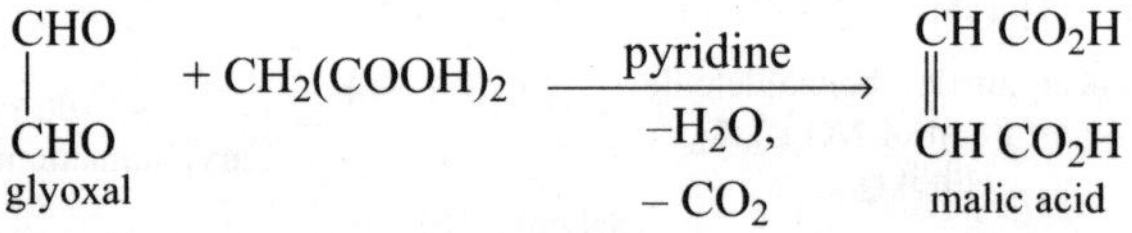

$$\underset{\substack{CHO \\ \text{glyoxal}}}{\overset{CHO}{|}} + CH_2(COOH)_2 \xrightarrow[\substack{-H_2O, \\ -CO_2}]{pyridine} \underset{\substack{CH\,CO_2H \\ \text{malic acid}}}{\overset{CH\,CO_2H}{||}}$$

(iv) Synthesis of substituted ethyl cinnamates

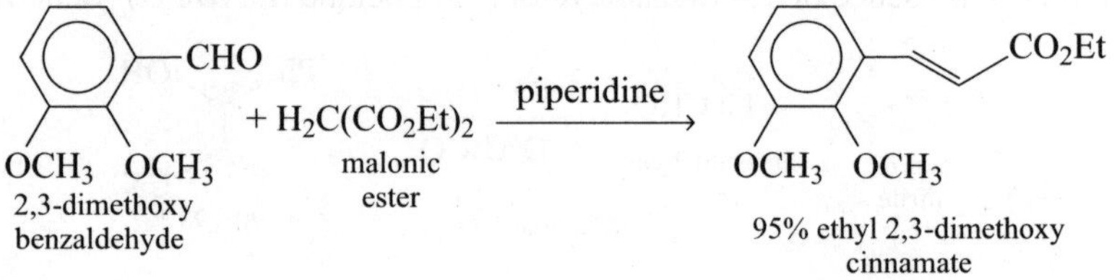

2,3-dimethoxy benzaldehyde (Ar—CHO) + H$_2$C(CO$_2$Et)$_2$ (malonic ester) $\xrightarrow{piperidine}$ 95% ethyl 2,3-dimethoxy cinnamate (Ar—CH=CH—CO$_2$Et)

(v) $Cl\text{—}C_6H_4\text{—}CHO + CH_2(CO_2Et)_2 \xrightarrow[\text{EtOH}]{Et_3N}$ $Cl\text{—}C_6H_4\text{—}\underset{OH}{\overset{}{CH}}\text{—}CH(CO_2Et)_2$

$\downarrow$ –H$_2$O

$Cl\text{—}C_6H_4\text{—}CH = C(CO_2Et)_2$

86 %

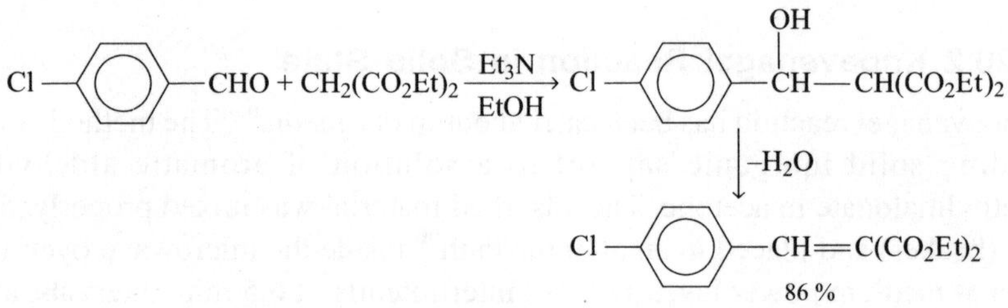

REFERENCES

1. F. Knoevenagel, Ber., 1896, 29, 172; J.R. Johnson, Org. Reactions, 1942, 1, 210.
2. O. Doebner, Ber., 1900, 33, 2140.
2a. F. Texier-Boullet and A. Foucand, Tetrahedron Lett., 1982, 4927.
3. H. Stobbe, Ber., 1893, 26, 2312.
4. K. Takai, C.H. Heathcock, J. Org. Chem., 1985, 50, 3247; A.E. Vougioakas and H.B. Kagan, Tetrahedron Lett., 1987, 28, 5513; K. Mikami, M. Terada and T. Nakai, J. Chem. Soc., Chem., Commun., 1993, 343 and references cited therein.

5. Paul A. Grieco Ed. in organic synthesis in water, Blackie Academic and Professional, London, 1998, p 255.

6. J. Aug., M. Lubin and A. Lubineau, Tetrahedron Lett., 1994, *35*, 7947.

6a. M. Kidwai, P. Sapra and K.R. Bhushan, J. Indian Chem. Soc., 2002, *79*, 596.

6b. G. Bram, A. Loupy and M. Majoub, Tetrahedron Lett., 1990, *46*, 5167.

7. C.G. Butler, R.K. Callow and N.C. Johnson, Proc. Royal Soc., 1961, *B,155*, 417.

8. Reported in the Review by Yakobson and N.E. Akhmetova, Synthesis, 1983, 173.

9. J.R. Harjani, S.J. Nara, M.M. Salunkhe, Tetrahedron Lett., 2002, *43*, 1127.

2.21 MICHAEL ADDITION

The base catalysed addition reaction between α, β-unsaturated carbonyl compounds (e.g., cinnamaldehyde, $C_6H_5CH=CH-CHO$; benzylidene acetone, $C_6H_5CH=CHCOCH_3$; mesityloxide, $(CH_3)_2C=CHCOCH_3$) etc and a compound with active methylene group (e.g., malonic ester, acetoacetic ester, cyanoacetic esters, nitroparalfins) is known as Michael addition[1]. The base usually employed is sodium ethoxide or a secondary amine (usually piperidine). Thus, methyl vinyl ketone reacts with diethyl malonate in presence of sodium ethoxide to give the addition product (Scheme-1).

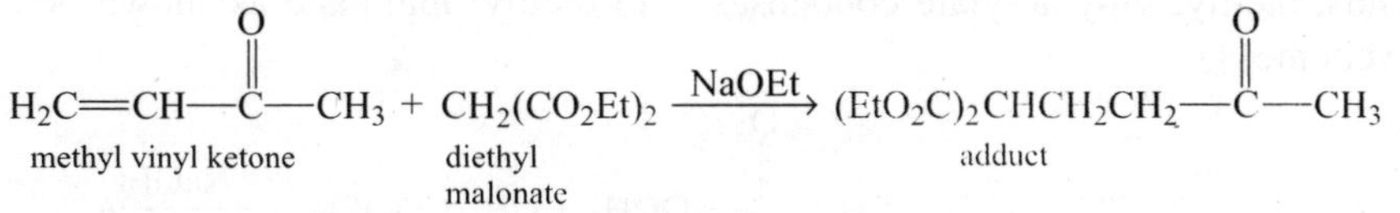

methyl vinyl ketone   diethyl malonate   adduct

(Scheme-1)

MECHANISM

Michael addition is regarded as nucleophilic addition of carbanions to α, β-unsaturated compounds. The base generates a carbanion (1) from active methylene compound, which then adds to the β-carbon of the α, β-unsaturated compound to give another anion (2), which in turn takes a proton from alcohol to produce an enol (3). The enol tautomerises to give the stable product, ketone. The reaction between diethylmalonate and benzylidene acetone is represented as shown below. (Scheme-2).

$$H_2C(CO_2Et)_2 + C_2H_5ONa \rightleftharpoons HC^-(CO_2Et)_2 + C_2H_5OH$$

(1)

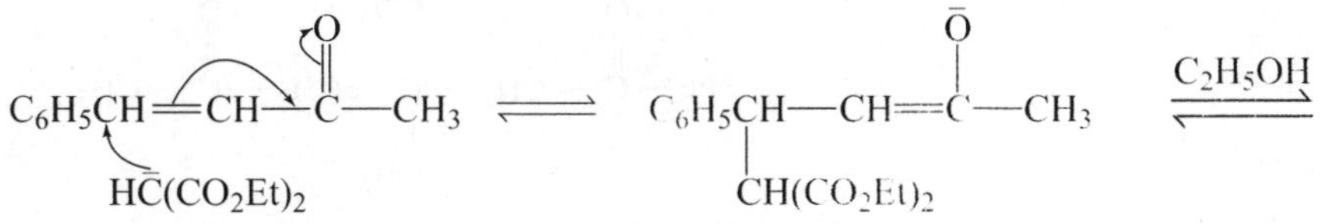

(2)

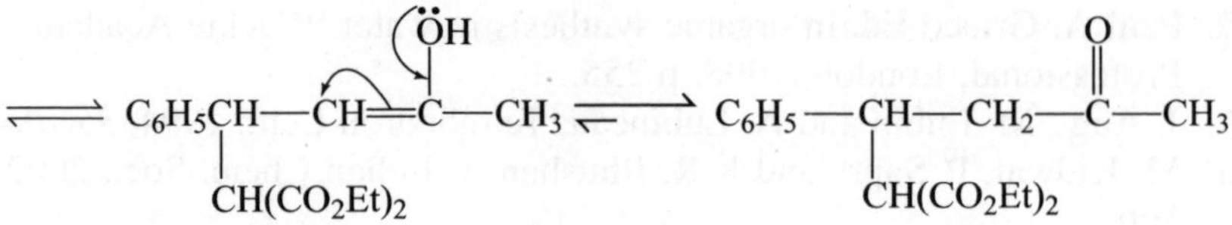

$$\rightleftharpoons C_6H_5CH\underset{\underset{CH(CO_2Et)_2}{|}}{-}CH\!\!=\!\!\overset{\overset{\ddot{O}H}{|}}{C}\!\!-\!\!CH_3 \rightleftharpoons C_6H_5\underset{\underset{CH(CO_2Et)_2}{|}}{-}CH\!\!-\!\!CH_2\!\!-\!\!\overset{\overset{O}{\|}}{C}\!\!-\!\!CH_3$$

(3)

(Scheme-2)

Michael additions take place with a variety of other reagents, e.g., acetylenic esters and α, β-unsaturated nitriles (Scheme-3).

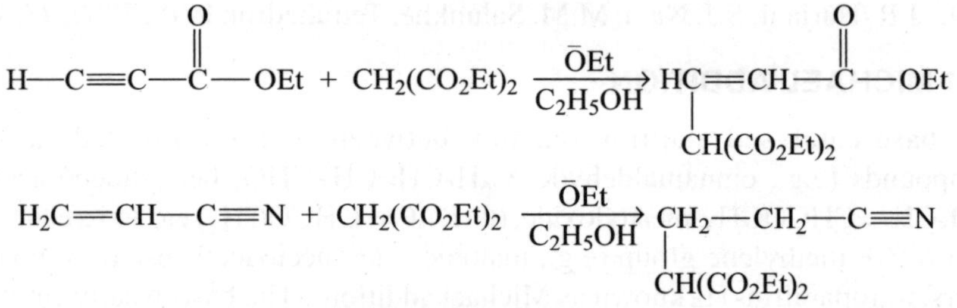

$$H\!\!-\!\!C\!\!\equiv\!\!C\!\!-\!\!\overset{\overset{O}{\|}}{C}\!\!-\!\!OEt + CH_2(CO_2Et)_2 \xrightarrow[C_2H_5OH]{\bar{O}Et} HC\!\!=\!\!\underset{\underset{CH(CO_2Et)_2}{|}}{CH}\!\!-\!\!\overset{\overset{O}{\|}}{C}\!\!-\!\!OEt$$

$$H_2C\!\!=\!\!CH\!\!-\!\!C\!\!\equiv\!\!N + CH_2(CO_2Et)_2 \xrightarrow[C_2H_5OH]{\bar{O}Et} \underset{\underset{CH(CO_2Et)_2}{|}}{CH_2}\!\!-\!\!CH_2\!\!-\!\!C\!\!\equiv\!\!N$$

(Scheme-3)

Compounds containing conjugated double bonds (conjugated with carbonyl group) react with active methylene compounds to give 1,6-addition products. Thus, methyl vinylacrylate condenses with diethyl malonate as shown below[2] (Scheme-4).

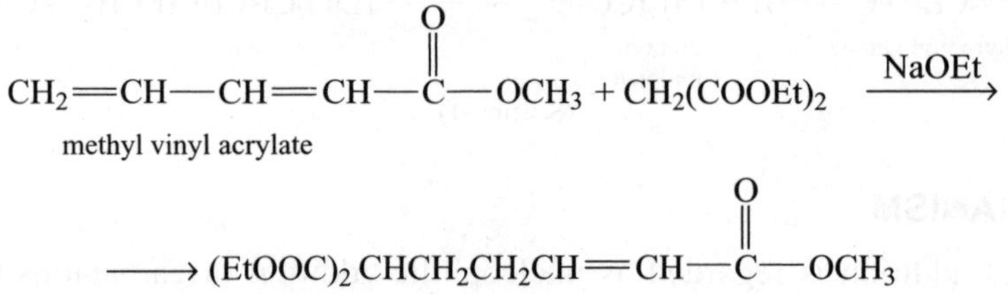

$$CH_2\!\!=\!\!CH\!\!-\!\!CH\!\!=\!\!CH\!\!-\!\!\overset{\overset{O}{\|}}{C}\!\!-\!\!OCH_3 + CH_2(COOEt)_2 \xrightarrow{NaOEt}$$

methyl vinyl acrylate

$$\longrightarrow (EtOOC)_2\,CHCH_2CH_2CH\!\!=\!\!CH\!\!-\!\!\overset{\overset{O}{\|}}{C}\!\!-\!\!OCH_3$$

(Scheme-4)

Michael addition products[3] are also obtained by the addition of silyl ethers of enols to α, β-unsaturated ketones and esters when catalysed by $TiCl_4$ (Scheme-5).

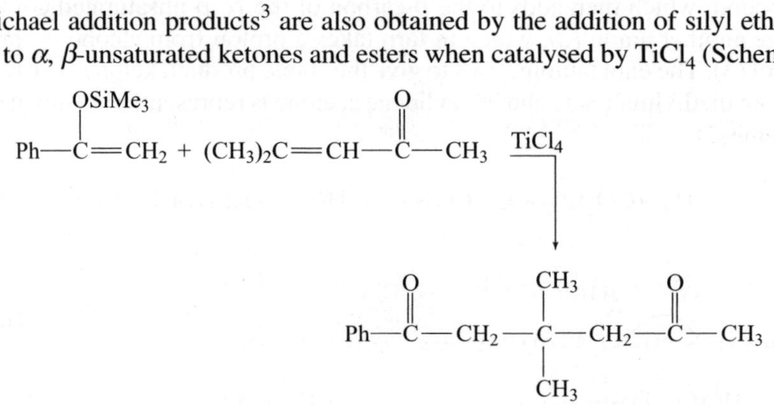

$$Ph\!\!-\!\!\overset{\overset{OSiMe_3}{|}}{C}\!\!=\!\!CH_2 + (CH_3)_2C\!\!=\!\!CH\!\!-\!\!\overset{\overset{O}{\|}}{C}\!\!-\!\!CH_3 \xrightarrow{TiCl_4}$$

$$Ph\!\!-\!\!\overset{\overset{O}{\|}}{C}\!\!-\!\!CH_2\!\!-\!\!\underset{\underset{CH_3}{|}}{\overset{\overset{CH_3}{|}}{C}}\!\!-\!\!CH_2\!\!-\!\!\overset{\overset{O}{\|}}{C}\!\!-\!\!CH_3$$

(Scheme-5)

2.21.1 Michael Addition Under PTC Conditions

The Michael addition of active nitriles to acetylenes can be catalysed[4,5] by the addition of quaternary ammonium chloride (Scheme-6).

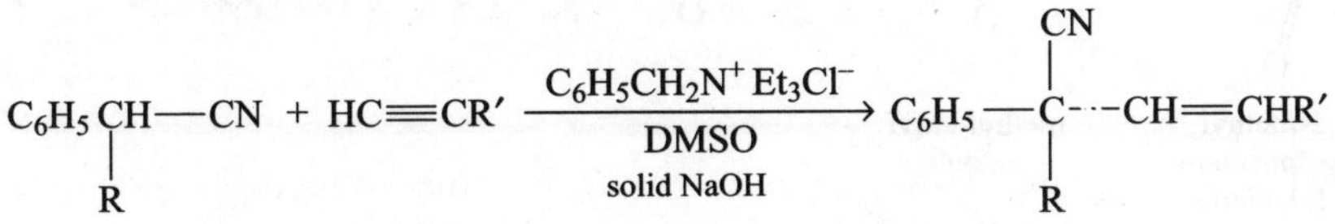

$$C_6H_5\,\underset{\underset{R}{|}}{CH}—CN\ +\ HC≡CR' \xrightarrow[\substack{DMSO \\ solid\ NaOH}]{C_6H_5CH_2N^+Et_3Cl^-} C_6H_5—\underset{\underset{R}{|}}{\overset{\overset{CN}{|}}{C}}\!-\!\cdots CH═CHR'$$

R–	R′	% yield
Me	H	83
Et	H	80
iso Pr	H	82
C_5H_{11}	H	88
Et	Ph	94
iso Pr	Ph	83
$PhCH_2$	Ph	98

(Scheme-6)

A remarkable use of phase transfer Michael reaction was reported[6] in 1975. The reaction of nitro sugar (1) or (2) with ethylmalonate in benzene in 0.2 N NaOH at room temperature in presence of hexadecyltributylphosphonium bromide gives 1,6-O-benzylidene-2,3-dideoxy-2-C-bis(ethoxycarbonyl) methyl-3-nitro-α-D-manno-pyranoside (3) in 92% yield.

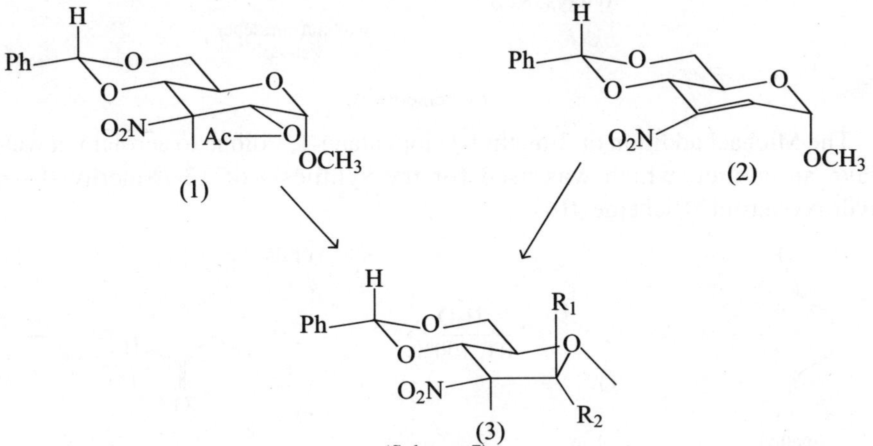

(Scheme-7)

2.21.2 Michael Addition in Aqueous Medium

Michael addition in aqueous phase was first reported in 1970s. 2-Methylcyclopentane-1,3-dione on reaction with methyl vinyl ketone in water gave an adduct with the use of a basic catalyst (pH > 7). The adduct further cyclises to give fused ring systems[7] (Scheme-8).

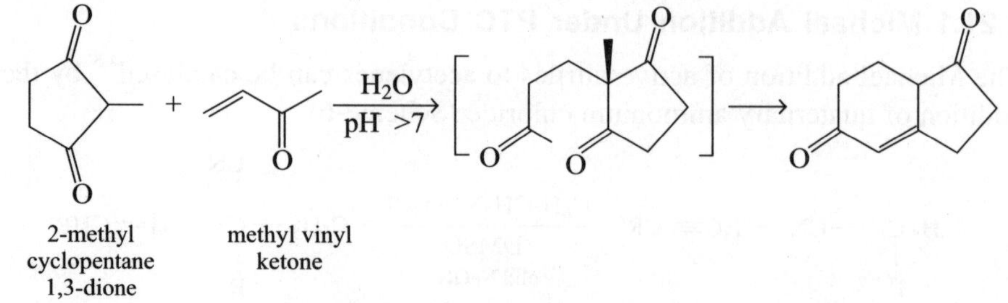

2-methyl cyclopentane 1,3-dione   methyl vinyl ketone

(Scheme-8)

Michael reaction of 2-Methyl cyclohexane-1,3-dione with methyl vinyl ketone gave optically pure Wieland-Miescher ketone[8] (Scheme-9).

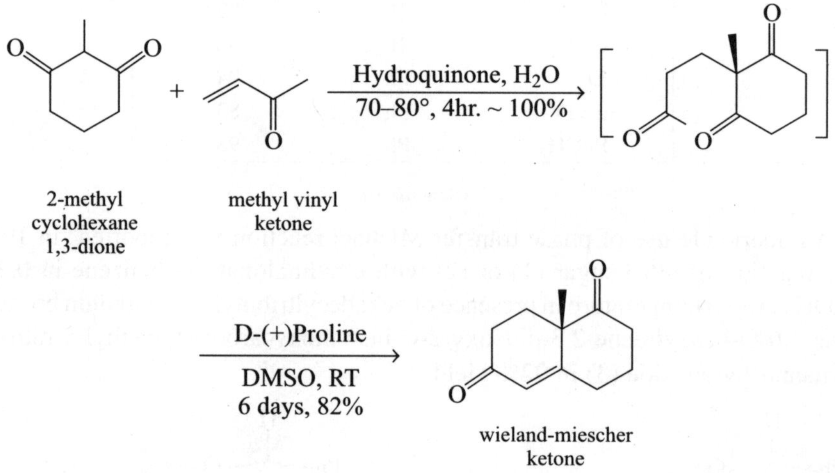

2-methyl cyclohexane 1,3-dione   methyl vinyl ketone

D-(+)Proline

DMSO, RT
6 days, 82%

wieland-miescher ketone

(Scheme-9)

The Michael addition of 2-methylcyclopentane-1,3-dione to acrolein in water gave an adduct, which was used for the synthesis of 13-α-methyl-14-α-hydroxysteroid[9] (Scheme-10).

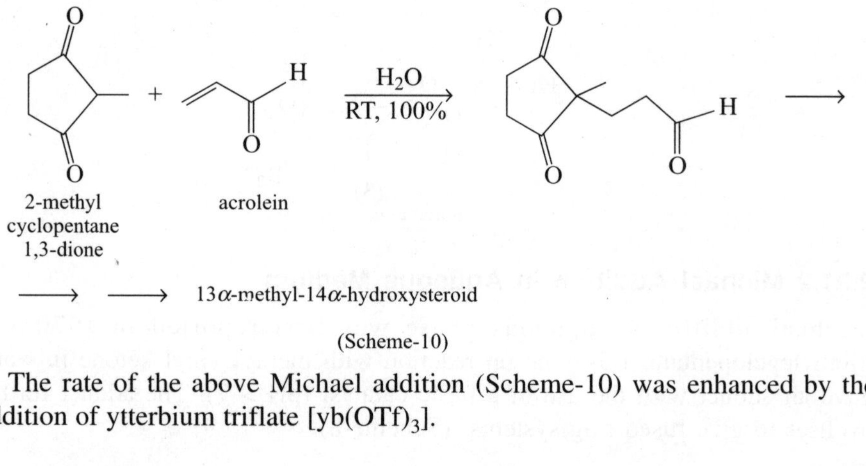

2-methyl cyclopentane 1,3-dione   acrolein

$\longrightarrow$ $\longrightarrow$ 13α-methyl-14α-hydroxysteroid

(Scheme-10)

The rate of the above Michael addition (Scheme-10) was enhanced by the addition of ytterbium triflate [yb(OTf)$_3$].

The Michael addition of nitromethane to methyl vinyl ketone in water (in absence of a catalyst) give[10] 4 : 1 mixture of adducts (A and B) (Scheme-11).

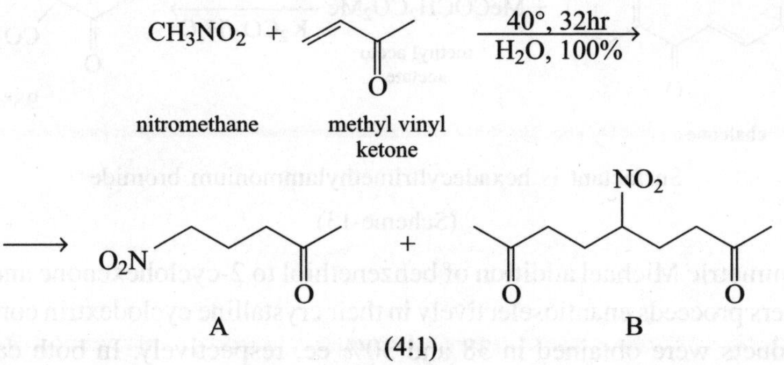

$$CH_3NO_2 \ + \ \diagup\!\!\!\diagdown\!\!\!\diagdown \quad \xrightarrow[\text{H}_2\text{O, 100\%}]{40°, 32hr}$$

nitromethane   methyl vinyl
            ketone

$$\longrightarrow \quad O_2N \diagdown\!\!\!\diagup\!\!\!\diagdown\!\!\!\diagup \quad + \quad$$

            A   B

                         (4:1)

(Scheme-11)

Use of methyl alcohol as solvent (in place of H_2O) gave 1:1 mixture of A and B. The above reaction does not occur in neat conditions or in solvents like THF, PhMe etc in the absence of a catalyst.

The Michael addition of cyclohexenone to ascorbic acid was carried out in water in presence of an inorganic acid[11] (rather than a base) (Scheme-12).

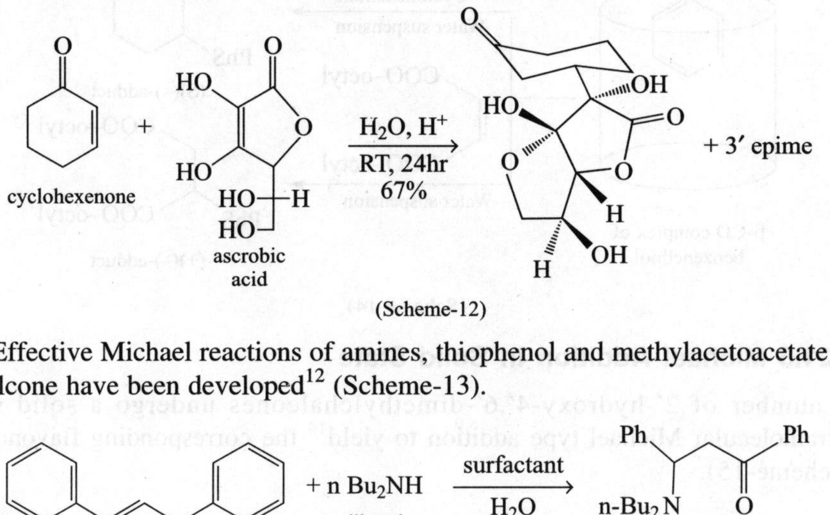

$$\xrightarrow[\text{RT, 24hr}]{\text{H}_2\text{O, H}^+} \quad + \ 3' \ \text{epime}$$

                         67%

cyclohexenone   ascrobic
                       acid

(Scheme-12)

Effective Michael reactions of amines, thiophenol and methylacetoacetate to chalcone have been developed[12] (Scheme-13).

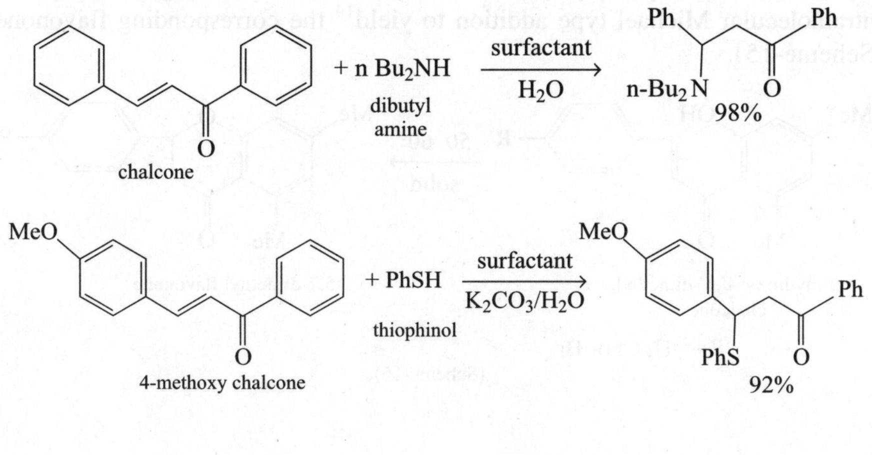

$$+ \text{ n Bu}_2\text{NH} \quad \xrightarrow[\text{H}_2\text{O}]{\text{surfactant}}$$

chalcone   dibutyl   n-Bu$_2$N   O
             amine   98%

$$+ \text{ PhSH} \quad \xrightarrow[\text{K}_2\text{CO}_3/\text{H}_2\text{O}]{\text{surfactant}}$$

4-methoxy chalcone   thiophinol   PhS   O
                                             92%

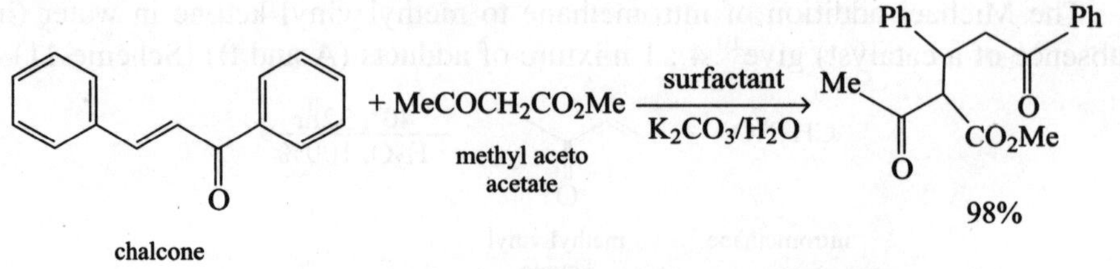

+ MeCOCH$_2$CO$_2$Me

methyl aceto
acetate

chalcone

$\xrightarrow[\text{K}_2\text{CO}_3/\text{H}_2\text{O}]{\text{surfactant}}$

98%

Surfactant is hexadecyltrimethylammonium bromide

(Scheme-13)

Asymmetric Michael addition of benzenethiol to 2-cyclohexenone and maleic acid esters proceeds enantioselectively in their crystalline cyclodextrin complexes. The adducts were obtained in 38 and 30% ee, respectively. In both cases, the reaction was carried out[13] in water suspension (Scheme-14). (See also Weiss-Cook Reaction, section 2.27)

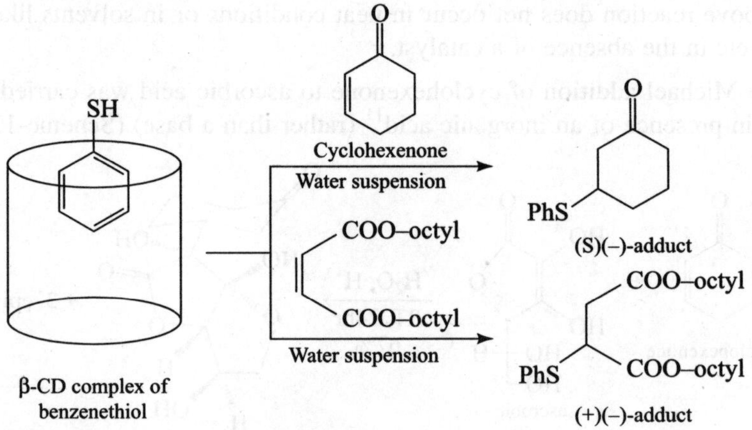

β-CD complex of
benzenethiol

Cyclohexenone
Water suspension

COO—octyl

COO—octyl

Water suspension

(S)(−)-adduct

(+)(−)-adduct

(Scheme-14)

2.21.3 Michael Addition in Solid State

A number of 2'-hydroxy-4',6'-dimethylchalcones undergo a solid sate intramolecular Michael type addition to yield[14] the corresponding flavonones (Scheme-15).

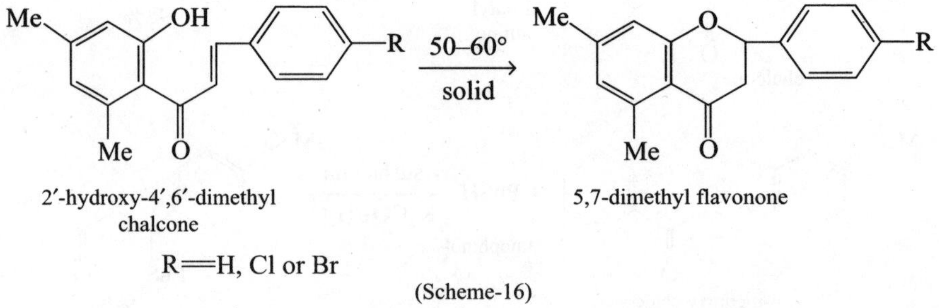

2'-hydroxy-4',6'-dimethyl
chalcone

$\xrightarrow[\text{solid}]{50-60°}$

5,7-dimethyl flavonone

R══H, Cl or Br

(Scheme-16)

The Michael addition of chalcone to 2-phenylcyclohexanone under PTC conditions gave[15] 2,6-disubstituted cyclohexanone derivatives in high distereoselectivity (90% ee) (Scheme-17).

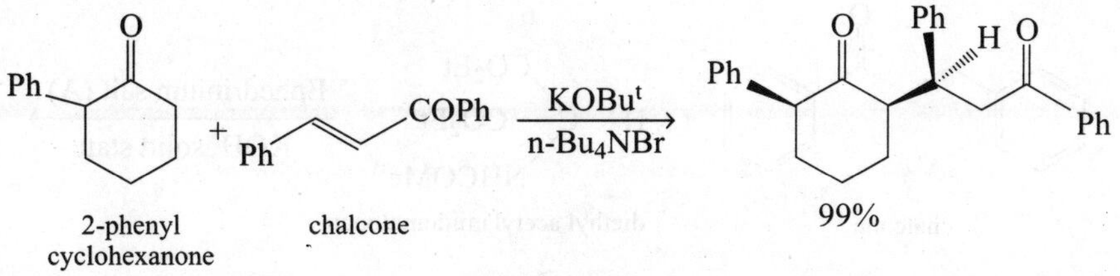

2-phenyl   chalcone   99%
cyclohexanone

(Scheme-17)

The enantioselective Michael addition of mercapto compounds with an optically active host compound derived from 1:1 inclusion complex of cyclohexanone with (−)–A derived from tartaric acid[16] and a catalytic amount of benzyltrimethyl ammonium hydroxide on irridation with ultrasound for 1 hr at room temperature gave the adduct in 50-78% yield with ee 75-80% (Scheme-18).

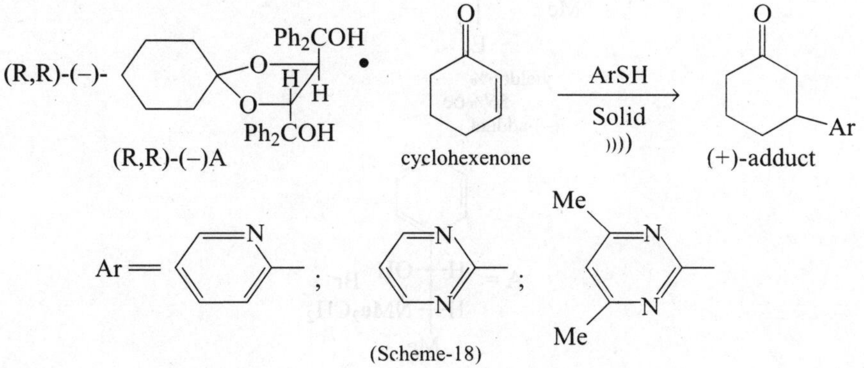

(R,R)-(−)A   cyclohexenone   (+)-adduct

Ar =

(Scheme-18)

The Michael addition of thiols to 3-methyl-3-buten-2-one in its inclusion crystal with (−)–A also occurred enantioselectively (Scheme-19).

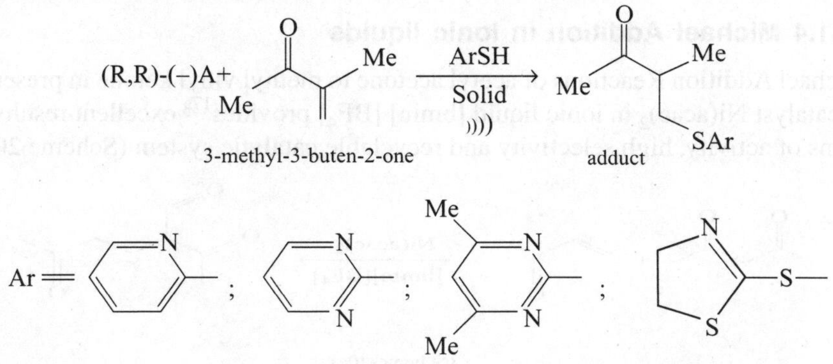

3-methyl-3-buten-2-one   adduct

Ar =

(Scheme-19)

Michael addition of diethyl (acetylamido) malonate to chalcone using asymmetric phase transfer catalyst (ephedrinium salt) in presence of KOH in the solid state gave[17] the adduct in 56% yield with ee of 60% (Scheme-20).

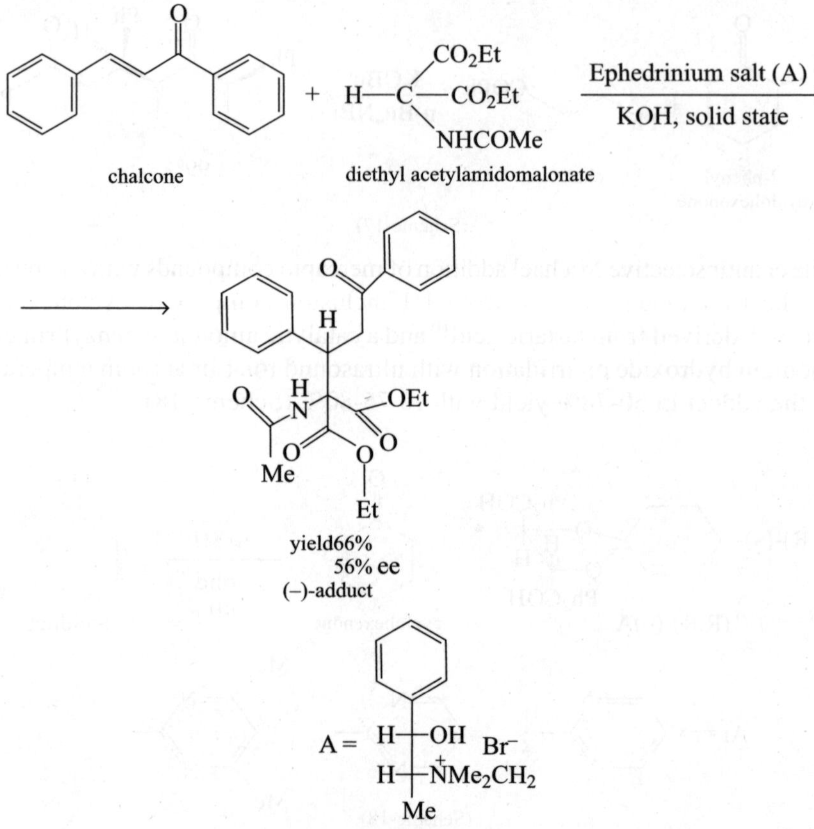

chalcone

diethyl acetylamidomalonate

Ephedrinium salt (A)

KOH, solid state

yield66%
56% ee
(−)-adduct

A =

(−)-N-methyl-N-benzylephedrinium bromide

(Scheme-20)

2.21.4 Michael Addition in Ionic liquids

Michael Addition Reactions of acetyl acetone to methyl vinyl ketone in presence of catalyst Ni(acac)$_2$ in ionic liquid [bmin] [BF$_4$] provides[17a] excellent results in terms of activity, high selectivity and recyclable catalytic system (Scheme 20a).

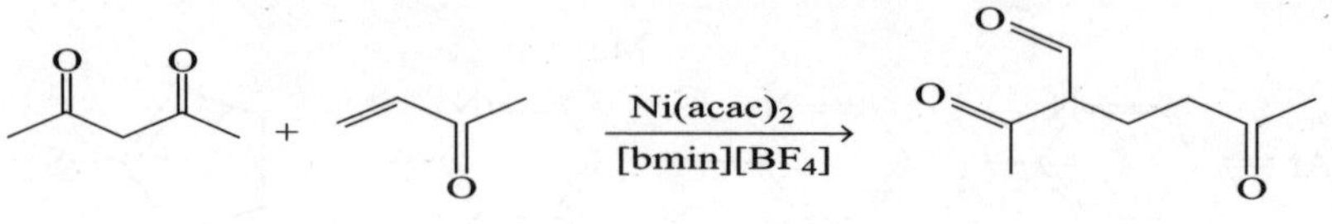

Ni(acac)$_2$

[bmin][BF$_4$]

(Scheme-20a)

2.21.5 Applications

In Michael addition a new carbon-carbon bond is produced. This procedure is of great synthetic importance since a variety of organic compounds can be synthesised. Some of the important applications are given below.

(i) Synthesis of dimidone

The Michael adduct obtained by the condensation of diethyl malonate with mesityl oxide in presence of sodium ethoxide undergoes **internal claisen condensation** to give dimidone.

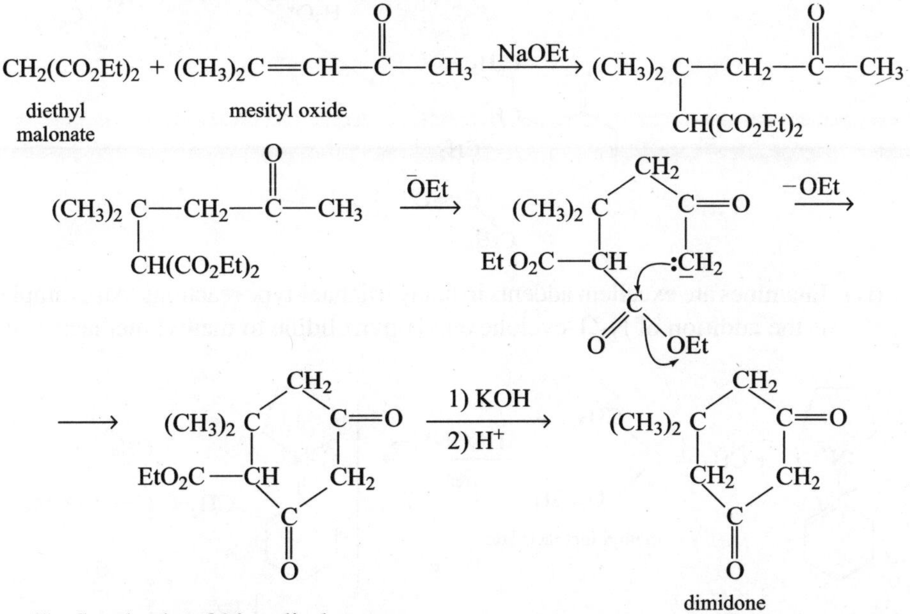

(ii) Synthesis of bicyclic ketones

Use of Michael addition followed by aldol condensation is an important routes for the synthesis of bicyclic ketones and is known as **Robinson annulation**. Thus Michael condensation of 2-methyl-1,3-cyclohexanedione with methyl vinyl ketone followed by aldol condensation give bicyclic ketone.

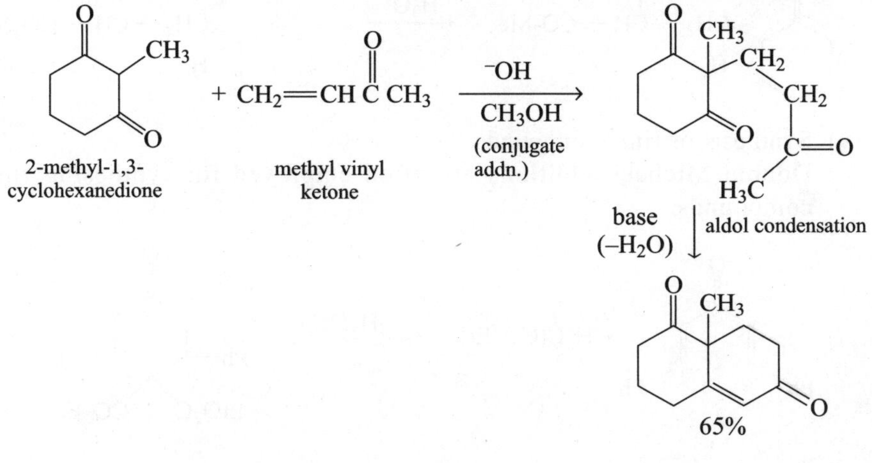

(iii) Synthesis of o-substituted cyclohexanone. Michael addition of cyclohexanone to chalcone give the 2-substituted cyclohexanone. See also (iv) given below.

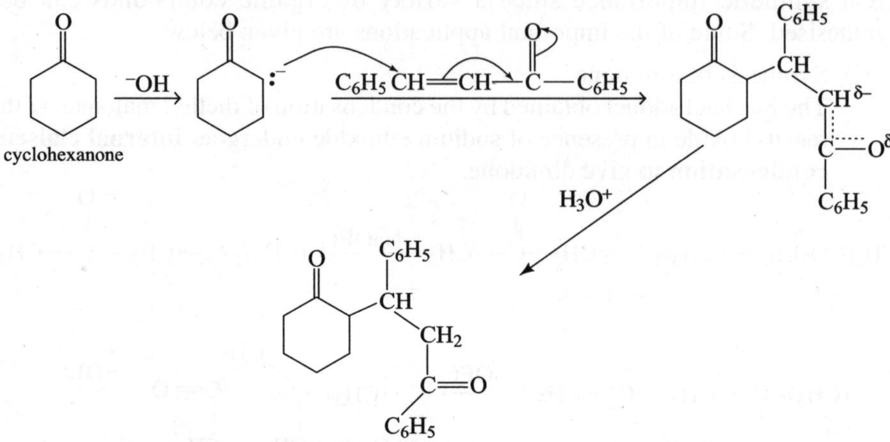

(iv) Enamines are excellent addents in many Michael-type reactions. An example is the addition of N-(1-cyclohexenyl)-pyrrolidine to methyl methacrylate.

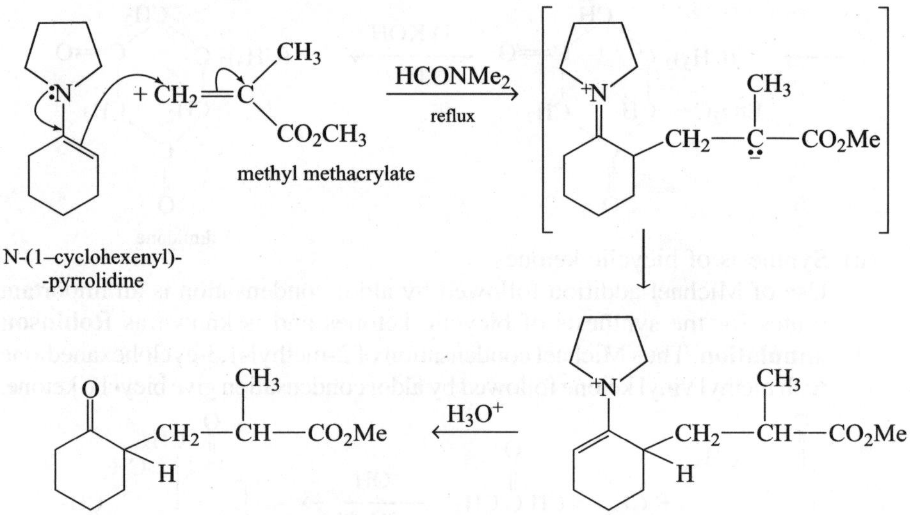

(v) Synthesis of ring compounds.
Double Michael additions are often employed for synthesing ring compounds.

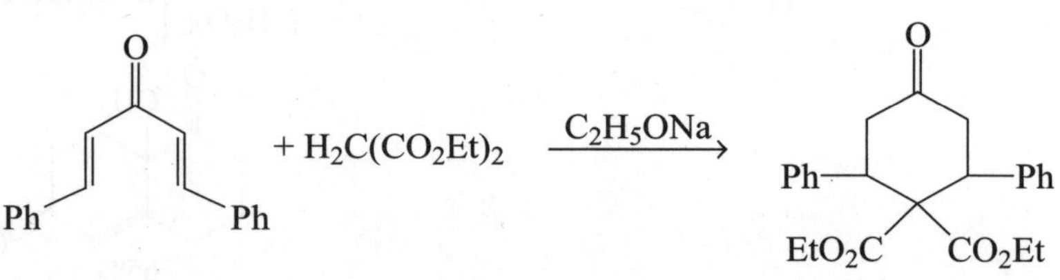

(vi) Synthesis of caronic acid, a cyclopropane derivative.

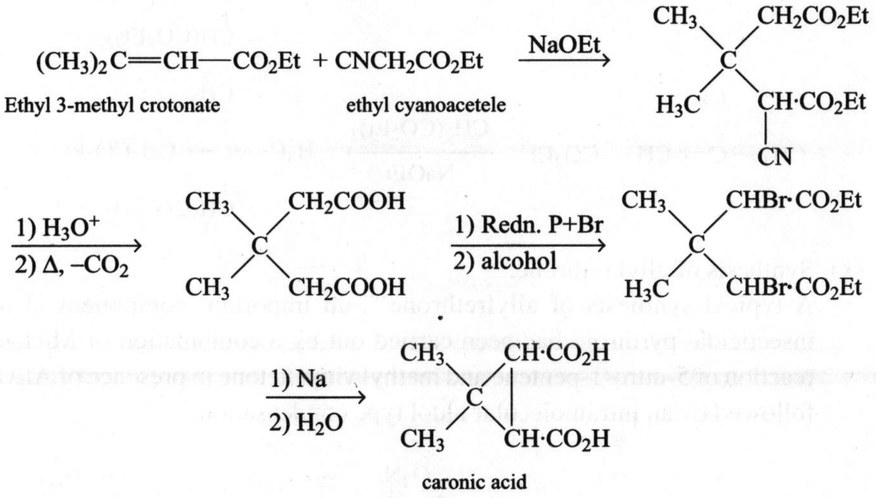

$(CH_3)_2C\!\!=\!\!CH\!\!-\!\!CO_2Et$ + $CNCH_2CO_2Et$ $\xrightarrow{\text{NaOEt}}$

Ethyl 3-methyl crotonate   ethyl cyanoacetele

$$
\begin{array}{c}
CH_3 \quad\; CH_2CO_2Et \\
\diagdown \;\; C \;\; \diagup \\
\diagup\;\;\;\; \diagdown \\
H_3C \quad\;\; CH\cdot CO_2Et \\
| \\
CN
\end{array}
$$

$\xrightarrow[\text{2) } \Delta, -CO_2]{\text{1) } H_3O^+}$

$$
\begin{array}{c}
CH_3 \quad\; CH_2COOH \\
\diagdown \;\; C \;\; \diagup \\
\diagup\;\;\;\; \diagdown \\
CH_3 \quad\; CH_2COOH
\end{array}
$$

$\xrightarrow[\text{2) alcohol}]{\text{1) Redn. P+Br}}$

$$
\begin{array}{c}
CH_3 \quad\; CHBr\cdot CO_2Et \\
\diagdown \;\; C \;\; \diagup \\
\diagup\;\;\;\; \diagdown \\
H_3C \quad\; CHBr\cdot CO_2Et
\end{array}
$$

$\xrightarrow[\text{2) } H_2O]{\text{1) Na}}$

$$
\begin{array}{c}
CH_3 \quad\;\; CH\cdot CO_2H \\
\diagdown \;\; C \;\; \diagup | \\
\diagup\;\;\;\; \diagdown | \\
CH_3 \quad\; CH\cdot CO_2H
\end{array}
$$

caronic acid

(vii) Synthesis of nitro and cyano compounds.

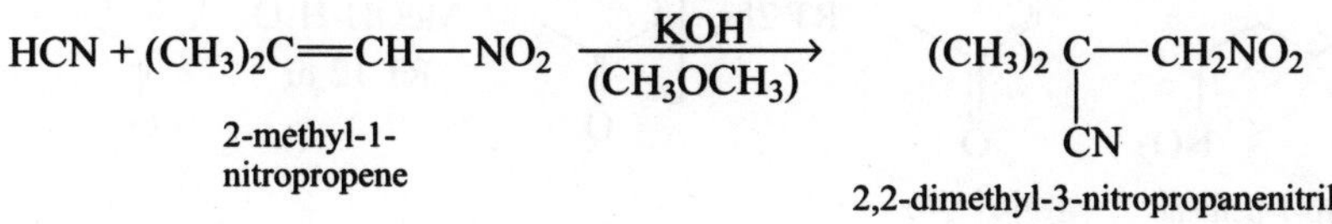

$HCN + (CH_3)_2C\!\!=\!\!CH\!\!-\!\!NO_2$ $\xrightarrow[(CH_3OCH_3)]{KOH}$

2-methyl-1-
nitropropene

$$
\begin{array}{c}
(CH_3)_2 \;\; C\!\!-\!\!CH_2NO_2 \\
| \\
CN
\end{array}
$$

2,2-dimethyl-3-nitropropanenitrile

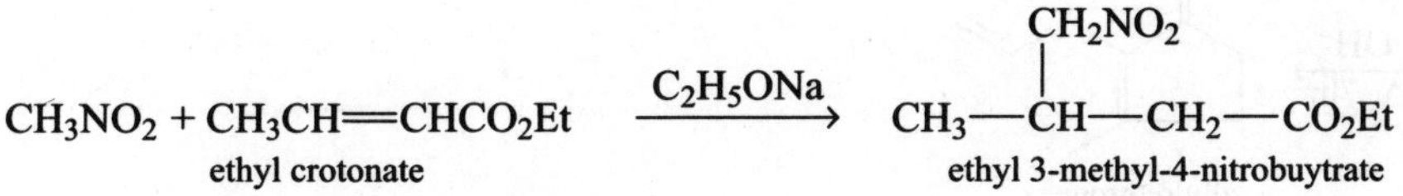

$CH_3NO_2 + CH_3CH\!\!=\!\!CHCO_2Et$ $\xrightarrow{C_2H_5ONa}$ $CH_3\!\!-\!\!\underset{\underset{CH_2NO_2}{|}}{CH}\!\!-\!\!CH_2\!\!-\!\!CO_2Et$

ethyl crotonate   ethyl 3-methyl-4-nitrobuytrate

(viii) Synthesis of aconitic acid.

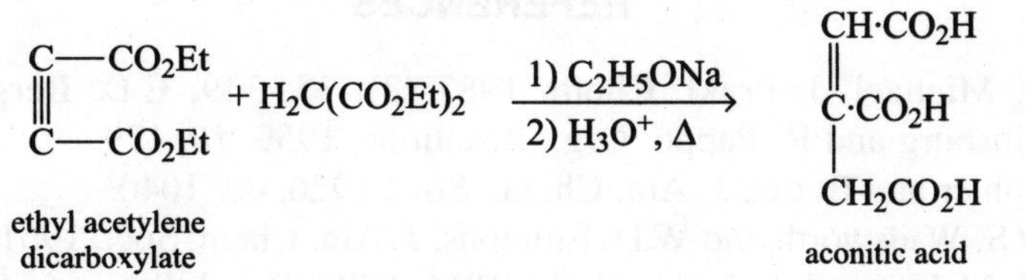

$$
\begin{array}{c}
C\!\!-\!\!CO_2Et \\
\|\| \\
C\!\!-\!\!CO_2Et
\end{array}
$$
+ $H_2C(CO_2Et)_2$ $\xrightarrow[\text{2) } H_3O^+, \Delta]{\text{1) } C_2H_5ONa}$ $$
\begin{array}{c}
CH\cdot CO_2H \\
\| \\
C\cdot CO_2H \\
| \\
CH_2CO_2H
\end{array}
$$

ethyl acetylene
dicarboxylate   aconitic acid

(ix) Some typical applications of Michael additions are given below:

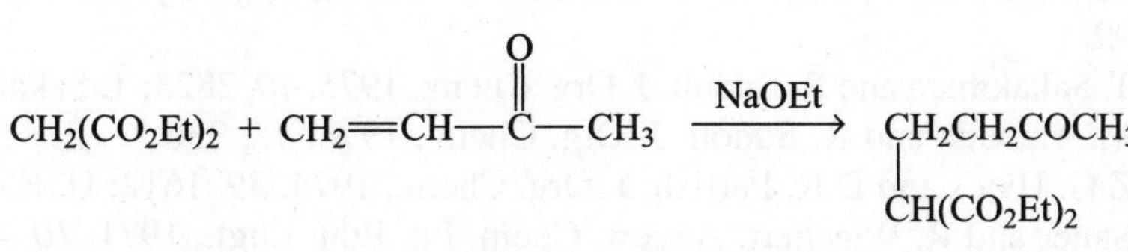

$CH_2(CO_2Et)_2 + CH_2\!\!=\!\!CH\!\!-\!\!\overset{\overset{\displaystyle O}{\|}}{C}\!\!-\!\!CH_3$ $\xrightarrow{\text{NaOEt}}$ $\underset{\underset{CH(CO_2Et)_2}{|}}{CH_2CH_2COCH_3}$

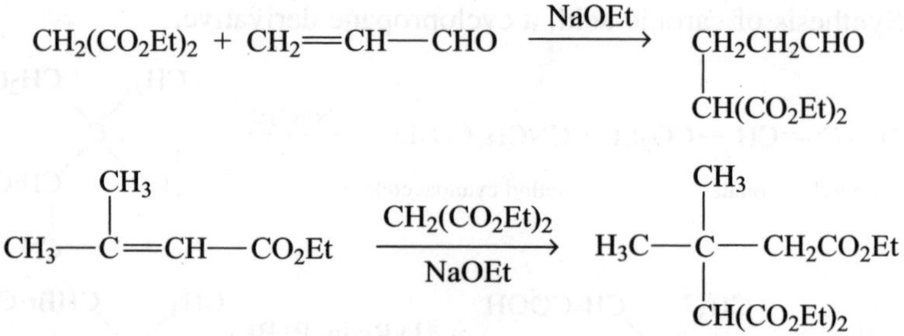

$$CH_2(CO_2Et)_2 + CH_2{=}CH{-}CHO \xrightarrow{NaOEt} \begin{array}{c} CH_2CH_2CHO \\ | \\ CH(CO_2Et)_2 \end{array}$$

$$\begin{array}{c} CH_3 \\ | \\ CH_3{-}C{=}CH{-}CO_2Et \end{array} \xrightarrow[NaOEt]{CH_2(CO_2Et)_2} \begin{array}{c} CH_3 \\ | \\ H_3C{-}C{-}CH_2CO_2Et \\ | \\ CH(CO_2Et)_2 \end{array}$$

(x) Synthesis of allyl rethrone.

A typical synthesis of allylrethrone[18], an important component of an insecticidal pyrthroid has been carried out by a combination of Michael reaction of 5-nitro-1-pentene and methyl vinyl ketone in presence of Al_2O_3 followed by an intramolecular aldol type condensation.

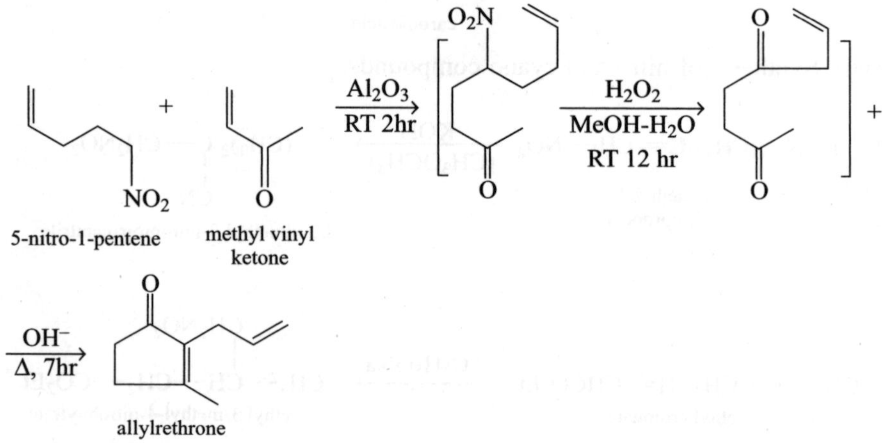

5-nitro-1-pentene   methyl vinyl ketone

allylrethrone

REFERENCES

1. A. Michael, J. Prakt. Chem. 1987 (2), 35, 349; E.D. Bergmann, D. Ginsburg and R. Pappo, Org., Reactions, 1959, *10*, 179.
2. Kohler and Butler. J. Am. Chem., Soc., 1926, *48*, 1040.
3. W.S. Wadsworth and W.D. Emmons, J. Am. Chem. Soc., 1961, *83*, 1733.
4. M. Makosza, Tetrahedron Lett., 1966, 5489; Polish Patent 55113 (1968); CA, 1969, *70*, 106006.
5. M. Makosza, J. Czyzewski and M. Jawdosiak, Org., Synth., 1976, *55*, 99.
6. T. Sakakibara and R. Sudoh, J. Org. Chem., 1975, *40*, 2823; T. Sakakibara, M. Yamada and R. Sudoh, J. Org. Chem., 1976, *41*, 736.
7. Z.G. Hajos and D.R. Parrish, J. Org. Chem., 1974, *39*, 1612; U. Elder, G. Sauer and R. Wiechert, Angew. Chem. Int. Edn. Engl., 1971, *10*, 496.

8. N. Harada, T. Sugioka, U. Uda and T. Kuriki, Synthesis, 1990, 53.
9. J.F. Lavelle and P. Deslongchamps, Tetrahedron Lett., 1988, *29*, 6033.
10. A. Lubineau and J. Auge, Tetrahedron Lett., 1992, *33*, 8073.
11. K. Sussang Karn, G. Fodor, I. Karle and C. George, Tetrahedron, 1988, *44*, 7047.
12. F. Toda, M. Takumi, M. Nagami and K. Tanaka, Heterocycles, 1988, *47*, 469.
13. H. Sakuraba, Y. Tanaka and F. Toda, J. Incl. Phenon, 1991, *11*, 195.
14. B. Satish, K. Panneesel-Vam, D. Zacharids and G.R. Desivaju, J. Chem. Soc. Perkin Trans., 1995, 2, 325.
15. E. Diez-Barra, A. de la Hoz, S. Merino and P. Sanchez-Verdu, Tetrahedron Lett., 1997, *38*, 2359.
16. F. Toda, K. Tanka and J. Sato, Tetrahedron Asymmetry, 1993, *4*, 1771.
17. A. Loupy, J. Sansoulet, A. Zaparucha and C. Merinne, Tetrahedron Lett., 1989, *30*, 333.
17a. M.M. Dell' Anna, V. Gallo, P. Mastrorilli, C. Francesco, G. Rommanazzi, G.P. Suranna, Chem., Commun., 2002, 434.
18. R.Ballini, synthesis, 1993-687.

2.22 MUKAIYAMA REACTION

It is a steroselective aldol condensation[1] of silyl enol ethers of ketones with aldehydes in presence of titanium tetrachloride. For example, condensation of silyl enol ether of 3-pentanone with 2-methylbutyraldehyde in presence of $TiCl_4$ gives the aldolate product, which on hydrolysis yield aldol product Manicone, an alarm pheromone (Scheme-1).

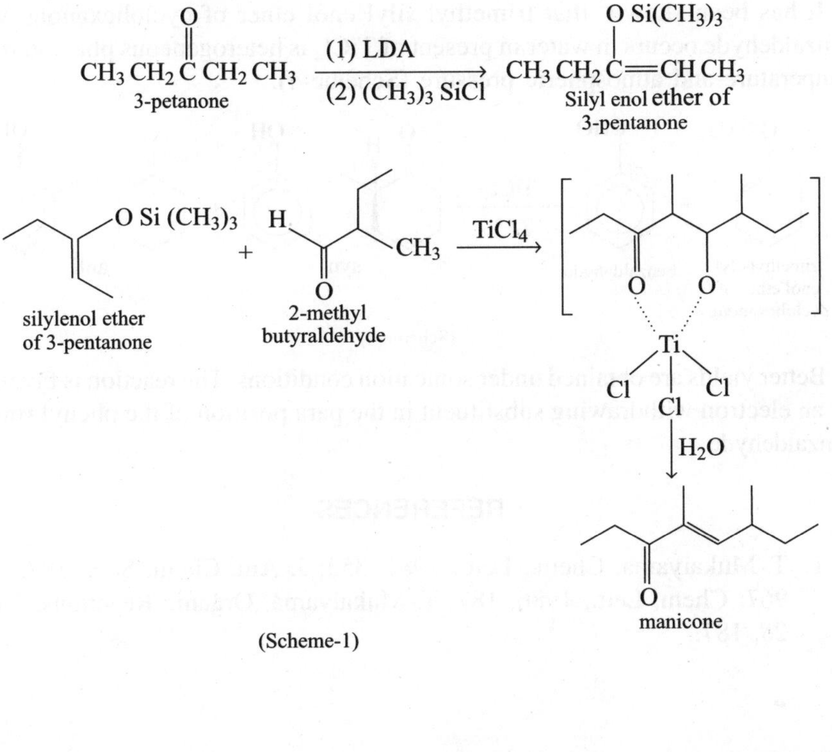

(Scheme-1)

In Mukaiyama reaction, other Lewis acids such as tin tetrachloride ($SnCl_4$) and boron trifluoride etherate ($BF_3 \cdot OEt_2$) can also be used.

2.22.1 Mukaiyama Reaction in Aqueous Phase

The Mukaiyama reactions are carried out under non-aqueous conditions. The first water promoted Mukaiyama reaction of silyl enol ethers with aldehydes was reported in 1986 (Scheme-2)[2].

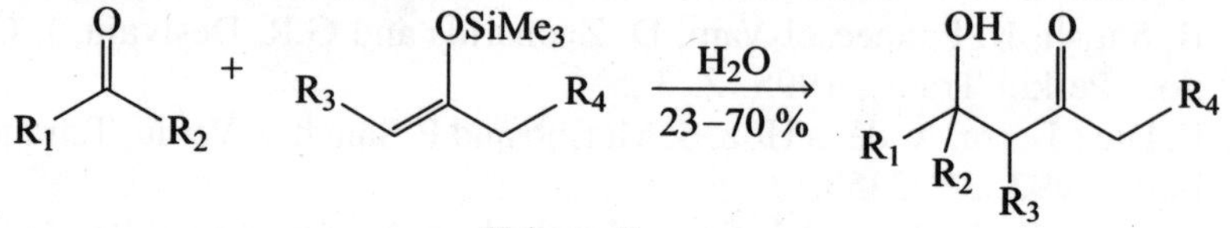

(Scheme-2)

The above reactions were carried out in aqueous medium without any acid catalyst, but the reaction, however, took several days for completion, since water serves as a weak Lewis acid. The addition of stronger Lewis acid (e.g., ytterbium triflate) greatly improved the yield and also the rate of the reaction[3] (Scheme-3).

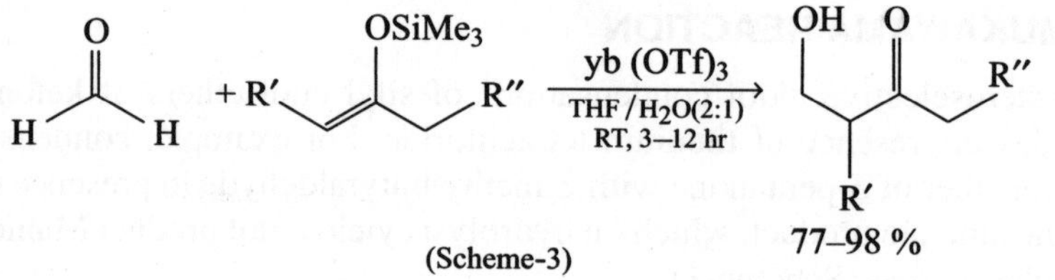

(Scheme-3)   77–98 %

It has been shown[2] that trimethyl silyl enol ether of cyclohexanone with benzaldehyde occurs in water in present of $TiCl_4$ is heterogeneous phase at room temperature and atmospheric pressure (Scheme-4).

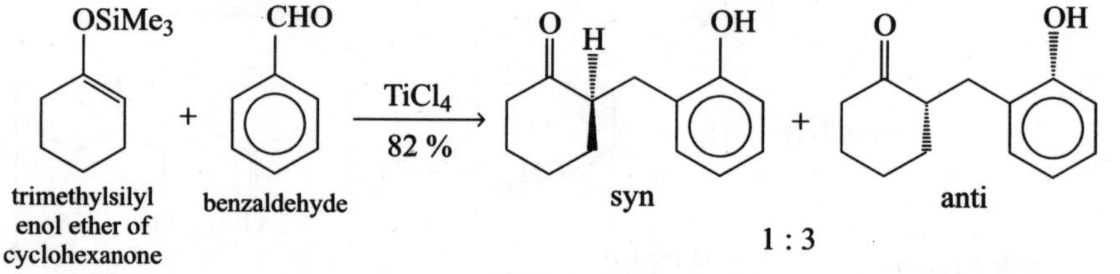

trimethylsilyl   benzaldehyde   syn   anti
enol ether of
cyclohexanone   1 : 3

(Scheme-4)

Better yields are obtained under sonication conditions. The reaction is favoured by an electron withdrawing substituent in the para position of the phenyl ring in benzaldehyde.

REFERENCES

1. T. Mukaiyama, Chem., Lett., 1982, 353; J. Am. Chem. Soc., 1973, *95*, 967; Chem. Lett., 1986, 187; T. Mukaiyama, Organic Reactions, 1982, *28*, 187.

2. A. Lubineau, J. Org. Chem., 1986, *51*, 2142; A. Lubineau, E. Meyer, Tetrahedron, 1988, *44*, 6065.

3. S. Kobayashi and I. Hachiya, J. Org., Chem., 1994, *59*, 3590. For a review on lanthanides catalysed organic reactions in aqueous media, see S. Kobayashi, Synlett, 1994, 589.

2.23 REFORMATSKY REACTION

The reaction of an α-halo-ester (usually an α-bromoester) with an aldehyde or ketone in presence of zinc metal in an inert solvent (ether-benzene) to produce β-hydroxy ester is known as Reformatsky Reaction[1]. (Scheme-1)

$$
\underset{\substack{\text{aldehyde} \\ \text{or} \\ \text{ketone}}}{\text{>C=O}} + \underset{\alpha\text{-bromoester}}{\text{Br}-\overset{|}{\underset{|}{C}}-CO_2R} \xrightarrow[\text{benzene}]{Zn} \underset{}{-\overset{\overset{BrZn\ O}{|}}{\underset{|}{C}}-\overset{|}{\underset{|}{C}}-CO_2R} \xrightarrow{H_3O^+} \underset{\beta\text{-hydroxyester}}{-\overset{\overset{HO}{\beta|}}{\underset{|}{C}}-\overset{\alpha|}{\underset{|}{C}}-CO_2R}
$$

(Scheme-1)

The Reformatsky reaction extends the carbon skeleton of an aldehyde or ketone and yields β-hydroxy esters. The initial product is a zinc alkoxide, which must be hydrolysed to yield the β-hydroxy ester.

MECHANISM

Zinc first reacts with the α-bromoester to form an organozinc intermediate (its formation is similar to that of the formation of grignard reagent). The formed organic zinc intermediate then adds to the carbonyl group of the aldehyde or ketone. Final hydrolysis gives β-hydroxy ester (Scheme-2).

$$
\underset{\alpha\text{-bromoester}}{Br-\overset{\overset{R}{|}}{\underset{|}{CH}}-CO_2Et} \xrightarrow[\text{benzene}]{Zn} \underset{\text{organo zinc intermediate}}{Br-Zn-\overset{\overset{R}{|}}{\underset{|}{CH}}-CO_2Et}
$$

$$
Br\ Zn \cdot \overset{\overset{R}{|}}{\underset{|}{CH}}-CO_2Et + {>}C{=}O \longrightarrow -\overset{\overset{BrZn\ O}{|}}{\underset{|}{C}}-\overset{\overset{R}{|}}{CH}\ CO_2Et
$$

$$
\downarrow HCl\ (H_2O)
$$

$$
-\overset{\overset{HO}{|}}{\underset{|}{C}}-\overset{\overset{R}{|}}{CH}\ CO_2Et
$$

(Scheme-2) β-hydroxy ester

Since the organo zinc reagent is less reactive than the organo magnesium reagent (grignard reagent), it does not add to the ester group. The β-hydroxy esters produced in the Reformatsky reaction are easily dehydrated to α, β-unsaturated ester since dehydration yields a system in which carbon-carbon double bond is in conjugation with the carbon-oxygen double bond of the ester (Scheme-3).

$$\underset{\beta\text{-hydroxy ester}}{-\overset{\overset{\displaystyle OH}{|}}{C}-\overset{\overset{\displaystyle R}{|}}{CH}-\overset{\overset{\displaystyle O}{\|}}{C}-OEt} \xrightarrow[-H_2O]{H_3O^+/heat} \underset{\alpha,\ \beta\text{-unsaturated ester}}{\overset{}{>}C=C\overset{R}{\underset{\overset{\displaystyle C}{\underset{\displaystyle \|}{O}}OEt}{<}}}$$

(Scheme-3)

In Reformatsky reaction, it is sometimes necessary to activate the zinc by addition of a small crystal of iodine, mercuric bromide or copper.

A modification of the Reformatsky reaction using nitriles (in place of aldehydes or ketones) is called the **Blaise Reaction** (Scheme-4).

$$\underset{\text{organo zinc intermediate}}{Br-Zn-\overset{\overset{\displaystyle R}{|}}{CH}-CO_2Et} + \underset{\text{nitrile}}{R'-C\equiv N} \longrightarrow \xrightarrow{H_2O} R'-\overset{\overset{\displaystyle R}{|}}{\underset{\overset{\displaystyle \|}{O}}{C}}-\overset{}{CH}-CO_2Et$$

(Scheme-4)

The Reformatsky reaction can also be applied to Schiff's bases to yield β-lactam (Scheme-5).

$$\underset{\substack{\text{schiff's base} \\ Ar = Ph,\ 4\text{-Me}-C_6H_4,\ 4Cl-C_6H_4}}{\overset{Ar}{\diagdown}\overset{}{\diagup}\overset{N}{\diagdown}Ar} + Br\,CH_2\,CO_2\,Et \xrightarrow[\text{dioxane}]{Zn/I_2} \underset{\beta\text{-lactams}}{\begin{array}{c}Ar \quad\quad Ar \\ \square\!=\!O\end{array}}$$

(Scheme-5)

2.23.1 Reformatsky Reaction Using Sonication

Excellent yields are obtained on sonication compared to more traditional methods, such as those employing activated zinc or trimethylborane as cosolvent[2]. Under optimal conditions quantitative yields of β-hydroxyester is obtained[3] (Scheme-6).

$$\underset{R'}{\overset{R}{\diagdown}}C=O \xrightarrow[\substack{\text{dioxane, rt, 5-30 min} \\ )))}]{Br\,CH_2\,CO_2\,Et/Zn/I_2} \underset{\substack{R' \quad OH \\ 97-98\ \%}}{\overset{R}{\diagdown}C\diagup}\overset{}{CO_2Et}$$

R = H; R' = Ph or $(CH_2)_2CH_3$

R–R' = $-(CH_2)_4^-$

(Scheme-6)

In the sonication procedure, it is necessary to activate zinc with iodine and to carry out the reaction in dioxane.

Even with nitriles, the Reformatsky reaction (known as **Blaise reaction**), sonication leads to amines which hydrolyse to give ketones—(Scheme-4). Also, the application of Reformatsky reaction to Schiffs bases to yield better yields of β-lactams (Scheme-5), but this modification is not of general applicability[3].

Using appropriate nitrile, keto-γ-butyrolactone is obtained in good yield[4] (Scheme-7).

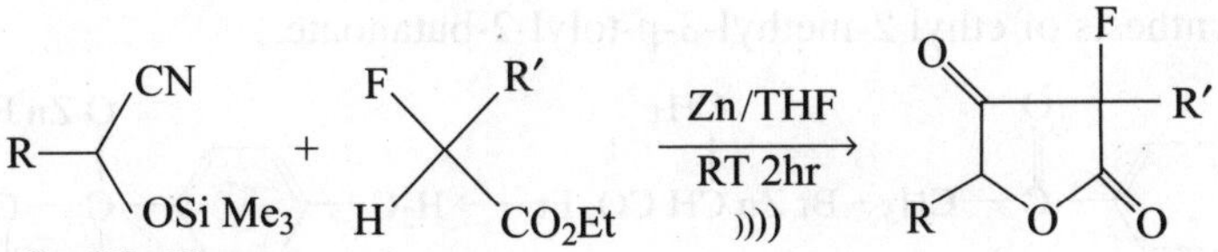

R= CH_3, Ph.;
R′= H, CH_3

(Scheme-7)

2.23.2 Reformatsky Reaction in Solid State

Treatment of aromatic aldehydes with ethylbromoacetate and Zn-NH_4Cl in the solid state give the corresponding Reformatsky products[5] (Scheme-8).

$$\text{R CHO} + \text{Br CH}_2\text{CO}_2\text{Et} \xrightarrow[\text{solid state, 3 hr}]{\text{Zn-NH}_4\text{Cl}} \text{R CH(OH) CH}_2\text{CO}_2\text{Et} \quad 80\text{–}90\ \%$$

R = Ph, p-BrC_6H_4, 3,4-methylenedioxyphenyl,

Ph — Ph —, 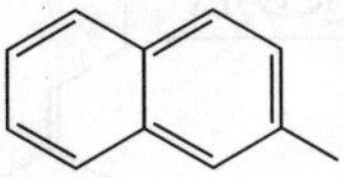

(Scheme-8)

2.23.3 Applications

(i) Synthesis of ethyl 3-phenyl-3-hydroxypropionate[6].

$$\underset{\text{benzaldehyde}}{C_6H_5CHO} + \underset{\text{ethyl bromo-acetate}}{BrCH_2CO_2Et} \xrightarrow[B(OEt)_3]{Zn} C_6H_5\text{—}CH(OH)CH_2CO_2Et$$

ethyl 3-phenyl-3-hydroxy-propionate
95%

(ii) Synthesis of α-phenyl-γ-fluorotetronic acid[4].

$$\text{Ph CH(O SiMe}_3\text{)—CN} + \text{Br CHF CO}_2\text{Et} \xrightarrow[\text{ultrasound}]{\text{Zn}} \left[\text{Ph CH(O SiMe}_3\text{)—C(=O)·CHF·CO}_2\text{Et}\right] \longrightarrow$$

trimethyl silyl ether of benzaldehyde cyanohydoin; α-fluobromoacetate

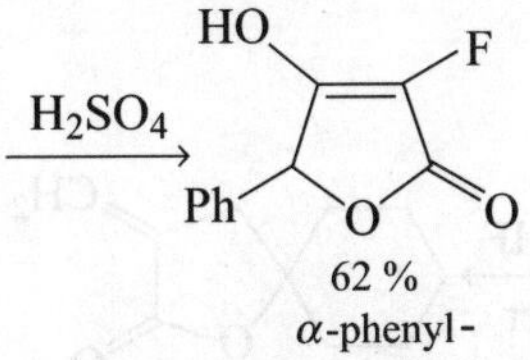

62 %
α-phenyl-
-Υ-fluorotetronic acid

(iii) Synthesis of ethyl 2-methyl-3-p-tolyl-2-butanoate.

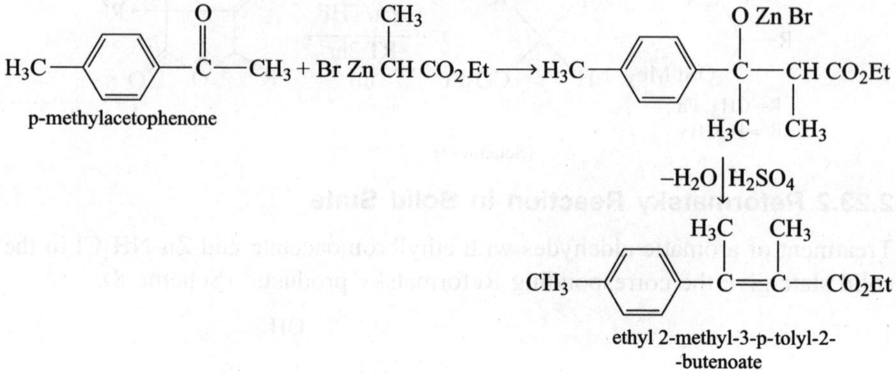

p-methylacetophenone

ethyl 2-methyl-3-p-tolyl-2-
-butenoate

(iv) Synthesis of citral.

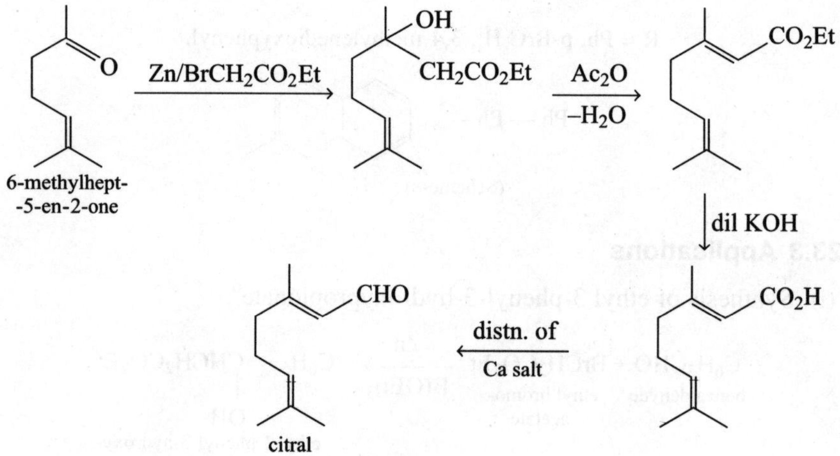

6-methylhept-
-5-en-2-one

$Zn/BrCH_2CO_2Et$

Ac_2O
$-H_2O$

dil KOH

distn. of
Ca salt

citral

(v) Reformatsky reaction of 6-methyl-2-heptanone[7].

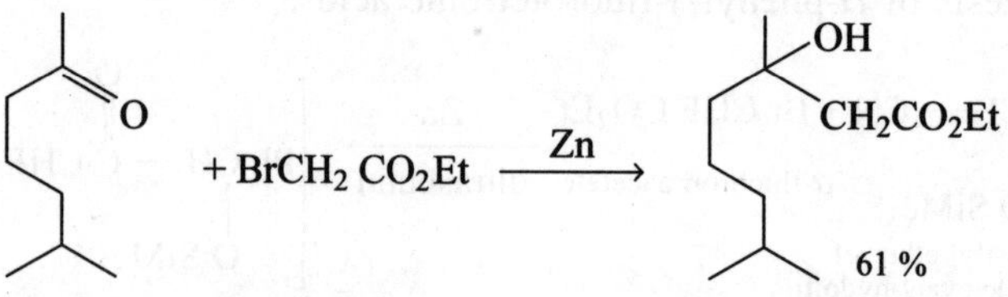

$+ BrCH_2 CO_2Et$ Zn

61 %

(vi) Synthesis of spiro compounds[8].

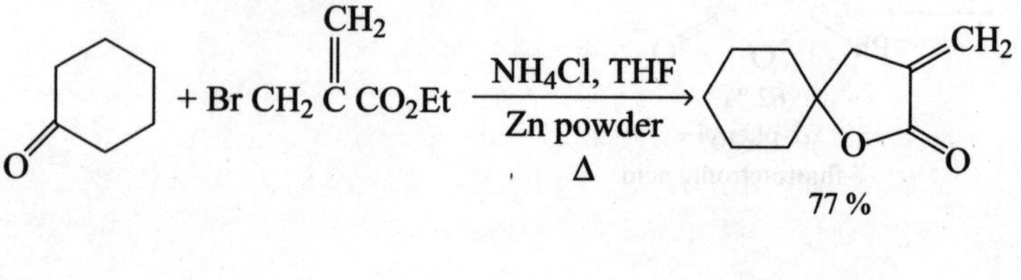

$+ Br CH_2 \overset{CH_2}{\underset{}{\overset{\|}{C}}} CO_2Et$ $\xrightarrow[\text{Zn powder}]{\text{NH}_4\text{Cl, THF}}$ Δ

77 %

(vii) Synthesis of citric acid.

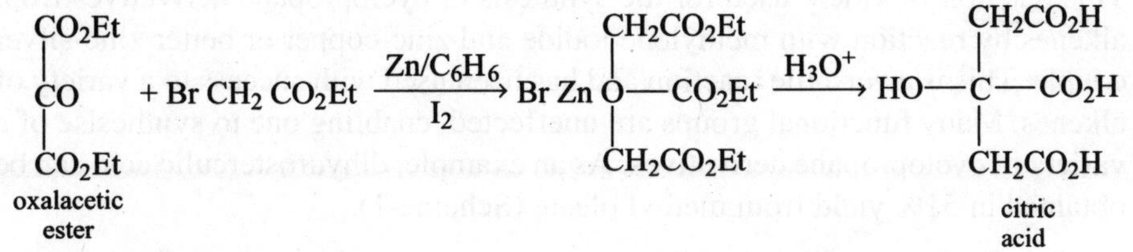

$$\begin{matrix} CO_2Et \\ | \\ CO \\ | \\ CO_2Et \end{matrix} + Br\, CH_2\, CO_2Et \xrightarrow[\;I_2\;]{Zn/C_6H_6} \begin{matrix} CH_2CO_2Et \\ | \\ Br\, Zn\, O\!-\!CO_2Et \\ | \\ CH_2CO_2Et \end{matrix} \xrightarrow{H_3O^+} \begin{matrix} CH_2CO_2H \\ | \\ HO\!-\!C\!-\!CO_2H \\ | \\ CH_2CO_2H \end{matrix}$$

oxalacetic ester

citric acid

(viii) Synthesis of vitamin A₁

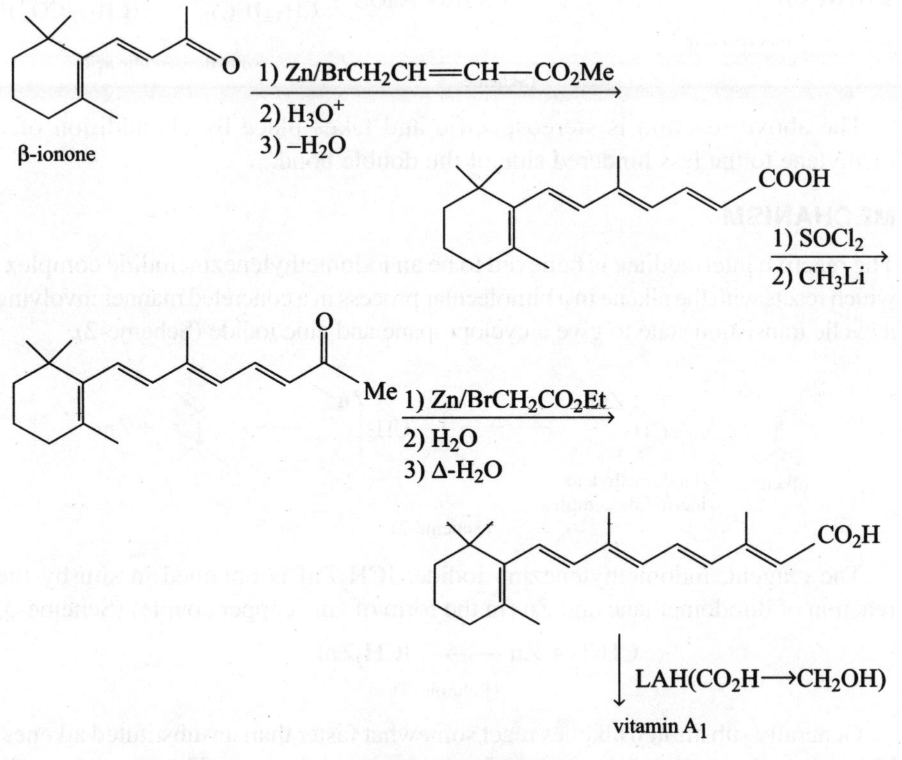

β-ionone

1) Zn/BrCH₂CH=CH—CO₂Me
2) H₃O⁺
3) –H₂O

1) SOCl₂
2) CH₃Li

1) Zn/BrCH₂CO₂Et
2) H₂O
3) Δ-H₂O

LAH(CO₂H→CH₂OH)

vitamin A₁

REFERENCES

1. S. Reformatsky, Ber., 1887, *20*, 1210; J. Russ. Chem. Soc., 1890, 22, 44.
2. B. Han, P. Boudjouk, J. Org. Chem., 1982, *47*, 5030.
3. A.K. Bose, K. Gupta and M.S. Manhas, J. Chem. Soc., Chem. Commun., 1984, 86; C. Petrier, L. Gemai and J. L. Luchi, Tetrahedron Lett., 1982, 3361.
4. T. Titazume, Synthesis, 1986, 855.
5. H. Tanka, S. Kishigami and F. Toda, J. Chem., Soc., 1991, *56*, 4333.
6. M.W. Rathke, J. Org. Chem., 1970, *35*, 3966.
7. R. Heilmann and R. Glenat, Bull. Soc. Chim. Fr., 1955, 1586.
8. H. Mattes and C. Benezra, Tetredron Lett., 1985, 5697.

2.24 SIMMONS–SMITH REACTION

This reaction is widely used for the synthesis of cyclopropane derivatives from alkenes by reaction with methylene iodide and zinc-copper or better zinc-silver couple. This is a versatile reaction and has been used with success to a variety of alkenes. Many functional groups are uneffected, enabling one to synthesise of a variety of cyclopropane derivatives. As an example, dihydrosterculic acid can be obtained in 51% yield from methyl oleate (Scheme-1).

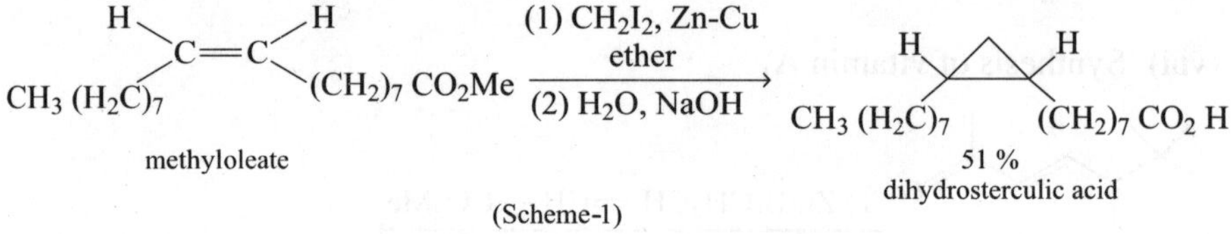

$$CH_3 (H_2C)_7 \quad C=C \quad (CH_2)_7 CO_2Me \xrightarrow[\text{(2) H}_2\text{O, NaOH}]{\substack{\text{(1) CH}_2\text{I}_2,\ \text{Zn-Cu} \\ \text{ether}}}$$

methyloleate

51%
dihydrosterculic acid

(Scheme-1)

The above reaction is stereospecific and takes place by cis addition of a methylene to the less hindered side of the double bond.

MECHANISM

The reactive intermediate is believed to be an iodomethylenezinciodide complex, which reacts with the alkene in a bimolecular process in a concreted manner involving a cyclic transistion state to give a cyclopropane and zinc iodide (Scheme-2).

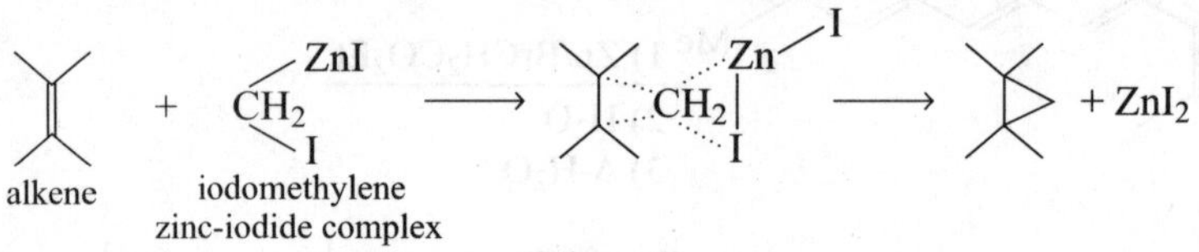

alkene   iodomethylene
zinc-iodide complex

(Scheme-2)

The reagent, iodomethylenezinc iodide, ICH_2ZnI is obtained in situ by the reaction of diiodomethane and Zn (in the form of zinc copper couple) (Scheme-3)

$$CH_2I_2 + Zn \longrightarrow ICH_2ZnI$$

(Scheme-3)

Generally substituted alkenes react somewhat faster than unsubstituted alkenes. Thus, 1-methylcyclohexene reacts faster than cyclohexene. The reaction, as in the case of other olefins (Scheme-1) occurs by the cis addition of methylene to the less hindered side of the double bond. For example, cis-and trans-3-hexene give pure cis-1,2-diethylcyclopropane and trans 1,2-diethylcyclopropane resp. (Scheme-4).

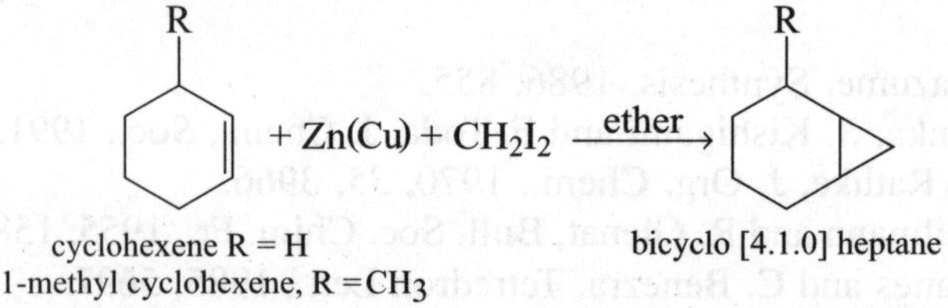

$$+ Zn(Cu) + CH_2I_2 \xrightarrow{\text{ether}}$$

cyclohexene R = H
1-methyl cyclohexene, R =CH₃

bicyclo [4.1.0] heptane

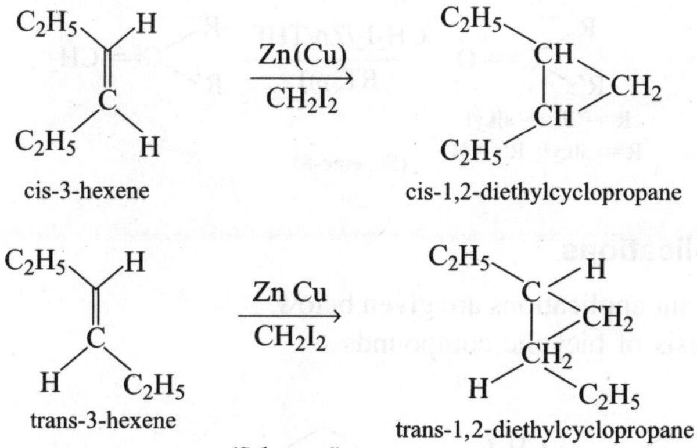

cis-3-hexene   cis-1,2-diethylcyclopropane

trans-3-hexene   trans-1,2-diethylcyclopropane

(Scheme-4)

Non-terminal acetylenes[2] also react with Simmons–Smit reagent to give the corresponding cyclopropenes but the yields are low (Scheme-5).

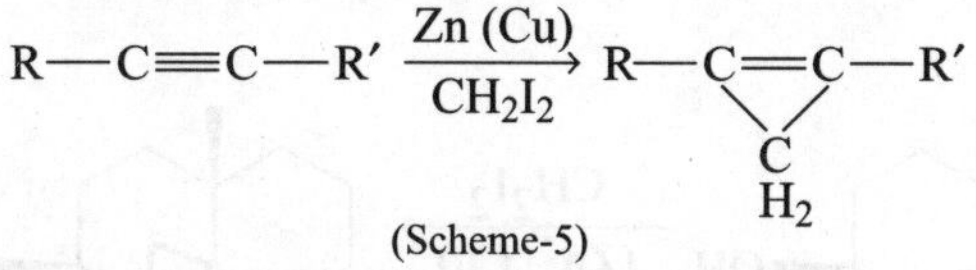

(Scheme-5)

Corey *et al*[3] observed that there is a pronounced effect of the neighbouring hydroxyl substituents. The oxygen atom of the starting alcohol coordinates with zinc followed by transfer of methylene to the nearer face of the adjacent double bond, thus increasing the rate and control of the stereochemistry of the adduct (Scheme-6).

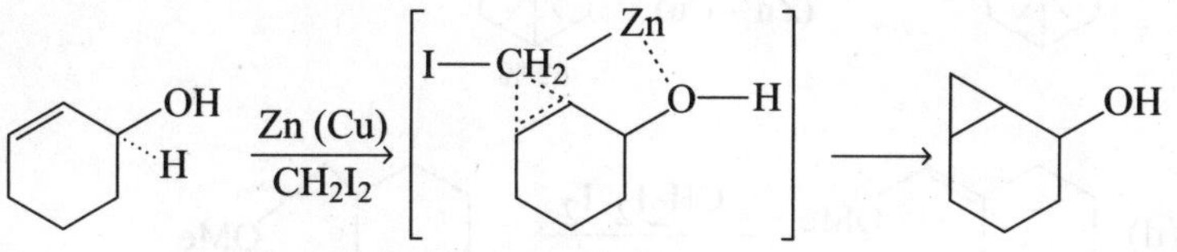

(Scheme-6)

2.24.1 Simmons–Smith Reaction Under Sonication

In this procedure, sonochemically activated zinc and methylene iodide are used[4]. The generated carbene adds on to the olefinic bond to give 91% yield of the cyclopropane derivative compared to 51% yield by the normal route (Scheme-7).

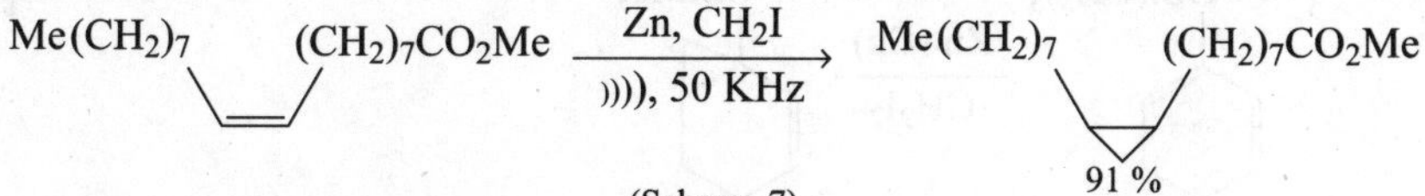

(Scheme-7)

The above procedure can be scaled up[5] and has several advantages. Ketones on reaction with Simmons-Smith reagent results in methylenation[6] of the carbonyl group (Scheme-8). Normally such methylenation of carbonyl group requires complex reagents; this can now be accomplished by sonication.

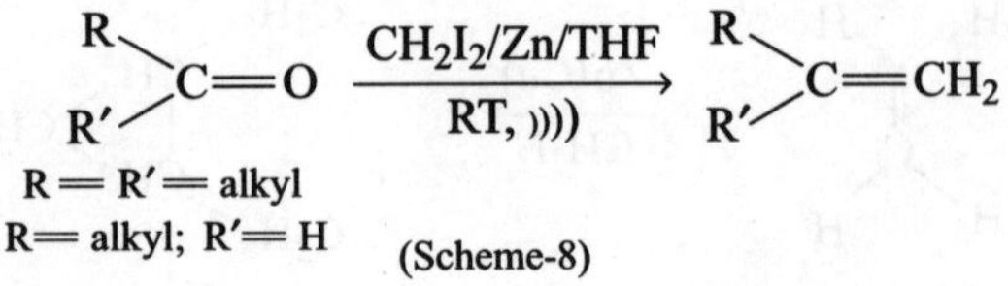

$$\underset{R'}{\overset{R}{\diagdown}}C{=}O \xrightarrow[RT,))))]{CH_2I_2/Zn/THF} \underset{R'}{\overset{R}{\diagdown}}C{=}CH_2$$

R = R′ = alkyl
R = alkyl; R′ = H   (Scheme-8)

2.24.2 Applications

Some important applications are given below.

1. Synthesis of bicyclic compounds

Ref.

(a) 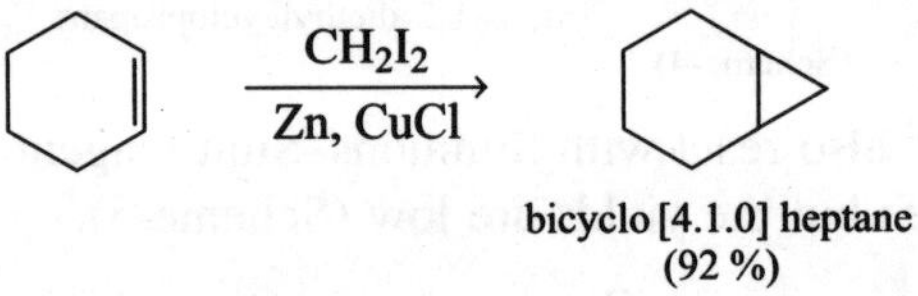 $\xrightarrow[Zn, CuCl]{CH_2I_2}$

bicyclo [4.1.0] heptane
(92 %)

7

(b) 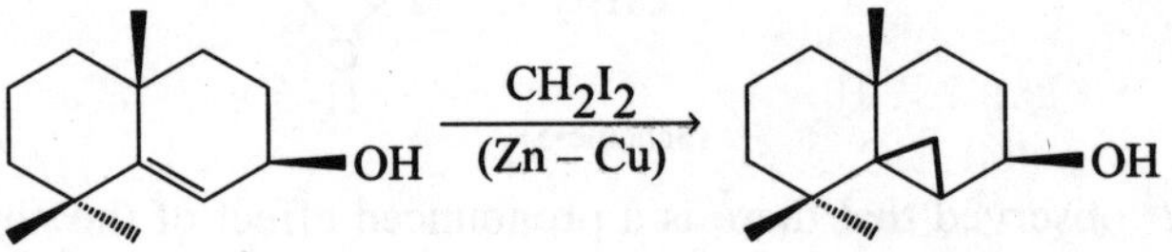 $\xrightarrow[(Zn-Cu)]{CH_2I_2}$ ···OH

8

(c) 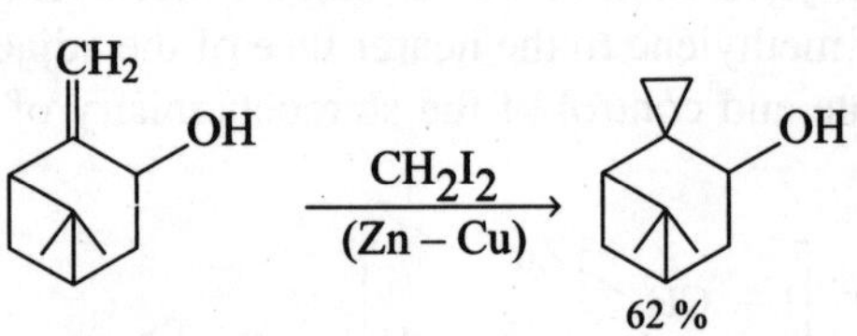 $\xrightarrow[(Zn-Cu)]{CH_2I_2}$

62 %

9

(d) 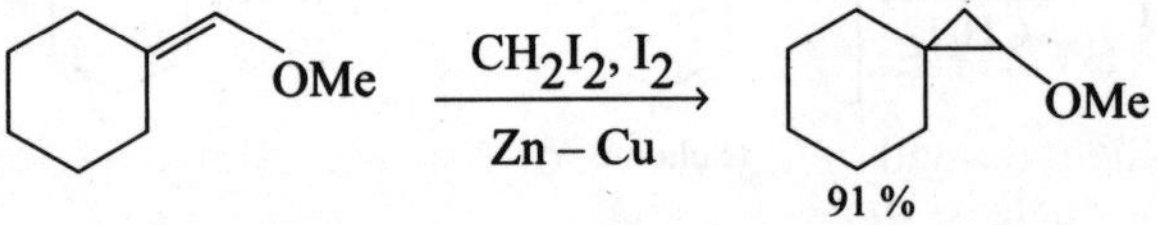 $\xrightarrow[Zn-Cu]{CH_2I_2, I_2}$ OMe

91 %

10

2. 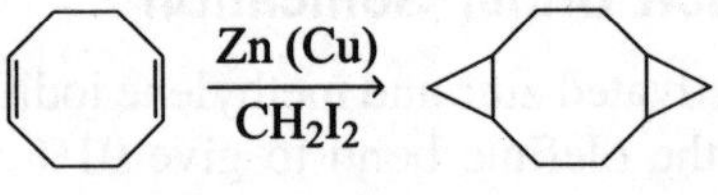 $\xrightarrow[CH_2I_2]{Zn (Cu)}$

11

3. 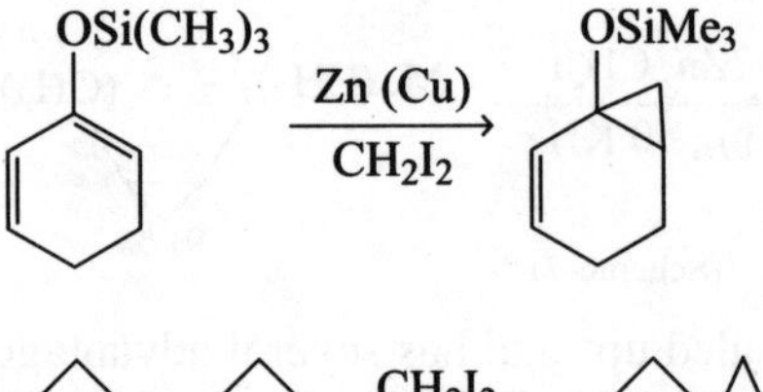 OSi(CH₃)₃ $\xrightarrow[CH_2I_2]{Zn (Cu)}$ OSiMe₃

12

4. 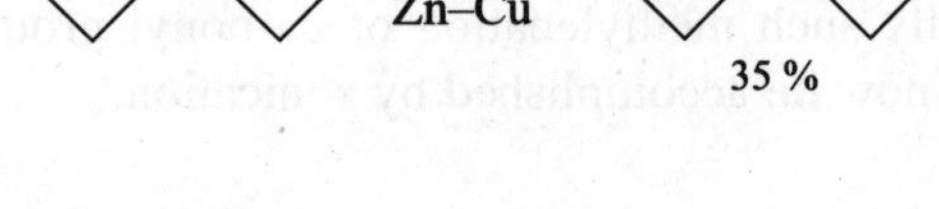 $\xrightarrow[Zn-Cu]{CH_2I_2}$

35 %

13

5.

$$\xrightarrow[\text{ether}]{\text{CH}_2\text{I}_2,\ \text{Zn–Cu}}$$

REFERENCES

1. H.E. Simmons and R.D. Smith. J. Am. Chem. Soc., 1958, *80*, 5323; H.E. Simmons, T.R. Carins, S.A. Vladuchick, and C.M. Hoiness, 1973, Organic Reactions, *29*, 1.
2. N.T. Castellucci and C.E. Griffin, J. Am., Chem., Soc., 1960, *82*, 4107.
3. E.J. Corey and H. Uda, J. Am. Chem., Soc., 1963, *85*, 1788.
4. O. Repic and S. Vogt, Tetrahedron Lett., 1982, *23*, 2729.
5. H. Tso, T. Chou and H. Hung, J. Chem. Soc. Chem. Commun., 1887, 1552.
6. C. Petrier, A.L. Gemal and J.L. Luck, Tetrahedron Lett., 1982, *23*, 3361.
7. R.J. Rawson and I.T. Harrison, J. Org. Chem., 1970, *35*, 2057.
8. W.G. Douben and A.C. Asheroft, J. Am. Chem. Soc., 1963, *85*, 3673.
9. C. Filliatre and C. Gueraud, C.R. Acad. Sci., *C*, 1971, *273*, 1186.
10. E. Wenkart, R.A. Mueller, E.J. Reardon, Jr., S.S. Sathe, D.J. Scharf and G. Tosi, J. Am. Chem. Soc., 1970, *92*, 7428.
11. S.D. Koch, J. Org. Chem., 1961, *26*, 3122.
12. S.D. Koch, R.M. Kliss, D.V. Lopiekes and R.J Wireman. J. Org. Chem., 1961, *26*, 3122.
13. L.K. Bee, J. Beeby, J.W. Everett and P.J. Garrat. J. Org. Chem., 1975, *40*, 2212.

2.25 STRECKER SYNTHESIS

Treatment of an aldehyde with ammonia and hydrogen cyanide produces an α-amino nitrile. Hydrolysis of the nitrile group of the α-aminonitrile converts the latter to an α-amino acid. This synthesis is called strecker synthesis[1] (Scheme-1).

$$\underset{}{\text{R—CH}} + \text{NH}_3 + \text{HCN} \longrightarrow \underset{\substack{| \\ \text{NH}_2 \\ \alpha\text{-amino} \\ \text{nitrile}}}{\text{RCH CN}} \xrightarrow[\text{H}_2\text{O}]{\text{H}_3\text{O}^+/\Delta} \underset{\substack{| \\ {}^+\text{NH}_3 \\ \alpha\text{-amino} \\ \text{acid}}}{\text{RCH CO}_2^-}$$

(with O double bonded to the first CH)

(Scheme-1)

In the **Erlenmeyer modification**[2] of the strecker synthesis, the aldehyde is treated with HCN and the formed cyanohydrin is reacted with ammonia (Scheme-2); the final step is same as given above.

$$\underset{\text{aldehyde}}{R-\overset{\overset{\displaystyle O}{\|}}{C}H} \xrightarrow{\text{HCN}} \underset{\text{cyanohydrin}}{R-\overset{\overset{\displaystyle OH}{|}}{C}H-CN} \xrightarrow{\text{NH}_3} R-\overset{\overset{\displaystyle NH_2}{|}}{C}H-CN \longrightarrow R-\overset{\underset{\displaystyle \overset{+}{N}H_3}{|}}{C}H\ CO_2^-$$

(Scheme-2)

A more convenient procedure is to treat the aldehyde in one step with ammonium chloride and sodium cyanide (this mixture is equivalent to ammonium cyanide, which in turn decomposes into ammonia and HCN). This procedure is referred to as the **Zelinsky–Stadnikoff modification**[3]. The final step is the hydrolysis of the intermediate α-aminonitrile under basic or acidic conditions to give the corresponding α-amino acid. The synthesis of phenylalanine from phenylacetaldehyde is given in Scheme-3.

$$\underset{\text{phenylacetaldehyde}}{Ph\ CH_2\ CHO} \xrightarrow{\text{NH}_3,\ \text{HCN}} Ph\ CH_2\ \overset{\overset{\displaystyle NH_2}{|}}{C}H\ CN \xrightarrow{\text{H}_3\text{O}^+} \underset{\text{phenylalanine}}{Ph\ CH_2\ \overset{\overset{\displaystyle \overset{+}{N}H_3}{|}}{C}H\ COO^-}$$

(Scheme-3)

MECHANISM

The first step is the formation of an imine from the aldehyde and ammonia. Subsequent addition of HCN on gives α-aminonitrile, which on hydrolysis gives α-aminoacid (Scheme-4).

$$R-CHO + NH_3 \longrightarrow \left[R\ \overset{\overset{\displaystyle OH}{|}}{\underset{\underset{\displaystyle NH_2}{|}}{C}}H \right] \xrightarrow{-H_2O} \underset{\text{imine}}{R\ CH=NH}$$

$$\underset{\text{imine}}{R\ CH=NH} \underset{}{\overset{CN^-}{\rightleftharpoons}} R\ \overset{\overset{\displaystyle}{}}{\underset{\underset{\displaystyle CN}{|}}{C}}H-\overset{-}{N}H \overset{H_3O^+}{\rightleftharpoons} \underset{\alpha\text{-aminonitrile}}{R\ \overset{}{\underset{\underset{\displaystyle CN}{|}}{C}}H-NH_2}$$

$$\underset{\alpha\text{-aminonitrile}}{R\ \overset{}{\underset{\underset{\displaystyle CN}{|}}{C}}H-NH_2} \xrightarrow[H_2O]{H_3O^+,\ \text{heat}} \underset{\alpha\text{-aminoacid}}{R-\overset{}{\underset{\underset{\displaystyle {}_+NH_3}{|}}{C}}H\ CO_2^-}$$

(Scheme-4)

2.25.1 Strecker Synthesis Under Sonication

Strecker synthesis of aminonitriles in much better yield is possible using ultrasonic acceleration[4] (Scheme-5).

$$\underset{\substack{\text{aldehyde} \\ \text{or ketone}}}{R_2\ CO} \xrightarrow[))))]{R'NH_2,KCN,AcOH} \underset{\substack{\alpha\text{-amino} \\ \text{nitriles}}}{R_2C \overset{\displaystyle \diagup CN}{\diagdown NH\ R'}}$$

(Scheme-5)

A modified Streker synthesis for the preparation of α-aminonitriles in excellent yields involves the adsorption of the reagent on the surface of a catalyst before the reaction. In fact, this technique is a combination of the 'support reagents' with sonochemical activation; the side reactions are suppressed[5] (Scheme-6).

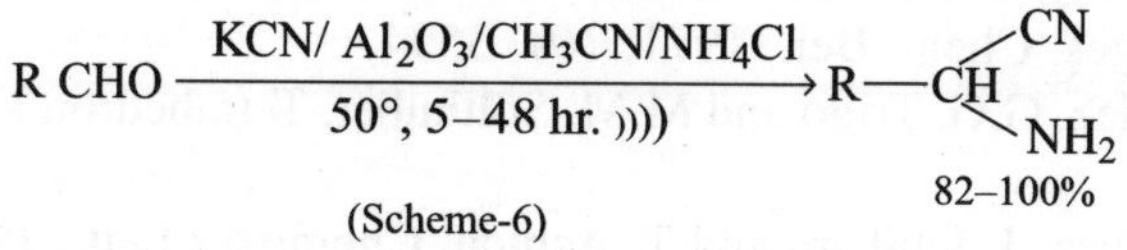

$$R\,CHO \xrightarrow[50°,\ 5-48\ hr.\ ))))]{KCN/\ Al_2O_3/CH_3CN/NH_4Cl} R-\underset{\underset{NH_2}{|}}{\overset{\overset{CN}{|}}{CH}}$$

82–100%

(Scheme-6)

Using simple ultrasonic cleaning baths, the reaction time for the synthesis of α-aminonitriles can be reduced from 12 days to 20–25 hr. and yields up to 60% are obtained[6] (Scheme-7).

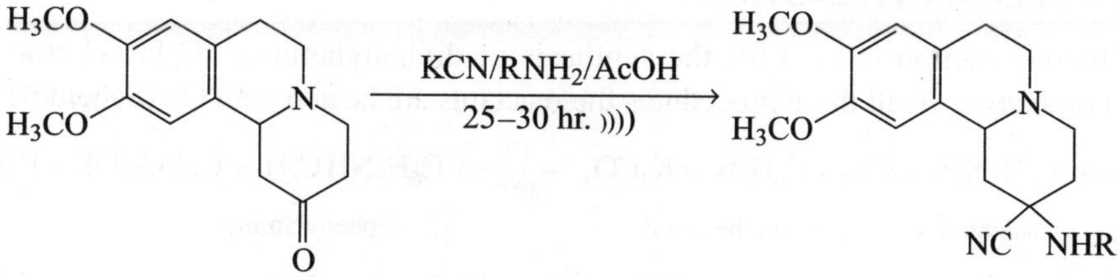

$$\xrightarrow[25-30\ hr.\ ))))]{KCN/RNH_2/AcOH}$$

R = H, Bu^n, Ph, PhCH$_2$-, 4-MeC$_6$H$_4$

(Seheme-7)

2.25.2 Applications

1. Strecker synthesis is mostly used for the synthesis of α-amino acids. Thus by using formaldehyde, acetaldehyde, 2-methylpropionaldehyde and phenylacetaldehyde, glycine, alanine, valine and phenylalanine respectively can be obtained. The synthesis of dl-Tyrosine is given below.

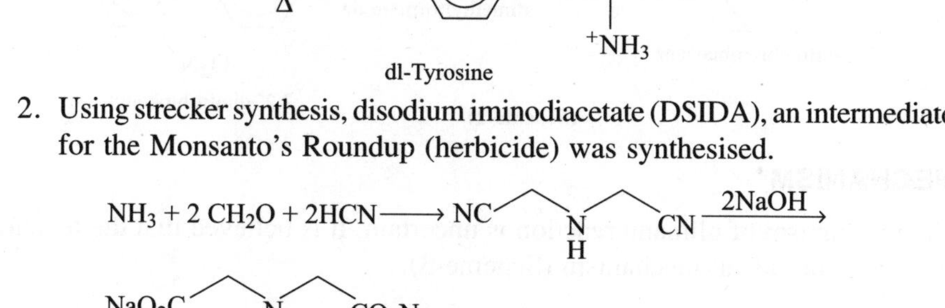

p-hydroxyphenyl-
acetaldehyde

dl-Tyrosine

2. Using strecker synthesis, disodium iminodiacetate (DSIDA), an intermediate for the Monsanto's Roundup (herbicide) was synthesised.

$$NH_3 + 2\ CH_2O + 2HCN \longrightarrow NC\overset{}{\underset{}{\diagup}}\underset{\underset{H}{|}}{N}\diagdown CN \xrightarrow{2NaOH}$$

DSIDA

REFERENCES

1. A. Strecker, Ann., 1850, *75*, 25; 1954, *91*, 349; D.T. Mowry, Chem., Rev., 1948, *42*, 236.
2. D.T. Mowry, Chem. Rev., 1948, *42*, 189.
3. K. Weinges, Chem. Ber., 1971, *104*, 3594.
4. J. Menedez, G.G. Trigo and M.M. Solthuber, Tetrahedron Lett., 1986, *27*, 3285.
5. T. Hanafusa, J. Ichihara and T. Ashida, Chemistry Lett., 1987, 687.
6. D.R. Borthakur and J.S. Sandhu, J. Chem., Soc., Chem. Commun., 1988, 1444.

2.26 ULLMANN REACTION

Ullmann reaction is used for the synthesis of diphenylamines, diphenyl ethers and diphenyls. In all these procedures the reactants are heated with Cu (Scheme-1).

(i) $C_6H_5NHCOCH_3 + C_6H_5Br + K_2CO_3 \xrightarrow[\text{reflux}]{\text{Cu}} C_6H_5NHC_6H_5 + CH_3COOK + KBr$

    acetanilide   bromobenzene   diphenylamine

(ii) $C_6H_5OH + C_6H_5Br + KOH \xrightarrow[\text{reflux}]{\text{Cu}} (C_6H_5)_2O + KBr + H_2O$

    phenol   bromobenzene   diphenyl ether

(iii) $2C_6H_5I + Cu \xrightarrow[\Delta]{C_6H_5NO_2} C_6H_5C_6H_5 + CuI_2$

    iodobenzene

(Scheme-1)

The last reaction used for the preparation of diaryls is called **ullmann coupling reaction**.

Aryl chlorides and bromides usually do not undergo this coupling reactions unless the halogen is activated by a suitable substituent (e.g. NO_2) in the ortho or para position. Thus o-nitrochlorobenzene on heating with copper powder gives 2,2′-dinitrobiphenyl (Scheme-2).

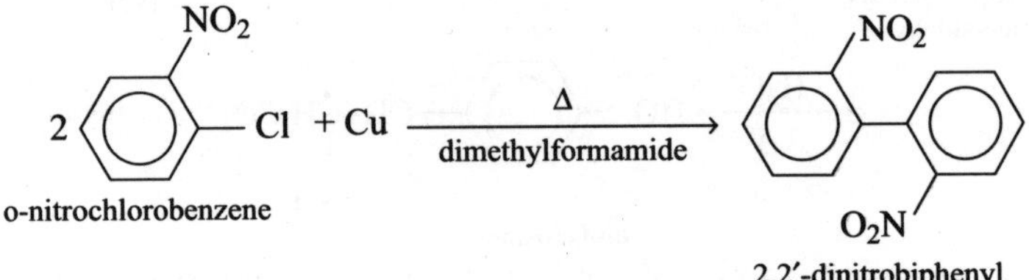

(Scheme-2)

MECHANISM

The mechanism of ullmann reaction is uncertain. It is believed that the reaction proceeds via radical mechanism (Scheme-3).

$$ArX + Cu \longrightarrow Ar\cdot \xrightarrow{Cu} ArCu \xrightarrow{ArX} Ar\text{–}Ar + CuX_2$$

(Scheme-3)

2.26.1 Ullmann Coupling Under Sonication

Under sonication, the size of the copper powder used in ullmann coupling in considerably reduced[2]. Breaking of the particles bring in contact with reactive solution fresh surface; the reactivity is not even hindered by the usual oxide layer on copper powder. The coupling is carried out in dimethylformamide (Scheme-4).

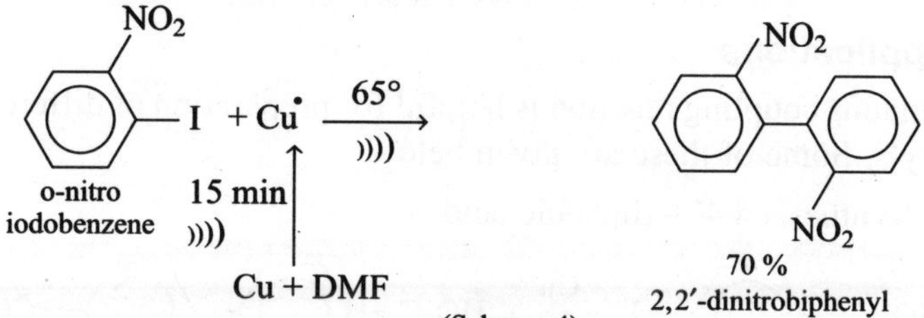

$$NO_2$$

o-nitro
iodobenzene

$Cu + DMF$

(Scheme-4)

2,2'-dinitrobiphenyl
70 %

The yield is much lower in decalin (20%) and toluene (5%).

Sonication of arylsulphonates in presence of in situ generated nickel (O) complex is an interesting Ulmann type coupling[3]. This method best works for triflates (R = CF$_3$). However the yields are low for tosylates (R = 4 – CH$_3$C$_6$H$_4$) (Scheme-5).

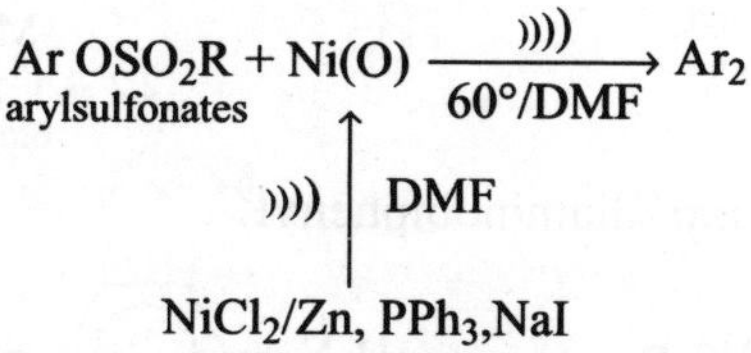

$$Ar\,OSO_2R + Ni(O) \xrightarrow[60°/DMF]{)))} Ar_2$$

arylsulfonates

))) DMF

NiCl$_2$/Zn, PPh$_3$,NaI
(Scheme-5)

The above Ullmann type coupling is useful in the formation of silicon bonds. Thus, a number of chlorosilanes can be coupled by sonication in the presence of lithium[4]. Using this method, highly hindered tetramesityldisilene can be prepared in good yield under sonication[5] (Scheme-6).

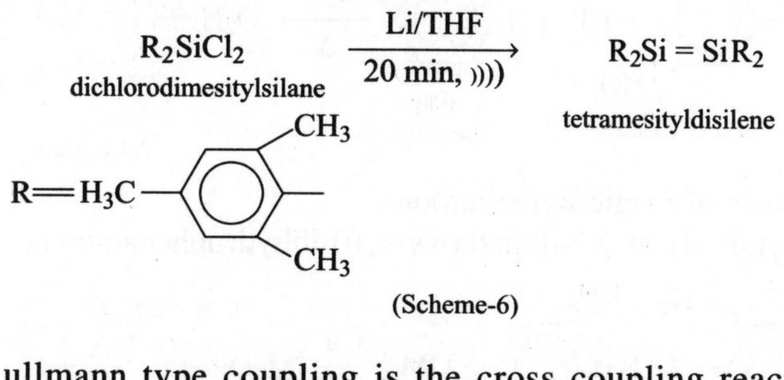

$$R_2SiCl_2 \xrightarrow[20\ min,\ )))]{Li/THF} R_2Si = SiR_2$$

dichlorodimesitylsilane

tetramesityldisilene

$$R = H_3C - \overset{CH_3}{\underset{CH_3}{\bigcirc}}$$

(Scheme-6)

Another ullmann type coupling is the cross coupling reactions[6,7,8] of perflouralkylzine reagents with vinyl, alkyl or aryl halides can be achieved by using a cleaning bath (35–45 KHz) (Scheme-7).

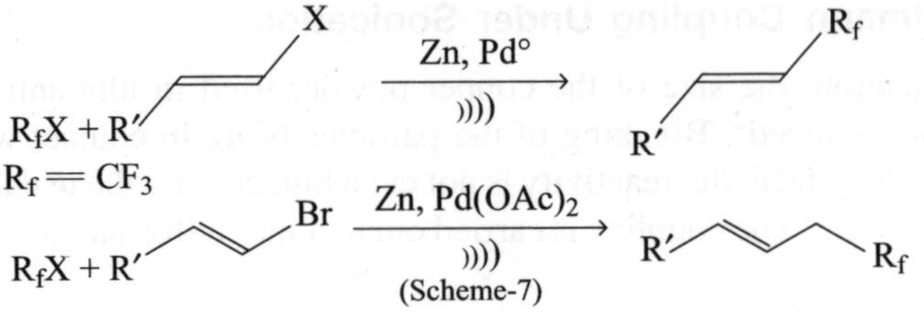

$$R_fX + R' \xrightarrow[\text{)))}]{\text{Zn, Pd}^\circ} \quad R'$$

$$R_f = CF_3$$

$$R_fX + R' \text{—Br} \xrightarrow[\text{(Scheme-7)}]{\text{Zn, Pd(OAc)}_2 \atop \text{)))}} R' \diagup\diagdown R_f$$

2.26.2 Applications

1. Ullmanns coupling reaction is helpful for preparation of different types of biaryls. Some of these are given below.

   (a) Synthesis 4,4′ – diphenic acid

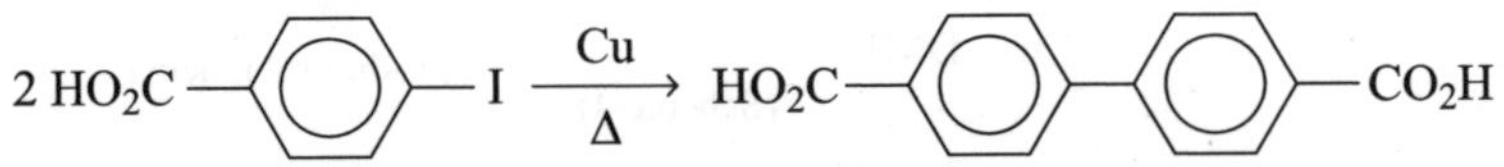

   $$2 \text{ HO}_2C \text{—}\bigcirc\text{—I} \xrightarrow[\Delta]{\text{Cu}} \text{HO}_2C\text{—}\bigcirc\text{—}\bigcirc\text{—CO}_2H$$

   p-iodobenzoic acid   4,4′-diphenic acid

   (b) Synthesis 2,2′,4,4′-tetramethylbiphenyl

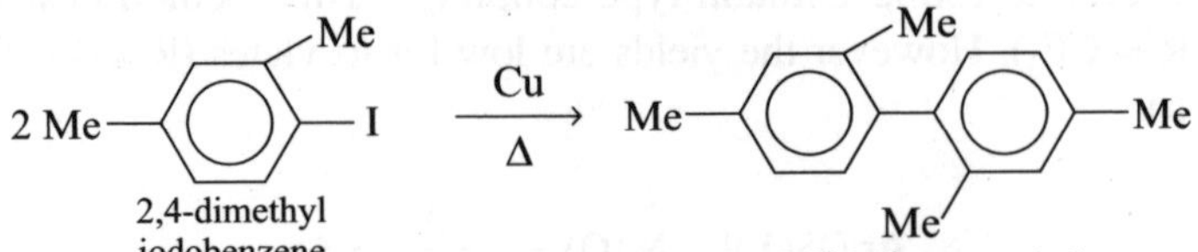

   2,4-dimethyl
   iodobenzene

   2,2′,4,4′-tetramethyl-
   biphenyl

   (c) Synthesis of p,p′-diaminobiphenyl[9]

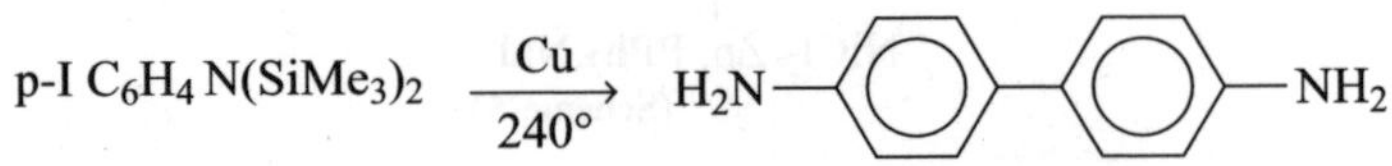

   $$\text{p-I C}_6\text{H}_4\text{N(SiMe}_3)_2 \xrightarrow[240^\circ]{\text{Cu}} \text{H}_2\text{N}\text{—}\bigcirc\text{—}\bigcirc\text{—NH}_2$$

   p-N,N-trimethylsilyl-
   iodobenzene

   60 %
   p,p′-diaminobiphenyl

   (d) Synthesis of 2,4,6-trinitrobiphenyl

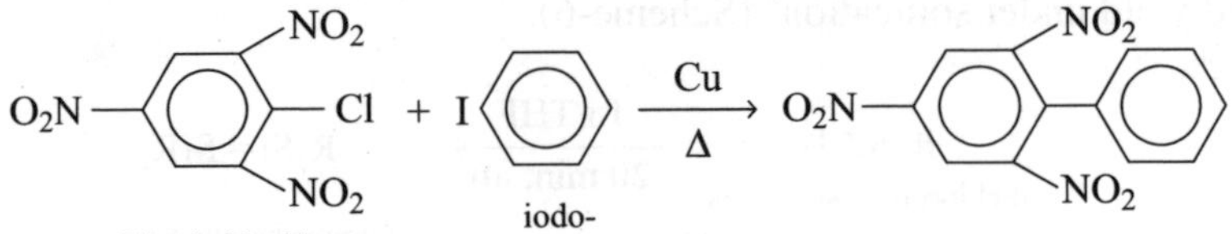

   Picryl chloride   iodobenzene

   2,4,6-trinitrobiphenyl

2. Synthesis of cyclic hydrocarbons

   (a) Synthesis of 2,7-dimethoxy-9,10-dihydrophenanthrene

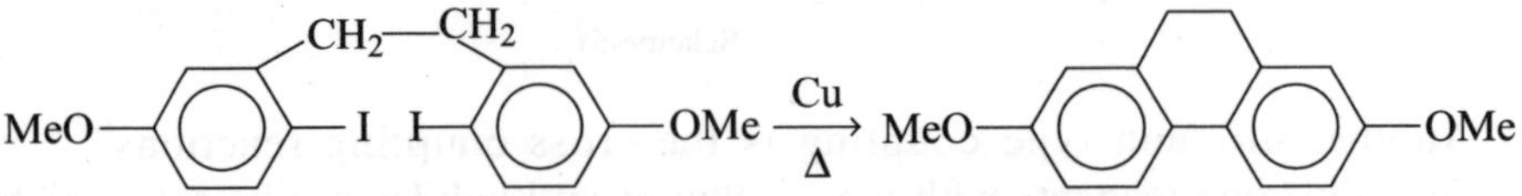

   2,2′-diiodo-5,5′-dimethoxy
   dibenzyl

   2,7-dimethoxy-9,10-
   dihydrophenanthrene

(b) Synthesis of perylene.

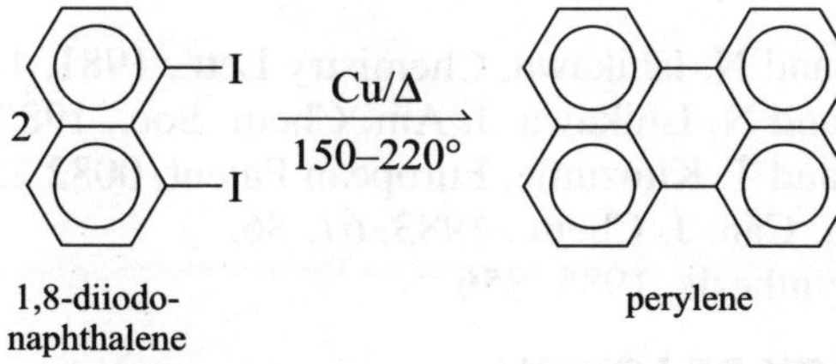

$$2 \text{ (1,8-diiodonaphthalene)} \xrightarrow[150–220°]{Cu/\Delta} \text{perylene}$$

1,8-diiodo-
naphthalene   perylene

(c) Synthesis of anthanthrone.

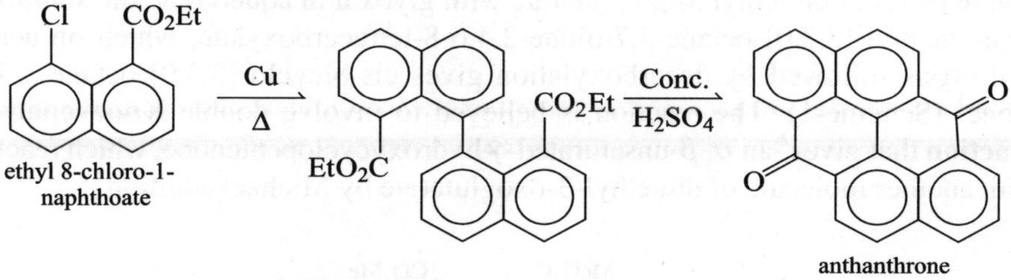

ethyl 8-chloro-1-
naphthoate $\xrightarrow{Cu, \Delta}$ $\xrightarrow[H_2SO_4]{Conc.}$ anthanthrone

3. Synthesis of diphenylamine[9].

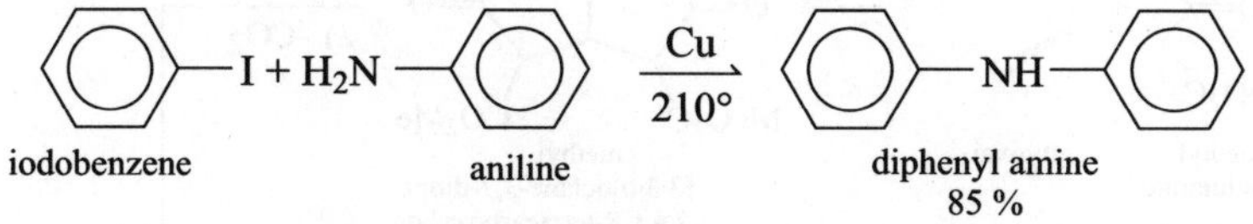

$$\text{iodobenzene} + H_2N\text{–(aniline)} \xrightarrow[210°]{Cu} \text{diphenyl amine } 85\%$$

iodobenzene aniline diphenyl amine
85 %

The is a commercial method for making diphenyamine

4. Synthesis of 2-oxo-1-phenyltetrahydropyrrole[10]

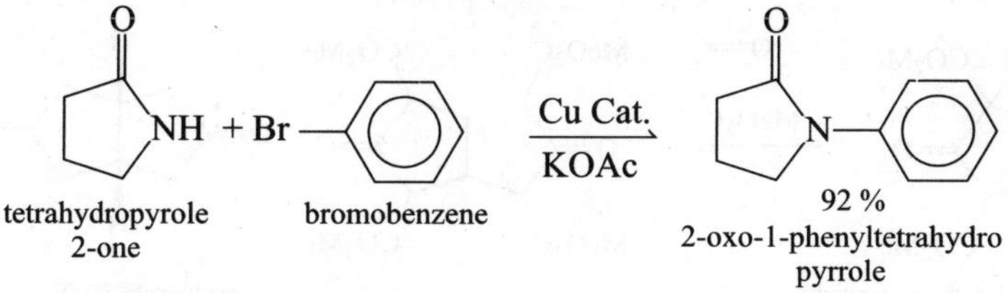

tetrahydropyrole
2-one bromobenzene $\xrightarrow[KOAc]{Cu\ Cat.}$ 92 %
2-oxo-1-phenyltetrahydro
pyrrole

5. Synthesis of diphenyl ether.

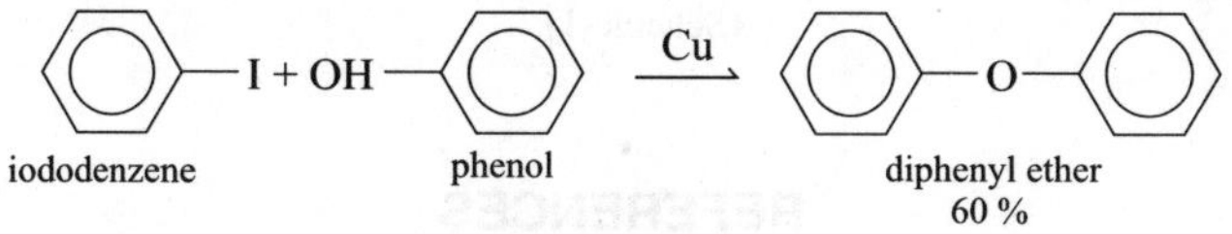

$$\text{iododenzene} + OH\text{–(phenol)} \xrightarrow{Cu} \text{diphenyl ether } 60\%$$

iododenzene phenol diphenyl ether
60 %

REFERENCES

1. F. Ullmann, Ann., 1904, *332*, 38; F. Ullmann and P. Sponagel, Ber., 1905, *38*, 2211; P.E. Fanta, Chem. Rev., 1946, *38*, 139.
2. J. Lindley, T.J. Manson and J.P. Lorimer, ultrasonics, 1987, *25*, 45.
3. T. Yamashita. Y. Inouse, T. Konodo and H. Hashimoto, Chem., Lett., 1986, 407.
4. P. Boudjouk and B. Hans. Tetrahedron Lett., 1981, *22*, 3813.

5. P. Boudjouk, B. Hans and K.R. Anderson, J. Am. Chem. Soc., 1982, *104*, 4992.
6. T. Kitazume and N. Ishikawa, Chemistry Lett., 1981, 1679.
7. T. Kitazumi and N. Ishikawa, J. Am. Chem. Soc., 1985, *107*, 5186.
8. N. Ishikawa and T. Kituzume, European Patent, 0082 252 Al, 1982.
9. T. Yamamoto, Can. J. Chem., 1983, *61*, 86.
10. B. Renger, Synthesis, 1985, 856.

2.27 WEISS—COOK REACTION

The reaction of dimethyl 3-oxyglutarate with glyoxal in aqueous acidic solution gives methyl [3.3.0] octane-3,7-dione-2,4,6-8-tetracarboxylate, which on acid hydrolysis followed by decarboxylation gives cis-bicyclo [3.3.0] octane-3-7-dione[1] (Scheme-1). The reaction is believed to involve double **Knoevenagal reaction** that gives[2] an α, β-unsaturated-γ-hydroxycyclopentenone, which reacts with another molecule of dimethyl-3-oxoglutarate by Michael addition.

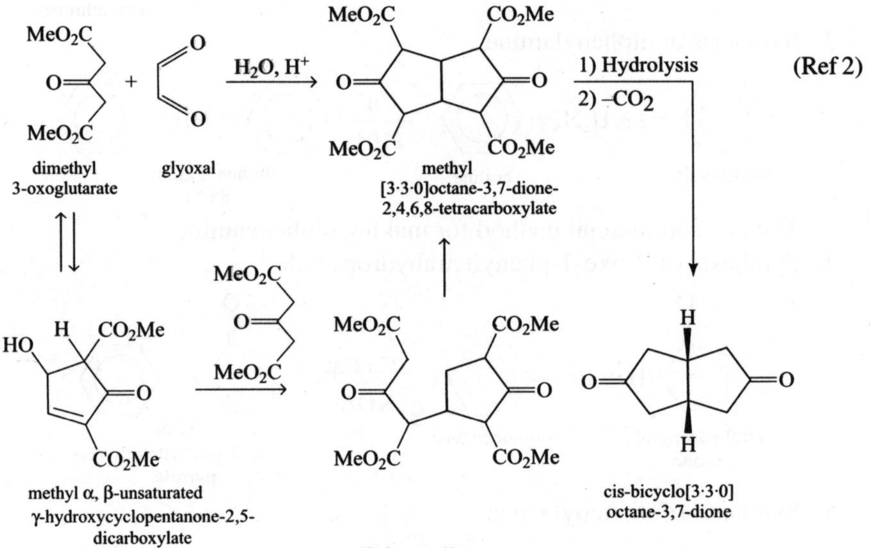

dimethyl 3-oxoglutarate   glyoxal

methyl [3·3·0]octane-3,7-dione-2,4,6,8-tetracarboxylate

methyl α, β-unsaturated γ-hydroxycyclopentanone-2,5-dicarboxylate

cis-bicyclo[3·3·0] octane-3,7-dione

(Scheme-1)

REFERENCES

1. W. Weiss and J.M. Edwards, Tetrahedron Lett., 1968, 4885.
2. C. Mannich and W. Krosche, Arch. Pharm. 1912, *250*, 647, F.E. Blike, Organic Reactions, 1942, *1*, 303.

2.28 WILLIAMSONS ETHER SYNTHESIS

It is an important useful procedure for the synthesis of unsymmetrical ethers. This synthesis consist of an SN$_2$ reaction of a sodium alkoxide with an alkyl halide, alkyl sulfonate or alkyl sulphate (Scheme-1).

$$R' - X + {}^-OR \longrightarrow R' - O - R' + X^-$$

$$X = I, -Br, -OSO_2R'' \text{ or } OSO_2OR''$$

(Scheme-1)

Since the secondary and tertiary alkyl halides undergos elimination reaction in the presence of a strong base such as alkoxide, it is important to use the alkoxide of the corresponding secondary or tertiary alcohol.

MECHANISM

The alkoxide ion reacts with the substrate in an SN_2 reaction resulting in the formation of an ether (Scheme-2). The substrate must have a good leaving group (as indicated in Scheme-1).

$$R - O^- Na^+ + R' - I \longrightarrow R - O - R' + NaI$$

sod. or potassium   alkyl halide   ether
    alkoxide   alkyl sulfonates
                         or alkyl sulphate

(Scheme-2)

2.28.1 Phase Transfer Catalysed Williamson Ether Synthesis

The phase transfer technique provides a simple and convenient method for conducting Williamson ether synthesis. It is found [2,3] that use of excess of alcohol or alkylhalide, lower temperature and larger alcohol (e.g., $C_8H_{17}OH$) give higher yield of ethers (Scheme-3).

$$C_8H_{17}OH + C_4H_9Cl \xrightarrow[\text{NaOH solution}]{\text{PTC}} C_8H_{17}OC_4H_9 + C_8H_{17}OC_8H_{17}$$

by product

(Scheme-3)

Use of five fold excess of aqueous sodium hydroxide (50%) over alcohol, excess alkyl chloride (also used as solvent) and tetrabutylammonium bisulfate (1–5 mole) as catalyst at 25–70° gave[3] optimum yields of ether. Primary alcohols require longer time or greater amount of catalyst.

Although dimethyl sulphate does not react with most alcohols in the presence of aqueous sodium hydroxide or even by the use of alkali metal alkoxides, the reaction proceeds easily[4] with tetrabutylammonium salts as catalyst. Activated alcohols and primary alcohols give high yield of ethers but secondary alcohols react very slowly and tertiary alcohols do not at all react.

In case of phenols, potassium carbonate is mostly used to scavenge protons. It is found that crown ethers can enhance their solubility as well as reactivity. A good example is the conversion of phenol into benzyl phenyl ether in quantative yield by using K_2CO_3 and a catalytic amount of 18-Crown 6.

The aromatic ethers are obtained by using a phenolate derivative with an alkyl halide. The Williamsons Ether synthesis is used for diaryl ethers (see ullmann reaction, section 2.26).

2.28.2 Applications

(i) Synthesis of ethyl propyl ether[5]

$$CH_3CH_2CH_2OH + NaH \longrightarrow CH_3CH_2CH_2O\overset{+}{N}a + H\!\!-\!\!H$$

propyl alcohol   sod. propoxide

$$\downarrow CH_3CH_2I$$

$$CH_3CH_2OCH_2CH_2CH_3 + N\overset{+}{a}I^-$$

ethyl propyl ether
70%

(ii) Synthesis of p-nitrophenyl butyl ether[6].

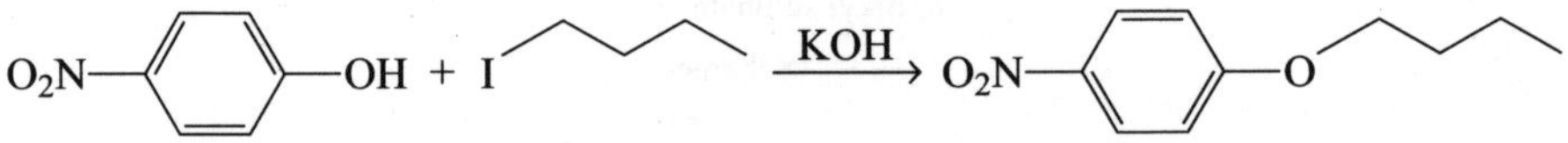

p-nitrophenol   butyl   55%
                       iodide

(iii) Synthesis of methylphenyl ether[7].

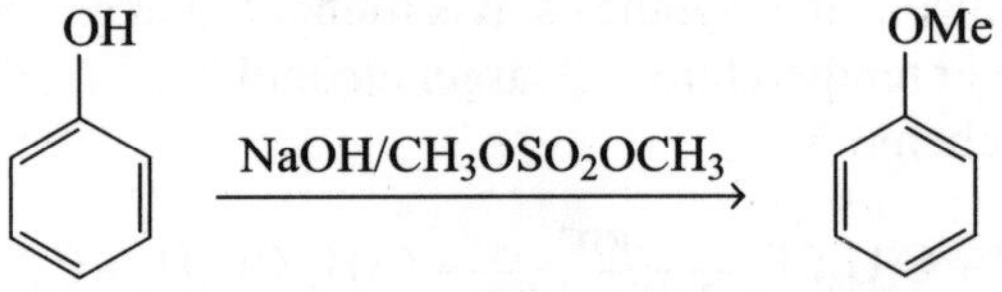

75%

(iv) Synthesis of isobutyl ethyl ether[8].

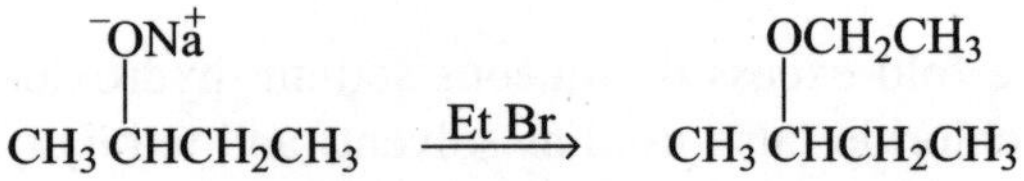

(v) A variation of Williamsons[1] ether synthesis is by using thallium (I) ethoxide. The method[6] is best for substrates containing an additional oxygen function such as —OH, —COOR, —CO—NR$_2$.

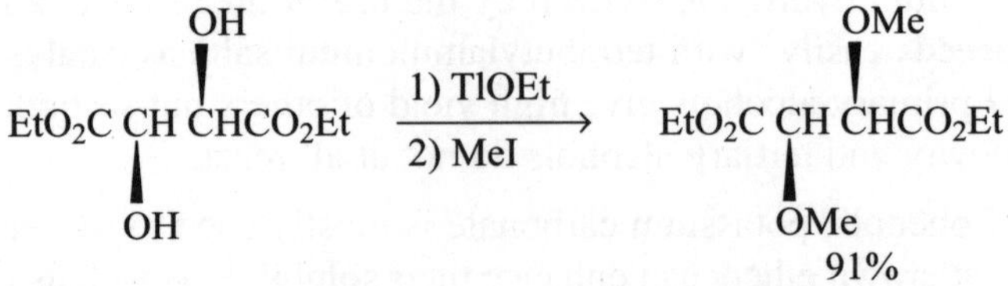

91%

(vi) The use of benzyl ether[9] as protecting group is very common, since the protecting group can be easily removed by catalytic reduction or by heating with conc. HCl and glacial acetic acid.

$$\underset{\substack{\text{CH}_2\text{OH}}}{\text{HO}}\quad\text{...}\quad \xrightarrow[\text{2) PhCH}_2\text{Cl}]{\text{1) NaH/dioxane}}$$

(Scheme structures: a disaccharide with CH₂OH, HO, OH groups and SPh, converting to a perbenzoylated disaccharide with CH₂OBz, BzO groups and SPh)

REFERENCES

1. A. W. Williamson, J. Chem. Soc., 1852, *4*, 229; O.C. Dermer, Chem. Revs., 1934, *14*, 409.
2. J. Jarrouse, C.R. Hebd, Scances Acad. Sci. Ser. C., 1951, *232*, 1424.
3. H.H. Freeman and R.A. Dubois, Tetrahedron Lett., 1975, 3251.
4. A. Merz, Angew Chem. Int. Ed. Engl, 1973, *12*, 846.
5. O. Gurney, J. Am. Chem. Soc., 1922, *44*, 1742.
6. H. Kaunowki, G. Cross and D. Seebach, Ber., 1981, *114*, 477.
7. G.S. Hiers and F.D. Hager, Org., Synth. 1941, *Coll. Vol. 1*, 58.
8. J.F. Norris and G. W. Rigby, J. Am. Chem. Soc., 1932, *54*, 2088.
9. R.C. Beier, B.P. Mundy and G.A. Strobel, Carbohyd. Res., 1983, *121*, 79.

2.29 WITTIG REACTION

The reaction of carbonyl compounds (aldehydes or ketones) with phosphorus ylides (or phosphorane) (commonly known as Wittig reagent) yield alkenes and triphenylphosphine oxide (Scheme-1).

$$\underset{\substack{\text{aldehyde} \\ \text{or ketone}}}{\overset{\text{R}}{\underset{\text{R}'}{>}}\text{C}=\text{O}} + \underset{\substack{\text{phosphorus ylide} \\ \text{(or phosphorane)}}}{(\text{C}_6\text{H}_5)_3\,\text{P}^+\!\!-\!\!\overset{..}{\text{C}}\overset{\text{R}''}{\underset{\text{R}'''}{<}}} \longrightarrow \underset{\substack{\text{alkene} \\ \text{(E and Z isomers)}}}{\overset{\text{R}}{\underset{\text{R}'}{>}}\text{C}=\text{C}\overset{\text{R}''}{\underset{\text{R}'''}{<}}} + \underset{\substack{\text{triphenyl} \\ \text{phosphene} \\ \text{oxide}}}{\text{O}=\text{P}(\text{C}_6\text{H}_5)_3}$$

(Scheme-1)

This reaction is known as the Wittig reaction[1] and is a very convenient method for the synthesis of alkenes (a mixture of (E) and (Z) isomers result). In this reaction there is absolutely no ambiguity as to the location of the double bond in the product, in contrast to E1 eliminations, which may yield multiple alkene products by rearrangement to more stable carbocation intermediate and both E1 and E2 elimination reactions may occur; this produces multiple products when different β-hydrogens are available for removal.

Wittig Reaction was discovered by George Wittig (and hence the name) in 1954 and was awarded Nobel Prize in Chemistry in 1979 due to its tremendous synthetic potentialities, being a valuable method for synthesising alkenes.

THE PHOSPHORUS YLIDES

The phosphorus ylides (commonly known as Wittig reagent) are obtained by the reaction of an alkyl halide with triphenylphosphine. The formed phosphonium salt is treated with a strong base like sodium hydride or phenyl lithium to give phosphorus ylide. These phosphorus ylides carry a positive and negative charge on adjacent atoms and can be represented by doubly bonded species called phosphoranes (Scheme-2).

$$Ph_3 P: + CH_2 \overset{R'}{-} X \longrightarrow Ph_3 \overset{+}{P} - CH_2 R' \ X^- \xrightarrow{base}$$

triphenyl phosphene   alkyl halide   phosphonium salt

$$\longrightarrow \left[Ph_3 \overset{+}{P} - \overset{-}{C}H R' \longleftrightarrow Ph_3 P = CH R' \right]$$

ylide   phosphorane

(Scheme-2)

In the above procedure (Scheme-2), the first step is a nucleophilic substitution reaction, triphenylphosphine being an excellent nucleophile and a weak base readily reacts with 1° and 2° alkyl halides by an SN_2 mechanism to displace an halide ion from the alkyl halide to give an alkyl triphenyl phosphonium salt. In the second step (which is an acid-base reaction), a strong base removes a proton from the carbon atom that is attached to phosphorus to give the ylide.

MECHANISM

The mechanism of Wittig reaction has been the subject of considerable study. It was earlier suggested the ylide, acting as a carbanion attacks the carbonyl carbon of the aldehyde or ketone to form an unstable intermediate with separated charge called betaine. In the subsequent step, the betaine being unstable gets converted into a four-membered cyclic system called an oxaphosphetane, which spontaneously looses a molecule of triphenyl phosphine oxide to form an alkene. Subsequent studies have shown that betanine is not an intermediate and that the oxaphosphetane is formed directly by a cycloaddition reaction. The driving force for the Wittig reaction is the formation of very strong phosphorus-oxygen bond in triphenylphosphine oxide (Scheme-3).

$$\begin{array}{c} R \\ R \end{array} C = O + Ph_3 \overset{+}{P} - \overset{-}{C}H R' \longrightarrow \left[\begin{array}{c} R \\ R \end{array} \overset{O^-}{\underset{|}{C}} - \overset{\overset{+}{PPh_3}}{\underset{\underset{H}{|}}{C}} - R' \right]$$

aldehyde or ketone   ylide   betaine (may not be formed)

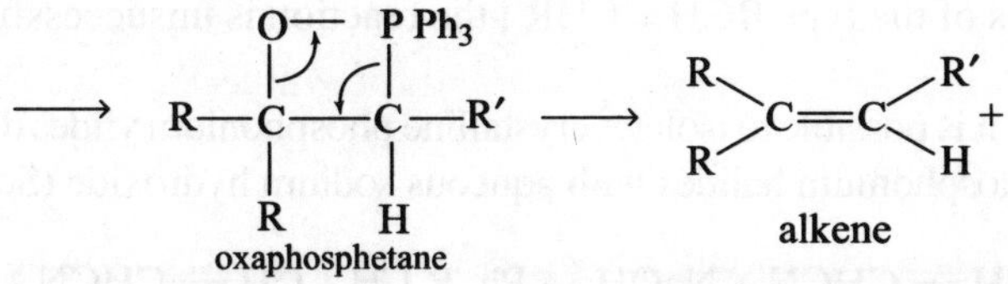

oxaphosphetane

alkene

(Scheme-3)

A typical example is given below (Scheme-4).

$$Ph_3P + CH_3Br \xrightarrow{C_6H_5 \text{ Li}} Ph_3P^+ - CH_3Br^- \xrightarrow{C_6H_5 \text{ Li}}$$

methyltriphenylphosphonium bromide 89%

$$\longrightarrow Ph_3\overset{+}{P} - \overset{-}{C}H_3 \leftrightarrow Ph_3P = CH_2 + C_6H_6 + LiBr$$

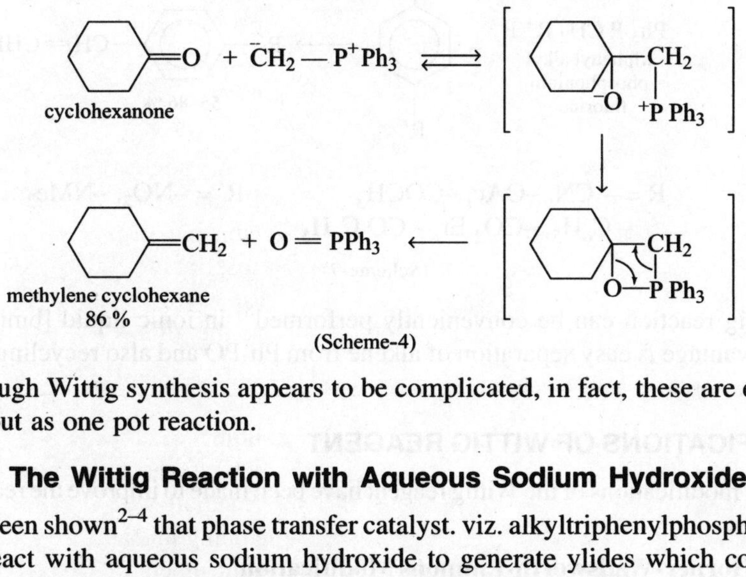

cyclohexanone

methylene cyclohexane
86 %

(Scheme-4)

Though Wittig synthesis appears to be complicated, in fact, these are easy to carry out as one pot reaction.

2.29.1 The Wittig Reaction with Aqueous Sodium Hydroxide

It has been shown[2-4] that phase transfer catalyst. viz. alkyltriphenylphosphonium salts react with aqueous sodium hydroxide to generate ylides which combine with organic phase aldehydes to produce olefins (Scheme-5).

$$Ph_3 \overset{+}{P} - CH_2 C_6 H_5 \overset{-}{Cl} + NaOH \text{ (org)} \xrightarrow[\text{solution}]{CH_2Cl_2} [Ph_3P = CH Ph]$$

$$\downarrow RCHO$$

$$RCH = CHPh + Ph_3PO$$

(Scheme-5)

In the above synthesis, the yield of the olefin increases[4] with the increase in concentration of alkali up to a maximum and then decreases. The yield depends on the alkyl group attached to the triphenylphosphonium salt. The quaternary phosphonium salts are better than the quaternary ammonium salts.

The PTC catalysed reaction is limited to only aldehydes and so is useful for preparation olefins of the type RCH = CHR'; the reaction is unsuccessful in case of ketones.

In certain cases, it is possible to isolate[5] crystalline phosphonium ylides directly by treatment of the phosphonium halides with aqueous sodium hydroxide (Scheme-6).

$$X^- Ph_3P^+CH_2CH = CHCN + NaOH \rightarrow Ph_3\overset{+}{P}\ \overset{-}{CH} - CH = CHCN + NaCl$$
(aq)
(Scheme-6)

Stable crystalline derivatives are also obtained when the phosphonium salts contains a $-CO_2Me$[6] or a $-CHO$[7] group instead of a cyano group.

It has been found[8] that triphenyl alkylphosphonium fluoride react with aldehydes to give olefins in good yield (Scheme-7)

$$Ph_3\ P\ CH_2\ R^+\ F^-$$
triphenyl alkyl
phosphonium
fluoride

55–86 %

$$R = - CN, -OAc, -COCH_3 \qquad R' = -NO_2, -NMe_2$$
$$- C_6H_5, -CO_2\ Et, - CO\ C_6H_5$$
(Scheme-7)

Wittig reaction can be conveniently performed[8a] in ionic liquid [bmin]BF_4. The advantage is easy separation of alkene from Ph_3PO and also recycling of the solvent.

MODIFICATIONS OF WITTIG REAGENT

Several modifications of the Wittig reagent have been made to improve the reactivity of ylides.

(i) **Horner-Wadsworth-Emmons Modification**

In this modification, the ylides are obtained from phosphonate ester (instead of a triphenylphosphonium salt), which in turn are readily available from alkyl halide and triethylphosphite via an **Arbuzov rearrangement**[9]. These ylides are more reactive than the corresponding phosphoranes and often react with ketones that are inert to phosphoranes. Thus the reaction of ethyl bromoacetate with triethylphosphite give phosphonate ester, which on treatment with base (NaH) and reaction with cyclohexanone gives α, β-unsaturated ester, ethyl cyclohexylideneacetate in 70% yield (with triphenylphosphorate only 20% yield is obtained). This method, commonly known as **Horner-Wadsworth Emmons modification** of the Wittig reaction (Scheme-8).

$$(EtO)_3P \quad + \; Br\,CH_2CO_2Et \longrightarrow (EtO)_3 \overset{\overset{\displaystyle O}{\|}}{P}\!\!-\!CH_2CO_2Et \longrightarrow$$

triethylphosphite   ethyl bromoacetate   phosphonate ester

$$\xrightarrow{\text{NaH}} (EtO)_3\overset{\overset{\displaystyle O}{\|}}{P}\!\!-\!\overset{-}{C}HCO_2Et$$

ethyl cyclohexylidene acetate

(Scheme-8)

The major product obtained in Horner-Wadsworth-Emmons modification is usually the (E) alkene isomer compared to (E) and (Z) isomeric mixture that is obtained in the usual Wittig reaction.

(ii) **The Wittig-Horner Reaction**

In this reaction, which is a modification of Wittig reaction, the readily available phosphine oxide $Ph_2\overset{\overset{\displaystyle O}{\|}}{P}CH_2R$ is used. Its lithio derivative is made to react with aldehydes and ketones to yield β-hydroxyphosphine oxides, which on treatment with sodium hydride smoothly eliminates water to give the corresponding alkene. This step is stereospecific, erythro hydroxy phosphine oxide gives the Z-alkene and the threo compound gives the E-alkene by preferential syn elimination. Various steps involved are given in Scheme-9.

$$Ph_3P \xrightarrow[\text{(2) hydrolysis}]{\substack{\text{(1) RCH}_2X \\ \text{quaternization}}} Ph_2\overset{\overset{\displaystyle O}{\|}}{P}\!\!\diagup^{R} \xrightarrow[\text{(2) R'CHO}]{\substack{\text{(1) BuLi} \\ \text{THF, }-78°}}$$

triphenyl
phosphene   phosphene
oxide

$$\longrightarrow \underset{\substack{\beta\text{-hydroxyphosphine}\\ \text{oxide}\\ \text{erythro : thero (9 : 1)}}}{Ph_2\overset{\overset{\displaystyle O}{\|}}{P}\cdots} \xrightarrow[\text{DMF}]{\text{NaH}} \underset{\text{z-alkene}}{\diagup^{R}} + Ph_2\overset{\overset{\displaystyle O}{\|}}{P}\!\!-\!ONa$$

(Scheme-9)

The phase transfer catalysed Wittig-Horner reaction using aqueous sodium hydroxide and either tetra-alkylammonium salts or crown ethers as catalysts give the olefins in 50–87% yield (Scheme-10).

$$(\text{EtO})_2\text{P(O)CH}_2\text{R} + \underset{\substack{\text{org.}}}{\underset{R''}{\overset{R'}{>}}}\text{C}{=}\text{O} + \underset{\substack{\text{aq.}}}{\text{NaOH}} \xrightarrow{\text{PTC}} \underset{\text{olefin}}{\text{RCH}{=}\text{C}\underset{R''}{\overset{R'}{<}}}$$

$$+ \ (\text{EtO})_2\text{PO}_2\text{Na}$$

(Scheme-10)

Following table gives the product obtained in phase-transfer catalysed Wittig Horner reaction.

R	R'	R''	Product	% yield	Ref.
–CN	C$_6$H$_5$	H	cinnamaldehyde	77	12 to 14
–CN	CH$_3$	H	crotonaldehyde	51	13
–CN	CH$_3$	CH$_3$	3-methyl-2-butenitrile	62	13
–CO$_2$Et	Ph	H	ethyl transcinnamate	56	13
–CO$_2$Et	Me	H	ethyl trans-crotonate	54	13
–CO$_2$Et	Ph	H	cinnamic acid	95	8
–COC$_6$H$_5$	Ph	H	C$_6$H$_5$CH=CHCOC$_6$H$_5$	55	12
–COC$_6$H$_5$	P–ClC$_6$H$_4$	H	P-Cl C$_6$H$_4$CH=CHCOC$_6$H$_5$	65	12
–COC$_6$H$_5$	p–BrC$_6$H$_4$	H	p-Br C$_6$H$_4$CH=CHCOC$_6$H$_5$	63	12
C$_6$H$_5$CH=CH–	2–pyridyl	H	C$_6$H$_5$ CH = CH — CH = CH—⟨pyridyl⟩	57	12–15
C$_6$H$_5$CH=CH	2–Furyl	H	C$_6$H$_5$ CH = CH —CH=⟨furyl⟩	84	16
⟨pyridyl⟩	C$_6$H$_5$	H	C$_6$H$_5$ CH = CH⟨pyridyl⟩	71	13

2.29.2 Wittig Reaction in Solid Phase

The well-known Wittig reaction has been reported[17] to occur in solid phase. In this procedure a 1:1 mixture of finely powdered inclusion compound of cyclohexanone or 4-methylcyclohexanone and (–)–B (derived from tartaric acid[18] and a catalytic amount of benzyltrimethylammonium hydroxide) was heated to 70° with Wittig reagent, carbethoxymethylene triphenylphosphorane to give optically active 1-(carbethoxymethylene) cyclohexane or the corresponding 4-methyl or 3,5-dimethyl compound (Scheme-11).

(R,R) -(–)— ⟨Ph$_2$COH inclusion compound structure⟩ (–)-B

+ ⟨cyclohexanone derivative⟩ $\xrightarrow[\text{solid, 70°, 4 hr}]{\text{Ph}_3\text{P} = \text{CHCO}_2\text{Et}}$ ⟨EtO$_2$C product structure⟩

R = CH$_3$; R' = H
R = H; R' = CH$_3$

(–)-product
45 % ee 73 % yield

(Scheme-11)

2.29.3 Wittig Reaction in Ionic Liquids

In the Wittig reaction, the separation of the alkene from the byproduct (Ph$_3$ PO) is a classical problem which is usually done by crystallisation or chromotography. The ionic liquid [bmin] BF$_4$ can be used as a medium to perform the reaction[19]. The advantage is easy separation of alkene from Ph$_3$ PO and also recycliny of the solvent.

2.29.4 Applications

Wittig reaction has tremendous applications for the synthesis of alkenes. It gives a great advantage over most other synthesis as there is no ambiguity as to the location of the double bond in the product. Some of the important applications are given below:

   (i) Synthesis of E-stilbene.

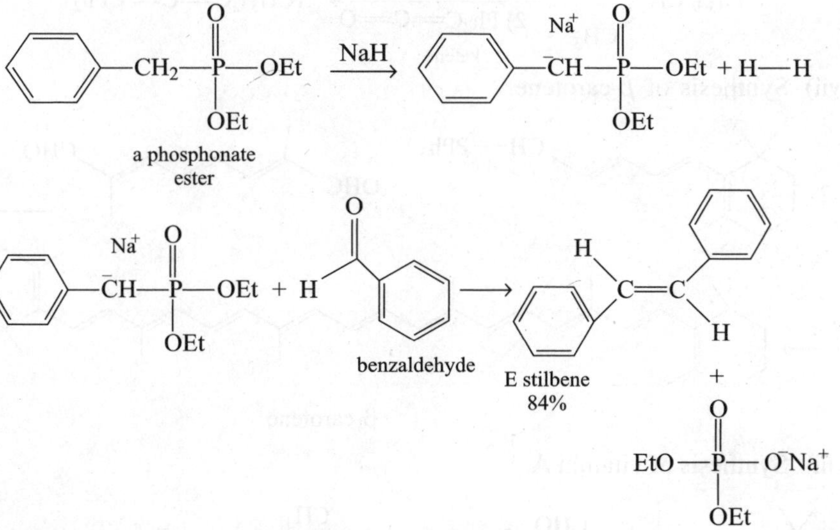

a phosphonate
ester

benzaldehyde   E stilbene
84%

   (ii) Synthesis of exocyclic methylene group compounds.

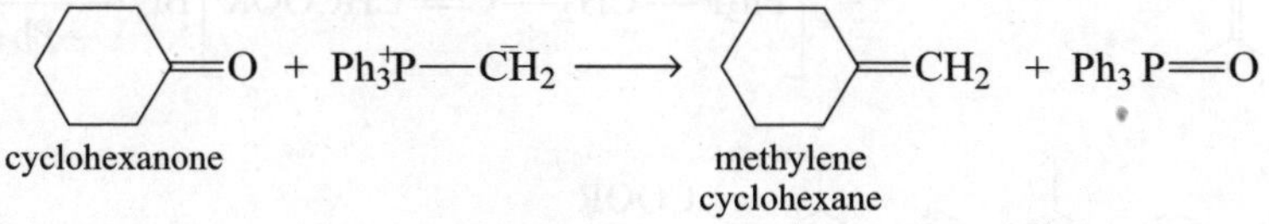

cyclohexanone   methylene
cyclohexane

   (iii) Synthesis of α, β-unsaturated esters.

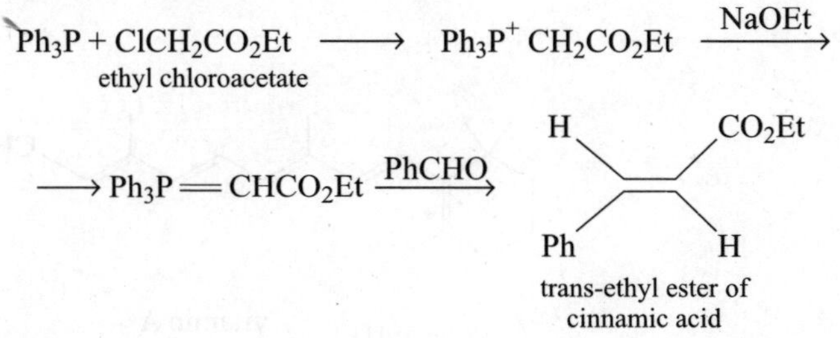

$$Ph_3P + ClCH_2CO_2Et \longrightarrow Ph_3P^+ CH_2CO_2Et \xrightarrow{NaOEt}$$

ethyl chloroacetate

$$\longrightarrow Ph_3P = CHCO_2Et \xrightarrow{PhCHO}$$

trans-ethyl ester of
cinnamic acid

(iv) Synthesis of polyzonimine, a natural insect repellant produced by millipedes.

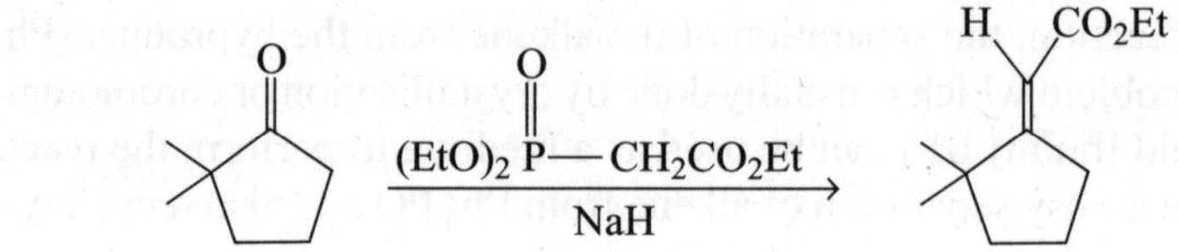

$$
\xrightarrow[\text{NaH}]{(EtO)_2 \overset{\overset{O}{\|}}{P}-CH_2CO_2Et}
$$

(v) Synthesis of dienes.

$$Ph_3P + ClCH_2CH = CHPh \longrightarrow Ph_3\overset{+}{P}-CH_2CH=CHPh$$

$$PhCHO + Ph_3\overset{+}{P}-CH_2CH=CHPh \xrightarrow{LiOEt} $$
$$PhCH=CH-CH=CHPh$$

(vi) Preparation of allenes

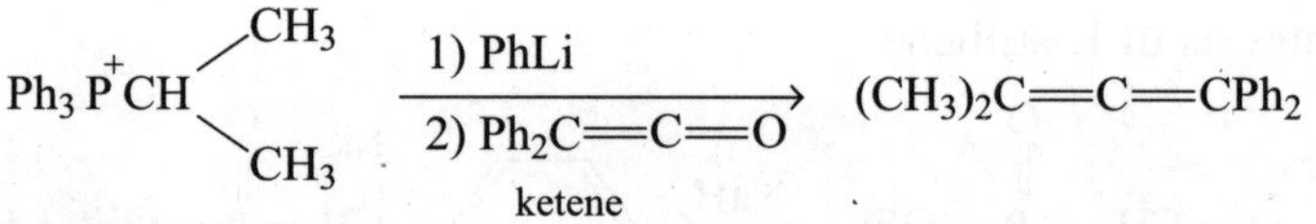

$$Ph_3\overset{+}{P}CH\overset{CH_3}{\underset{CH_3}{<}} \xrightarrow[\underset{\text{ketene}}{2)\ Ph_2C=C=O}]{1)\ PhLi} (CH_3)_2C=C=CPh_2$$

(vii) Synthesis of β-carotene.

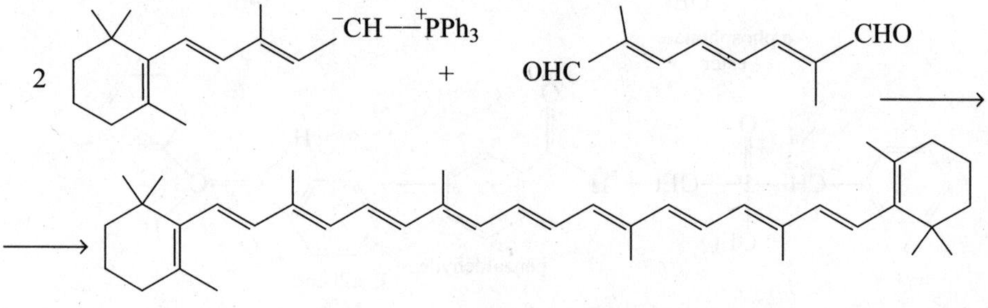

β-carotene

(viii) Synthesis of vitamin A.

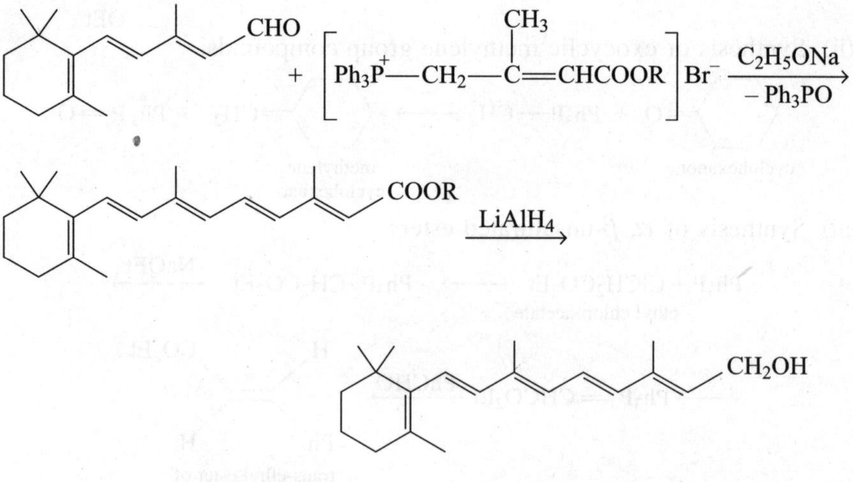

vitamin A

(ix) Synthesis of heterocyclic compounds[5].

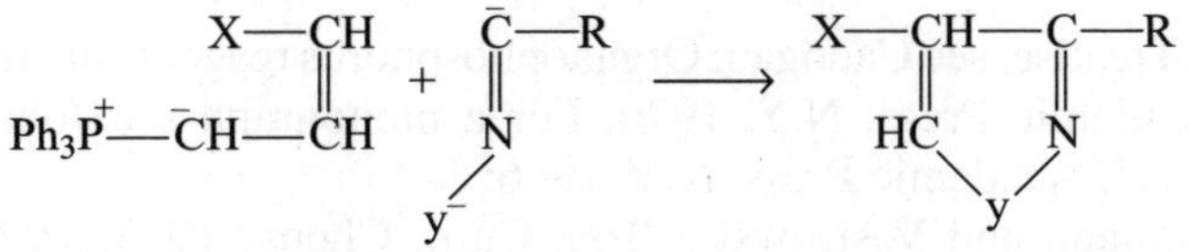

$$
\begin{array}{c}
X\!-\!CH \\
\| \\
Ph_3\overset{+}{P}\!-\!\bar{C}H\!-\!CH
\end{array}
+
\begin{array}{c}
\bar{C}\!-\!R \\
\| \\
N \\
\diagdown_y
\end{array}
\longrightarrow
\begin{array}{c}
X\!-\!CH\!-\!C\!-\!R \\
\| \quad\quad \| \\
HC \quad\quad N \\
\diagdown_y
\end{array}
$$

$x = -CN, -CHO, -CO_2Me$ $R = Ph, -CO_2CH_3, p\text{-}NO_2C_6H_4$

$y = O, -NPh,$

(x) Synthesis of bicyclic compounds[20].

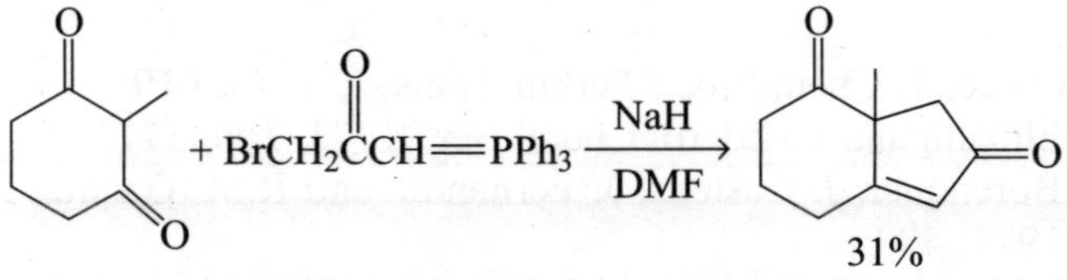

$$+ \; BrCH_2\overset{O}{\overset{\|}{C}}CH\!=\!PPh_3 \quad \xrightarrow[\text{DMF}]{\text{NaH}}$$

31%

(xi) The Wittig reaction in aqueous alkaline medium provides a convenient synthesis of olefines as per equation.

$$Ph_3P^+CH_2C_6H_5Cl \; + \; NaOH \,(aq) \longrightarrow \left[Ph_3P\!=\!CHPh \right]$$

$$\downarrow RCHO$$

$$RCH\!=\!CHPh + (C_6H_5)_2CHO$$

Following are given some olefines obtained

aldehyde RCHO	R' in phosphonium salt $Ph_3 P^+ R' X^-$	yield[a] % R—CH=CH—R'	Ref.
OHC—CHO	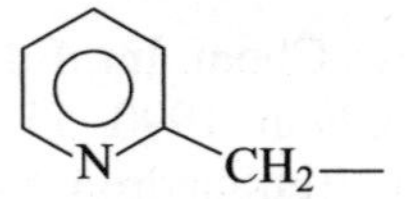	48[b]	3
OHC—CHO	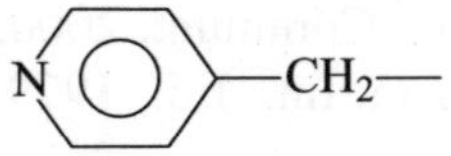	23–29[b]	3
OHC—CHO	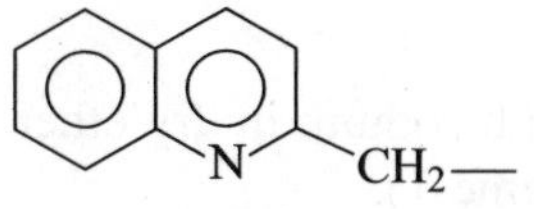	13[b]	3
C₆H₅CHO	CH₃–	99	4
C₆H₅CHO		0	3
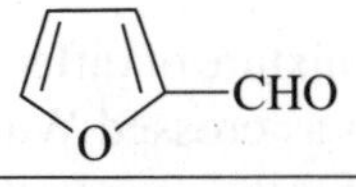 —CHO	CH₃⁻	63	4

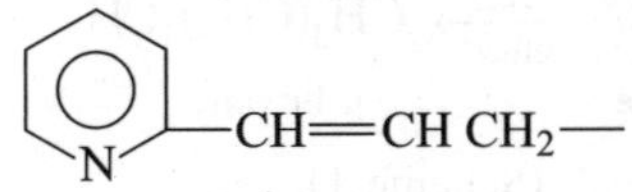

a   Ratio of cis and trans is approx. 1:1, when $R' \neq H$; b Product is $R'(CH = CH)_2 R'$

REFERENCES

1. For a general Treatise, see Cadogan: Organophosphorus reagents in organic synthesis, Academic Press, N.Y. 1970. For a monograph see Johnson; Ylid Chemistry', Academic Press, N.Y. 1966;
   Reviews Bestmann and Vostrowsky, Top. Curr., Chem., 1983, *109*, 85, 164.
2. G. Markl and A. Merz, Synthesis, 1975, 295.
3. S. Hung and I. Stemmler, Tetrahedron Lett., 1974, 3151.
4. W. Tagaki, I. Inouse, Y. Yano and T. Okonoge; Tetrahedron Lett., 1974, 2587.
5. P.D. Croce, J. Chem. Soc., Perkin Trans, I, 1976, 619.
6. F. Bohlmann and C. Zdero. Chem. Ber., 1973, *106*, 3779.
7. M.J. Borenquer, J. Castells, J. Fernandiz and R.M. Galard, Tetrahedron lett., 1971, 493.
8. G.P. Schiemenz, J. Becker and J. Stoeckigt, Chem. Ber., 1970, *103*, 2077.
8a. V.L. Boulaire, K.N. West, C.L. Liotta, C.A. Eckert., Chem. Commun., 2001, 887.
9. Prbuzov, Pune Applied Chem., 1964, *9*, 307.
10. J. Boutagy, R. Thomas, Chem. Rev., 1974, *74*, 87, W.S. Wadsworth and W.D. Emmons, J. Am. Chem. Soc., 19061, *83*, 1733.
11. Y. LeBigot, N. Hajjali, I. Rico, A. Lattes, M. Delmas and A. Gaset, Syn. Commun., 1985, 495.
12. M. Mikolajezyk, S. Grzejszezk, W. Miclura and A. Zatorski, Synthesis, 1976, 396.
13. C. Piechucki Synthesis, 1974, 869.
14. L.D. Incan and J. Seyden-Penne, Synthesis, 1975, 516.
15. M. Mikokajezuk, S. Grzejszezak, W. Modura and A. Zatorshi, Synthesis, 1975, 278.
16. A. Merz and G. Markl, Angew. Chem. Int. Ed. Engl. 1973, *12*, 845.
17. F. Toda and H. Akai, J. Org. Chem. 1990, *55*, 3446.
18. F. Toda, K. Tanka and J. Sato, Tetrahedron Asymmetry, 1993, *4*, 1771.
19. V.L. Boulaire, R. Gree, Chem. Commun., 2000, 2/95.
20. H.J. Altenbach, Angew., Chem. Int., Ed., 1979, *18*, 940.

2.30 WURTZ REACTION

The coupling of alkyl halides with sodium in dry ether to give hydrocarbons is known as Wurtz reaction[1] (Scheme-1).

$$CH_3CH_2CH_2Br \xrightarrow[\text{ether}]{\text{Na}} CH_3(CH_2)_4CH_3$$

n-propyl bromide   *n*-hexane

(Scheme-1)

Using two different alkyl halides (like RX and R′X) will give a mixture of different alkanes, viz. R–R + R′–R′ + R–R′ and so this method known as crossed Wurtz reaction is not of much importance. So this reaction is useful for the preparation of alkanes containing even number of carbon atoms (symmetrical). However, the hydrocarbons prepared by this method contain small amounts of olefins as byproducts.

Coupling of alkyl halides or sulfonates with grignard reagents or RLi in presence of Cu(I) salts is also a modified form of Wurtz reaction. Thus, alkyl halides react with grignard reagent in presence of catalytic amount of cobaltous chloride to give alkanes. Coupling products are also obtained by simply adding cobaltous chloride to a solution of grignard reagent (Scheme-2).

$$RMgX + R'X \xrightarrow{\ CoCl_2\ } R\text{--}R' + MgX_2$$

$$2RMgX \xrightarrow{\ CoCl_2\ } R\text{--}R$$

(Scheme-2)

Another variation of Wurtz reaction is known as **Wurtz-Fittig reaction**[2]. In this procedure, an alkyl halide and an aryl halide couple to form alkylated aromatic ring. For example, bromobenzene and butyl bromide react with sodium give *n*-butylbenzene (Scheme-3).

| bromobenzere | n-butyl bromide | | n-butyl benzene | |

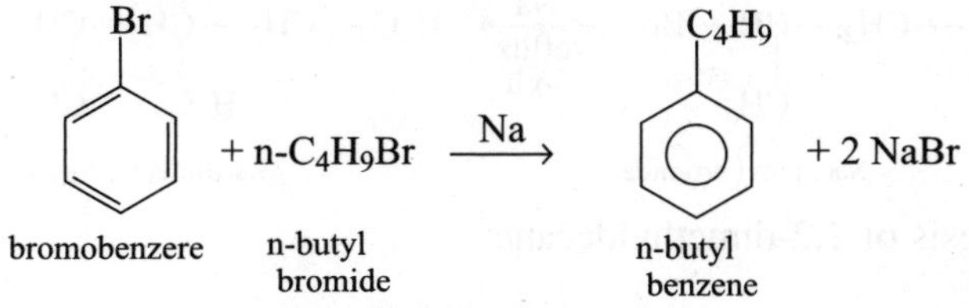

(Scheme-3)

The byproducts in this reaction are R-R and Ar-Ar, which can be separated easily.

MECHANISM

Two mechanisms have been proposed.

(i) Free radical mechanism. It operates in vapour state.

$$R\text{--}X + \dot{N}a \longrightarrow \dot{R} + NaX$$

$$\dot{R} + \dot{R} \longrightarrow R\text{--}R$$

(ii) Ionic mechanism. It operates in solutions. Initially sodium atom reacts with alkyl halide to form alkyl sodium having carbanion character. Final step is an S_N^2 type displacement on another molecule of alkyl halide.

$$RX + 2\dot{N}a \longrightarrow \bar{R} \overset{+}{N}a + NaX$$

$$R + R\text{--}X \xrightarrow{\ S_N^2\ } R\text{--}R + X^-$$

2.30.1 Wurtz Reaction Under Sonication

Wurtz Reaction carried out under sonication gives much better yields.

2.30.2 Wurtz Reaction in Water

It has been shown[3] that Wurtz coupling can also be carried out by Zn/H_2O (Scheme-4).

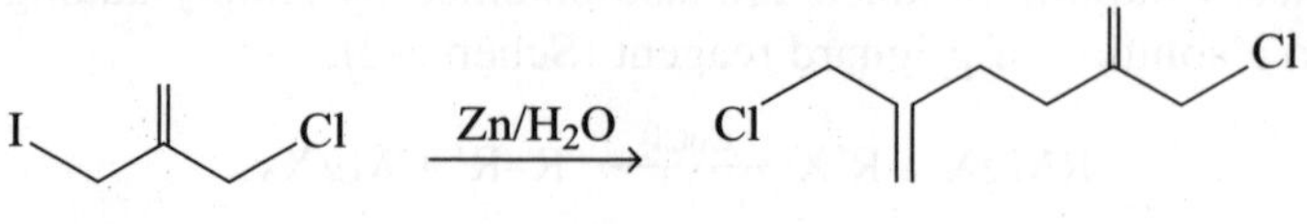

(Scheme-4)

Ullmann coupling reaction is also considered to be Wurtz-Type coupling (see ullmann reaction, section 2.26).

2.30.3 Applications

Some important applications are given below.

1. Synthesis of 3,4-dimethylhexane[4]

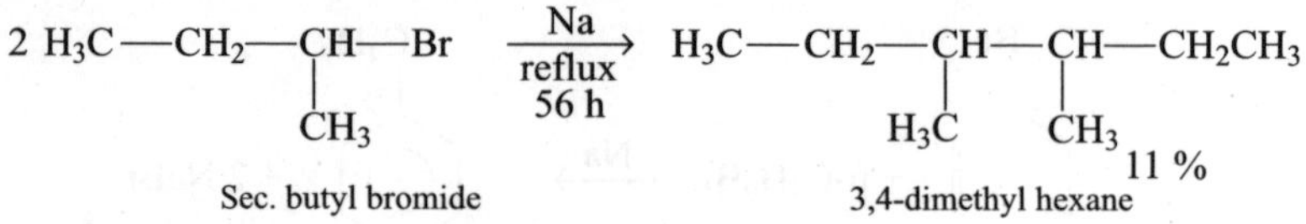

Sec. butyl bromide   3,4-dimethyl hexane

2. Synthesis of 2,2-dimethyldecane[5]

$$n\text{-}C_8H_{17}\text{-}OTS + tBu\ Br \xrightarrow{Li_2CuCl_4} n\text{-}C_8H_{17}\text{-}CMe_3$$

Octyl tosylate   ter. butyl   75%
                   bromide   2,2-dimethyldecane

3. Synthesis of bicyclic compounds

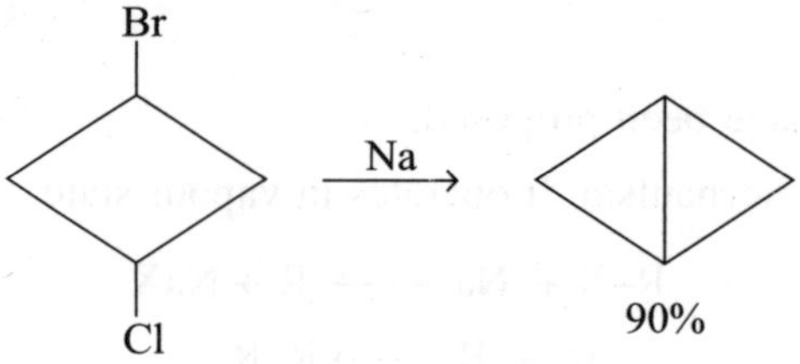

REFERENCES

1. A. Wurtz, Ann. Chem. Phys., 1855, *(3) 44*, 275; Ann., 1855, *96*, 364; R.E. Buntrock, Chem. Rev., 1968, *68*, 209.
2. B. Tollens and R. Fittig. Ann., 1864, *131*, 303; R. Fittig and J. Koneg., Ann., 1867, *144*, 277.
3. C.J. Li and T.H. Chan., Organometallics, 1991, *10*, 2548.
4. W.J. Balley, J. Org. Chem., 1962, *27*, 3088.

3
Green Preparation

In this part, some organic preparation, which are considered green synthesis and can be performed in a chemical laboratory are described. Few examples of the following types of reaction have been described.

1. Aqueous phase reactions
2. solid state (solventless) reactions
3. Photochemical reactions
4. PTC catalysed reactions
5. Rearrangement reactions
6. Microwave induced reactions
7. Enzymatic transformations
8. Sonication reactions
9. Esterification
10. Enamine reaction
11. Reactions in ionic liquids.

3.1 AQUOUS PHASE REACTIONS

3.1.1 Hydrolysis of Methyl Salicylate with Alkali

Alkaline hydrolysis of esters is called **saponification** and is an irreversible process.

Mechanism

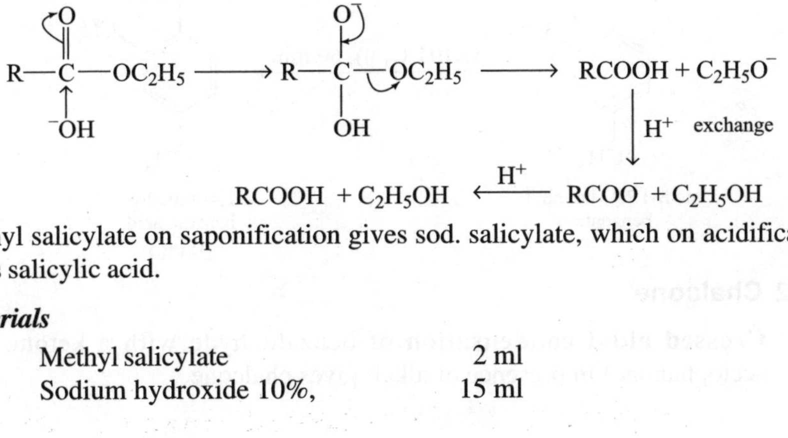

$$R-\overset{\overset{\displaystyle O}{\|}}{\underset{\underset{\displaystyle ^-OH}{\uparrow}}{C}}-OC_2H_5 \longrightarrow R-\overset{\overset{\displaystyle O^-}{|}}{\underset{\underset{\displaystyle OH}{|}}{C}}-OC_2H_5 \longrightarrow RCOOH + C_2H_5O^-$$

$$\Big\downarrow H^+ \quad exchange$$

$$RCOOH + C_2H_5OH \overset{H^+}{\longleftarrow} RCOO^- + C_2H_5OH$$

Methyl salicylate on saponification gives sod. salicylate, which on acidification gives salicylic acid.

Materials

Methyl salicylate	2 ml
Sodium hydroxide 10%,	15 ml

Procedure

Reflux a mixture of methyl salicylate (2 ml) with sodium hydroxide solution (15 ml, 10%) in a round bottomed flask using a reflux condenser or a sand bath (temp. 90–100°) for about 30 min. till the ester layer disappears. Cool the solution and acidify with hydrochloric acid. Cool the resultant solution (ice bath). Filter the separated salicylic acid and recrystallise from hot water yield 1.2 g m.p.158–159°.

Note

1. Use of a crown ether, viz. [18] crown-6 in small amount for saponification gives quantitative yield. The special feature of using crown ether, is that even the sterically hindered esters, which are difficult to saponify with alkali can be saponified conveniently by using [18] crown 6. (C.J. Pedersen and K. K. Friensdorff, Angew Chem. Int. Engl., 1972, 11, 16; C. J. Pedersen, J. Am. Chem.. Soc, 1967, 89, 2485, 7017; 1970, 92, 386, 391).

2. Saponification of esters can be brought about by using microwave under solid–liquid PTC conditions without solvent. The reaction can be completed in few minutes. For example,

$$C_6H_5 - \overset{\overset{\displaystyle O}{\|}}{C} - OCH_3 \xrightarrow[\substack{7\ min \\ microwave \\ aliquat\ 336}]{KOH} C_6H_5 - \overset{\overset{\displaystyle O}{\|}}{C} - OK$$
$$94\%$$

(A. Loupy, P. Pigeon, M. Ramdani, P. Jaequault, Synthetic Common., 1994, 24(2), 159).

3. Ester hydrolysis or saponification can also be conducted under milder conditions when sonication in used (S. Moon, L. Duclin, J. V. Croney, Tetrahedron Lett., 1979, 3971)

   In this case rate increase in attributed the emulsifying effect. A typical example is

methyl 2,4-dimethyl benzoate

$^-$OH/H$_2$O,))), 60 min

2,4-dimethyl benzoic acid (94%)

3.1.2 Chalcone

Crossed aldol condensation of benzaldehyde with a ketone (e.g. acetophenone) in presence of alkali gives chalcone.

$$C_6H_5CHO + CH_3\!-\!\overset{\overset{\displaystyle O}{\|}}{C}\!-\!C_6H_5 \xrightarrow{\;^-OH\;} \left[C_6H_5\!-\!\overset{\overset{\displaystyle H}{|}}{\underset{\underset{\displaystyle OH}{|}}{C}}\!-\!CH_2\!-\!\overset{\overset{\displaystyle O}{\|}}{C}\!-\!C_6H_5 \right]$$

Benzaldehyde   acetophenone

intermediate

$$\downarrow -H_2O$$

$$C_6H_5\,CH\!=\!CH\!-\!\overset{\overset{\displaystyle O}{\|}}{C}\!-\!C_6H_5$$

Chalcone
(Benzal acetophenone)

Materials

Benzaldehyde	2.2 ml
Acetophenone	2.5 ml
Rectified spirit	7 ml

Sodium hydroxide solution 1.1 g in 10 ml. H_2O

Procedure

Place benzaldehyde (2.2 ml) into a conical flask (100 ml capacity) and add acetophenone (2.5 ml) and rectified spirit (7 ml). Stir the mixture so as to obtain a homogeneous solution. If necessary, the mixture may be warmed to get a clear solution. To the clear solution at room temperature add sodium hydroxide solution (1.1 g NaOH in 10 ml H_2O). Stir the mixture. Keep the mixture over night in a refrigerator. Filter the separated chalcone, wash with cold water and recrystallise from alcohol. Yield 3.35 g (86.6 %), m.p. 56–57°.

Note

1. Using the same procedure and using veratraldehyde (3, 4-methylenedioxy benzaldehyde), p-anisaldehyde and 3-nitrobenzaldehyde in place of benzaldehyde prepare, 3,4-methylenedioxy chalcone (m.p. 112°), 4-methoxychalcone (m.p. 74°) and 3-nitrochalcone (m.p. 146°) respectively.

2. Chalcones are very conveniently obtained by the use of microwave ovens. There is considerable reaction rate enhancement and the reaction could be completed in 30 sec to 2 min time (R. Gupta, A. K. Gupta, S. Paul, P. L. Kachroo, Indian J chem., 1995, 34B, 61). In this procedure a solution of ketone and aromatic aldehyde in dry ethanol and catalytic amount of sodium hydroxide (1–2 pellet) is heated in a microwave oven for 30 sec to 2 minutes (at 210 watts. 30% microwave power). The reaction mixture is cooled and the chalcone filtered and washed with ethanol.

3.1.3 6-Ethoxycarbonyl-3,5-diphenyl-2-cyclohexenone

The α, β-unsaturated ketone, 6-ethoxycarbonyl-3,5-diphenyl-2-cyclohexenone is prepared (note 1) by **michael addition reaction** (sodium hydroxide catalysed conjugate addition of ethyl acetoacetate to trans-chalcone). In this case NaOH serves as a source of $^-$OH to catalyse the reaction.The Michael adduct obtained above on base-catalysed **aldol condensation reaction** gives (note 2) stable 6-membered ring compound (6-ethoxycarbonyl-3,5-diphenyl-3-hydroxy-cyclohexanone).

Finally the aldol intermediate is dehydrated to the required 6-ethoxycarbonyl-3.5-diphenyl-2-cyclohexenone. Various steps are given below.

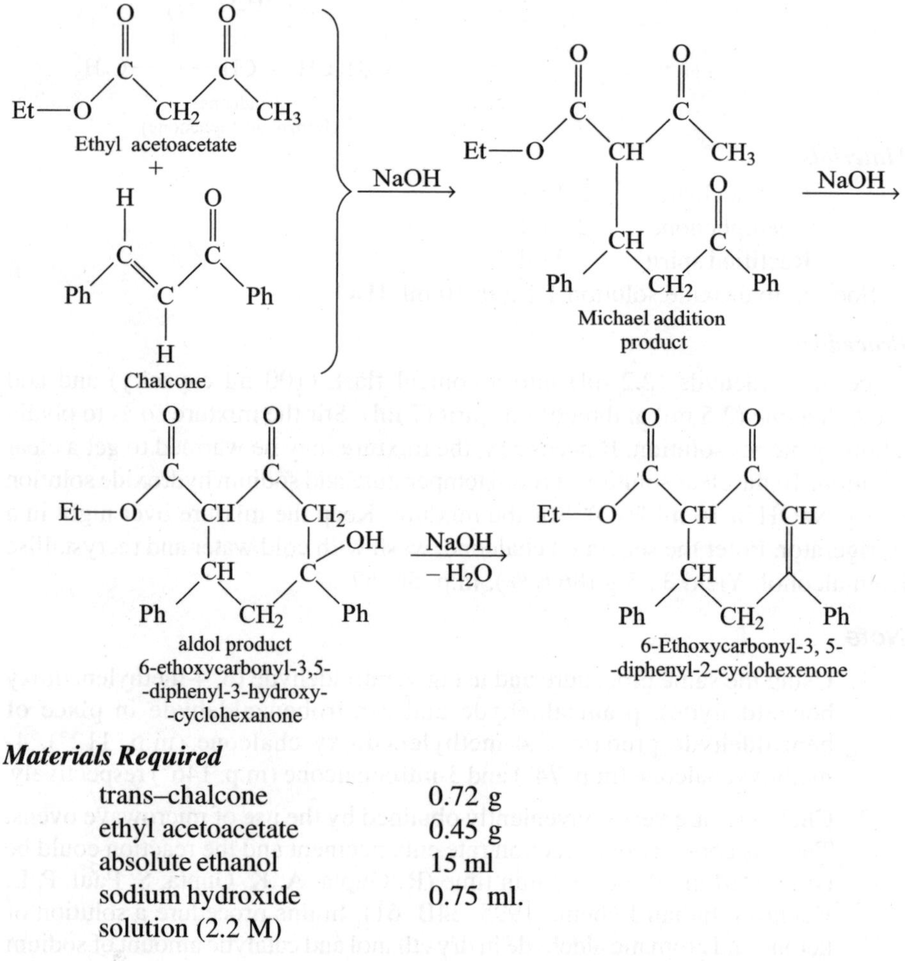

Materials Required

trans–chalcone	0.72 g
ethyl acetoacetate	0.45 g
absolute ethanol	15 ml
sodium hydroxide solution (2.2 M)	0.75 ml.

Procedure

Add finely powdered trans-chalcone (0.72 g) to a round bottomed flask (50 ml capacity), followed by the addition of ethyl acetoacetate (0.45 g) and absolute ethanol (15 ml). Swirl the flask until the solid almost dissolves. Add sodium hydroxide solution (2.2 M, 0.75 ml) and heat the flask which had been fitted with a reflux condenser on a hot plate. After the mixture starts boiling gently, continue reflux for 1 hour when the mixture becomes turbid. Cool the reaction mixture to room temperature, scratch the sides of the flask with a stirring rod to induce crystallisation. Cool the mixture in an ice-bath for $\frac{1}{2}$ hours. Filter the separated product using ice-cold water (2-3 ml) for transfer. Dry the crystals in the air (over night) or at 75-80° for 30 min.

To the separated solid in a test tube add acetone (4-5 ml) and stir the mixture. Most of the solid dissolves leaving behind some alkali. Separate the clear solution by centrifugation or filtration using filter acid. Evaporate the acetone from the clear solution. The oil produced solidifies on scratching with a glass rod. Crystallise from ethanol (4–5 ml). Weigh the formed 6-ethoxycarbonyl-3,5-diphenyl-2-cyclohexenone. m.p. 111-112°.

If possible record 1R spectra. You should observe absorbance at 1734 and 1660 cm^{-1} for the ester carbonyl and enone group respectively

Notes

1. The method is described by A. Garcia Ruso, J. Garcia–Raso, J. V. Sinisterra and R. Mestres, Michael Addition and Aldol condensation. A simple teaching model for organic laboratory" Journal of Chemical Education, 1986, *63*, 443.
2. From the Michael addition products, the methyl group looses a proton in presence of base and the resulting carbanion attacks the carbonyl group to give a stable 6-membered ring; in this case ethanol supplies a proton to yield the aldol intermediate product.
3. Since the reaction is carried out in aqueous phase, it is regarded as a Green Synthesis.

3.1.4 $\Delta^{1,9}$-2-octalone

It is obtained by **Michael addition** of methyl vinyl ketone to cyclohexanone followed by **aldol condensation** of the formed adduct.

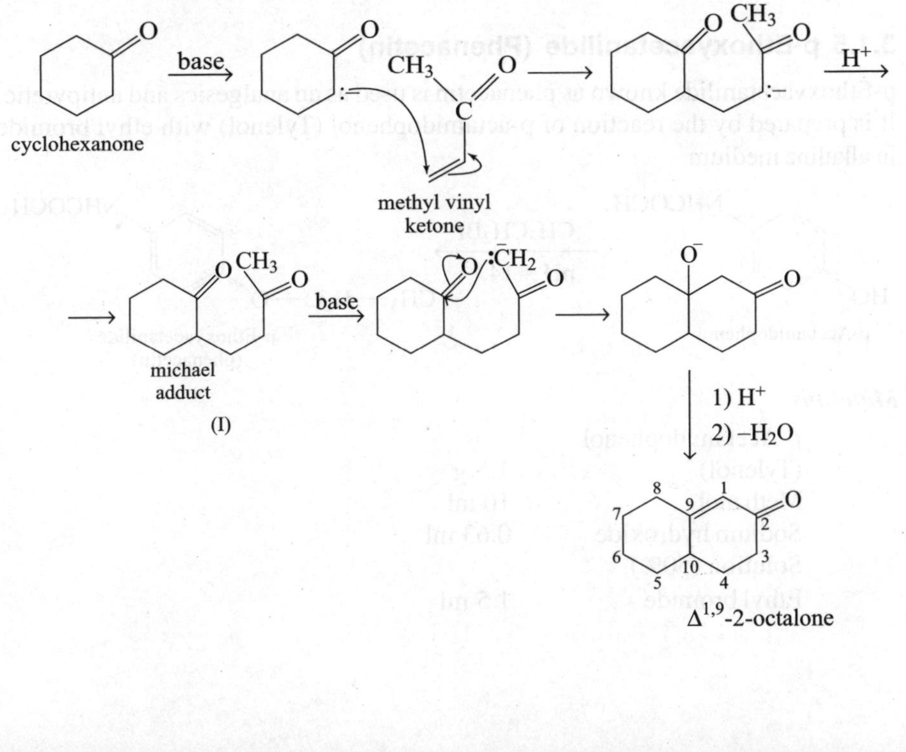

$\Delta^{1,9}$-2-octalone

The procedure is similar to that used for the preparation of 6-ethoxycarbonyl-3,5-diphenyl-2-cydohexenone (see 3.1.3).

The intermediate (Michael adduct) can be obtained more conveniently by the Michael addition of methyl vinyl ketone with the enamine of cyclohexanone. Subsequent aldol condensation of the michael adduct gives the required $\Delta^{1,9}$-2-octalone. (1)

For method of preparation of enamine at cyclohexanone see preparation of 2-acetyl cyclohexanone (sec. 3.10.1).

The overall procedure is similar to that used for the preparation of 6-ethoxycarbonyl-3,5-diphenyl-2-cyclohexenone.

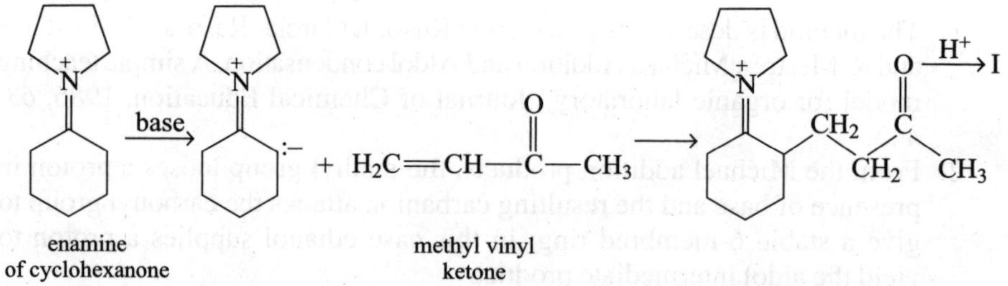

enamine
of cyclohexanone

methyl vinyl
ketone

Notes

1. Reactions that combine the Michael addition reaction and aldol condensation (as in the above case) to give a six-member ring fused on another ring are mostly used in steroid field and are known as **Robinson annelation reactions**.

3.1.5 p-Ethoxyacetanilide (Phenacetin)

p-Ethoxyacetanilide known as phenacetin is used as an analgesics and antipyretic. It is prepared by the reaction of p-actamidophenol (Tylenol) with ethyl bromide in alkaline medium

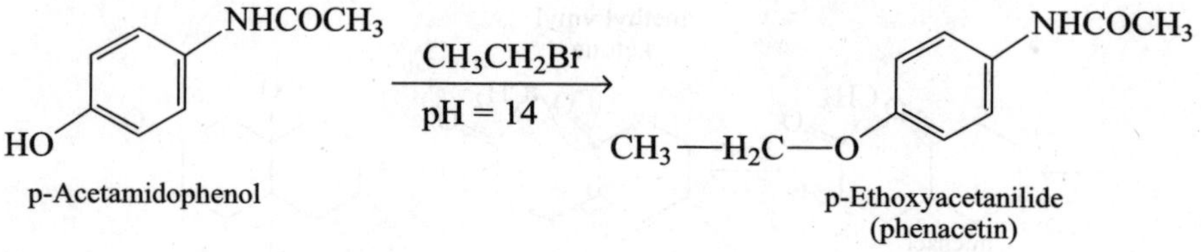

p-Acetamidophenol

p-Ethoxyacetanilide
(phenacetin)

Materials

p-Acetamidophenol (Tylenol)	1.5 g
Methanol	10 ml
Sodium hydroxide Solution (50%)	0.63 ml
Ethyl bromide	1.5 ml

Procedure

To a mixture of p-acetamidophenol (1.5 g) and methanol (10 ml) in a round bottomed flask add sodium hydroxide solution (50 %, 0.63 ml). Shake the mixture to get a clear solution. Add ethyl bromide (1.5 ml) and reflux the mixture for about 2 hours using a reflux condenser. Add hot water (20 ml). Leave the reaction mixture over night, cool (cold water) and filter the separated product. Yield 3 g (about 80%), m.p. 137–138°.

Note

p-Ethoxyacetanilide can also be prepared by first ethylation of p-aminophenol followed by acetylation with acetic anhydride at pH about 5.

3.1.6 p-Acetamidophenol (Tylenol)

p-Acetamidophenol, known as Tylenol is used as an analgesics. It is prepared by the acetylation of p-aminophenol with acetic anhydride at a pH of about 5.

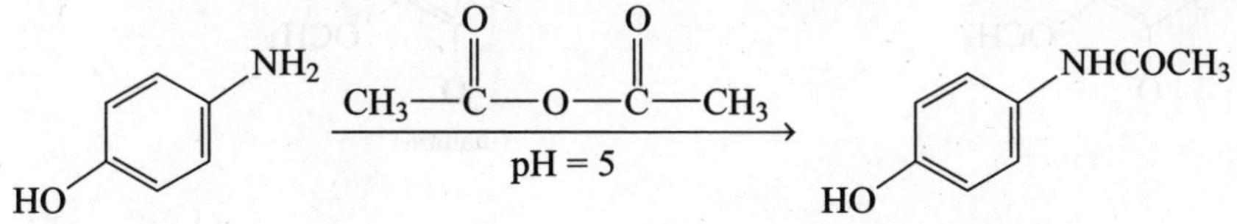

p- Aminophenol   p- acetamidophenol (Tylenol)

Materials

p-Aminophenol	1.1 g
Acetic anhydride	1.1 ml
Conc. hydrochloric acid	0.9 ml
Sodium acetate anhydrous	1g

Procedure

To a solution of p-aminophenol (1.1 g), in water (10 ml) containing conc. hydrochloric acid (0.9 ml) in a conical flask, add acetic anhydride (1.1 ml). Shake the mixture so that all the acetic anhydride dissolves. Finally add a solution of anhydrous sodium acetate (1 g) in water (6 ml), stirr the solution and leave it to stand overnight. Cool the solution in ice bath, filter the separated product. Crystallise from hot water. Yield 1.1 g (75%), m.p. 169°.

Notes

1. p-Aminophenol should be pure. If it is dark coloured, suspend it in methanol and filter.

2. Anhydrous sodium acetate is used since it is easily soluble in water than the trihydrate
3. Tylenol can also be prepared directly by the acetylation of p-aminophenol with acetic anhydride in aqueous medium i.e., the use of HCl and NaOAc can be avoided.

3.1.7 Vanillideneactone

It is obtained by the **Claisen–Schmidt reaction** of vanillin with acetone under basic conditions.

Mechanisn

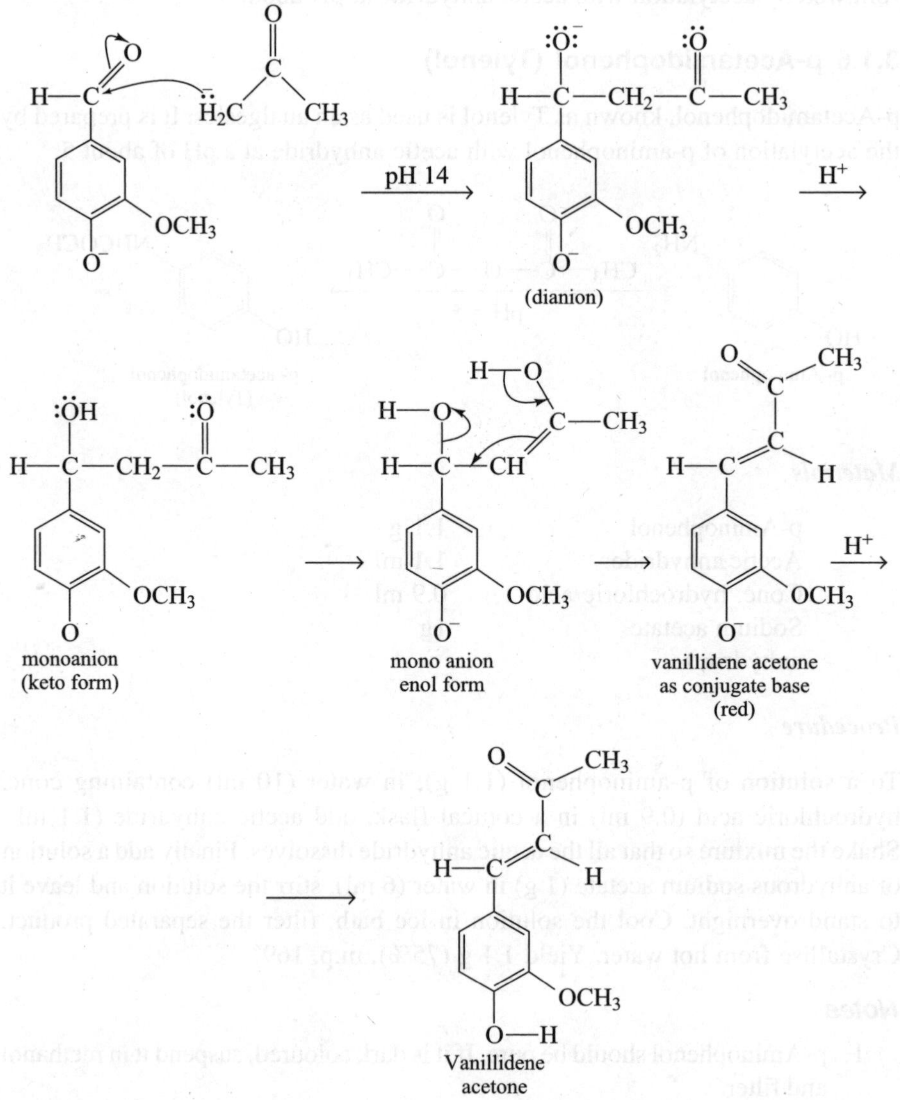

	(dianion)	

| monoanion (keto form) | mono anion enol form | vanillidene acetone as conjugate base (red) |

Vanillidene acetone

Materials

Vanillin	3.04 g
Acetone	12 ml
Sodium hydroxide	9 ml
Solution (10 %)	

Procedure

Dissolve vanillin (3.04 g) in acetone (12 ml) in a conical flask. To this solution add sodium hydroxide solution (10%, 9 ml). The solution which was yellow in the beginning, gradually darkens and becomes deep red in colour. Keep the reaction mixture for 72 hours with occasional shaking. Add water (50 ml) while stirring vigorously and then acidify by adding hydrochloric acid (10%, 15 ml). Filter the resulting product, wash with water and dry. Recrystallise form 1:1 ethanol-water (10–12 ml). Yield 3.3 g (85%) Record its m.p.

3.1.8 2,4-Dihydroxybenzoic aicd (β–Resorcylic acid)

It is prepared from resorcinol, pottasium bicarbonate and carbon dioxide gas.

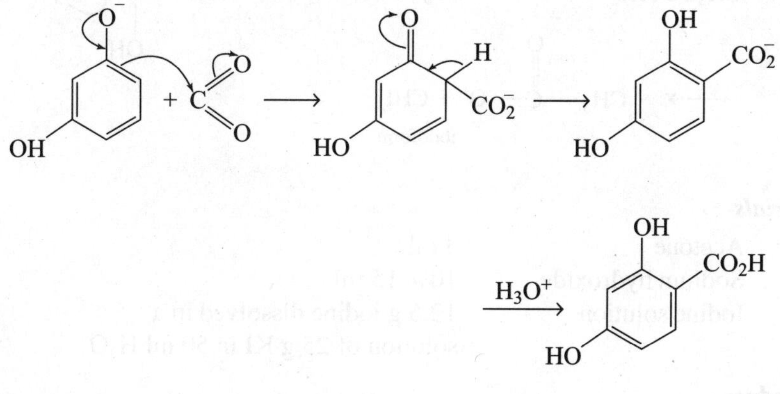

Materials

Resorcinol	15 g
Potassium bicarbonate	75 g
Water	150 ml

Procedure[1]

Heat the mixture of resorcinol (15 g), potassium bicarbonate (75 g) and water (150 ml) in a three necked round bottomed flask (250 ml capacity) fitted with a reflux condenser and a gas inlet tube. Heat the mixture gently on a steam bath for 3 hour and reflux for 30 minutes while passing a rapid stream of carbon dioxide gas through the solution. Acidify the hot solution by adding concentrated hydrochloric acid (45 ml) from a separatory funnel with a long tube delivering the acid to the bottom of the flask. Cool the flask to room temperature and then in ice bath. Filter the separated β-resorcyclic acid. Yield 9 g (64%) M.P. 216–17°.

Note

1. Nierensltein, Clibbens, Org. syn. Coll. Vol II, 1943, 557.

3.1.9 Iodoform

It is prepared by the well known **haloform reaction**, which consist in the cleavage of methyl ketones (CH_3–CO–R), acetaldehyde, ethanol and secondary methyl carbinols (CH_3CHOR) with halogens (mostly iodine) and a base to give haloform (iodoform if iodine is used) and carboxylic acids.

$$RCOCH_3 + 3NaOI \longrightarrow RCOI_3 + 3NaOH$$

$$RCOI_3 + NaOH \longrightarrow CHI_3 + RCOONa$$

From Acetone

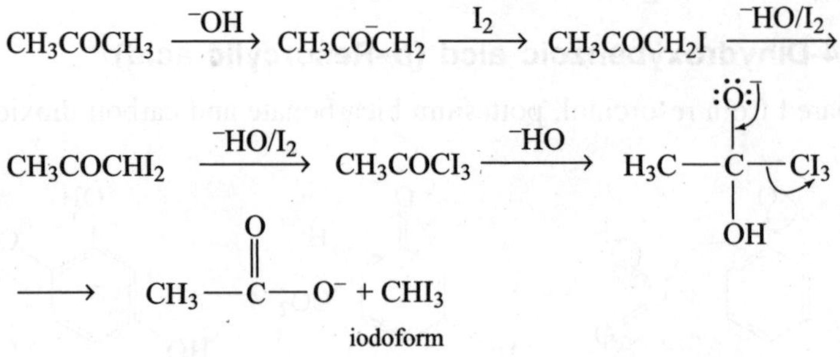

$$CH_3COCH_3 \xrightarrow{^-OH} CH_3COCH_2 \xrightarrow{I_2} CH_3COCH_2I \xrightarrow{^-HO/I_2}$$

$$CH_3COCHI_2 \xrightarrow{^-HO/I_2} CH_3COCI_3 \xrightarrow{^-HO} H_3C\!-\!\overset{\displaystyle :\ddot{O}:^-}{\underset{\displaystyle OH}{\overset{|}{\underset{|}{C}}}}\!\!-\!CI_3$$

$$\longrightarrow \quad CH_3\!-\!\overset{O}{\overset{\|}{C}}\!-\!O^- + CHI_3$$
iodoform

Materials

Acetone	3 ml
Sodium hydroxide	10% 15 ml
Iodine solution	12.5 g iodine dissolved in a solution of 25 g KI in 50 ml H_2O

Procedure

To a solution of acetone (3 ml) in water (30 ml) and sodium hydroxide solution (10%, 15 ml) add iodine solution (12.5 g iodine dissolve in a solution of 25 g potassium iodide in 50 ml H_2O). Heat the reaction mixture at 60° (water bath). Continue heating till the precipitate of iodoform settles down. Filter the yellow crystals of iodoform and crystallise from dilute methanol. Yield 5 g (31.2%). M.P. 119°.

Note

1. Iodoform can also be conveniently obtained from ethyl alcohol.

3.1.10 Endo-cis-1,4-endoxo–Δ^5-cyclohexene-2,3-dicarboxylic acid

It is obtained by **Diels–Alder Reaction** of furan with maleic acid or maleic anhydride in water at room temperature.

Maleic anhydride

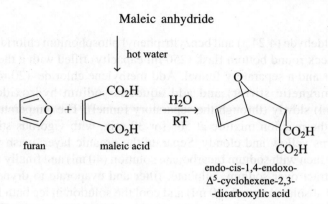

furan   maleic acid

endo-cis-1,4-endoxo-
Δ^5-cyclohexene-2,3-
-dicarboxylic acid

Materials

Furan	2.5g
Maleic acid	5g

Procedure[1]

Stir furan (2.5 g) with maleic acid (5 g) in water (25 ml) for 2–3 hr at room temperature. Filter the separated adduct, wash with water. Yield yield 7.2 g. Record its m.p.

Note

1. R.B. Woodward and Harold Baer, J. Am. Chem. Soc., 1948, *70*, 1161, R. Breslow and D. Rideout, J. Am. chem.. Soc., 1980, *102* , 7816.

3.1.11 Trans stilbene

Trans stilbene was earlier prepared (G. Markl and A. Merz, synthesis, 1973, 295) by the reaction of benzaldehyde with benzyltriphenyl phosphonium chloride.

$$C_6H_5-CH_2Cl + (C_6H_5)_3 P \xrightarrow{\ ^-OEt\ } (C_6H_5)_2 P^+ - CH_2C_6H_5Cl^-$$

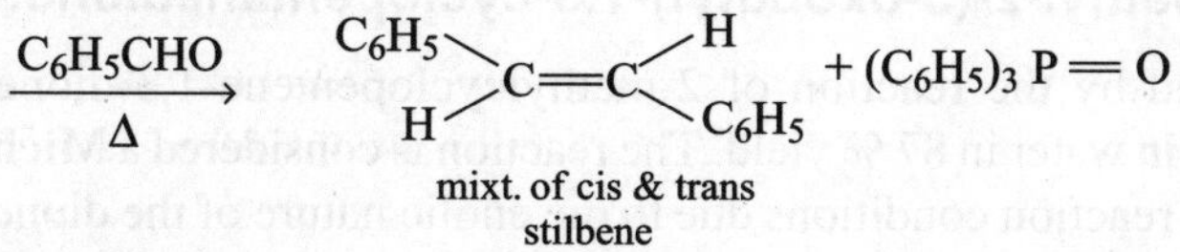

$$\xrightarrow[\Delta]{C_6H_5CHO} \underset{H}{\overset{C_6H_5}{>}}C=C\underset{C_6H_5}{\overset{H}{<}} + (C_6H_5)_3 P = O$$

mixt. of cis & trans
stilbene

It has now been possible to carry out the above reaction in presence of aqueous sodium hydroxide (John C. Warner, Paul T. Anastas and Jean–Pierre Anselme, J. Chem. Education 1985, *62*, 346).

Materials

Benzaldehyde	4.24 g
Benzyltriphenyl phosphonium chloride (see note 1)	15.72 g
Methylene chloride	20 ml
Sodium hydroxide solution	50%, 20 ml

Procedure

Take benzaldehyde (4.24 g) and benzyltriphenyl phosphonium chloride (15.72 g) in a three neck round bottom flask (250 ml capacity) filled with a thermometer, a condenser and a separatory funnel. Add methylene chloride. (20 ml), stir the mixture (magnetic stirrer) and add aqueous sodium hydroxide solution (50%, 20 ml) slowly (through the separatory funnel). The temperature rises to 50°. Keep the reaction mixture at 50° for 30 min. with vigorous stirring. The mixture turns yellow and cloudy. Separate the organic layer, wash with water (25 ml) and then with sodium bicarbonate solution (40 ml) and finally with water. Dry the extract over sodium, sulphate, filter and evaporate to dryness. To the residue add absolute ethanol (30 ml) and cool the solution in ice bath for 15 min. Filter the sepparated trans-stilbene (2 g), m.p. 122–123°.

Evaporate the filtrate, add petroleum ether (b.b 40–60°, 40 ml). Filter the precipitated triphenyl phosphine oxide (10 g), m.p. 146–47°. Evaporate the filtrate in vacuo at room temperature. The residual product consisted of cis–stilbene (3 g); it solidified at –5°, b.p. 135°.

Note

1. Benzyl triphenyl phosphonium chloride is prepared by refluxing a mixture of benzyl chloride (4.4 ml), triphenyl phosphine (14.3 g) and xylene (70 ml). The separated product is filtered, m.p. 310–11°. It is washed with xylene and dried in vacuum dessicator.

2. If possible take the nmr spectra of cis and trans stilbene and characterise both compounds.

3. The above procedure of Wittig condensation can also be performed with trans-cinnamaldehyde, p-metyl-, p-methoxy-, p-chloro-, m- and p-nitrobenzaldehydes and 9-anthracenealdehyde. The product is obtained in 65–85% yield (G. Markl and A. Merz, synthesis, 1973, 295).

3.1.12 2-Methyl-2-(3-oxobutyl)-1,3-cyclopentanedione.

It is prepared by the reaction of 2-methylcyclopentene-1,3-dione and methyl vinyl ketone in water in 87 % yield. The reaction is considered a **Michael addition** under acidic reaction conditions due to the enolic nature of the dione (Z.G. Hojas and D.R. Parrish, J. Org. Chem., 1974, *12*, 39).

2-methylcyclopentane-1,3-dione + methyl vinyl ketone $\xrightarrow{H_2O}$ 2-methyl-2-(3-oxybutyl)-1,3-cyclopentanedione

Materials

2-methylcyclopentane-1,3-dione	6.5 g
Methyl vinyl ketone	9.6 ml

Procedure

To a suspension of 2-methylcyclopentane-1,3-dione (6.5 g) in water (14 ml) add methyl vinyl ketone (9.6 ml). Stir the mixture under nitrogen at 20° for 5 days. Extract with benzene, treat with Na_2SO_4, charcoal and $MgSO_4$. Filter the solids, extract with hot benzene (20 ml) and evaporate the combined benzene extract. The oily product (10 g) consisted of the required 2-methyl-2-(3-oxobutyl)-1,3-cyclopentanedione. Characterise the product on the basis of IR spectra (1770 and 1725 cm^{-1}) and NMR spectral data [δ 1.12 (s, 3, 2 –CH$_3$), 2.22 (s, 3, CH$_2$CO$^-$) 2.82 (m, 4,–COCH$_2$CH$_2$CO–)].

Note

1. 2-methyl-2-(3-oxobutyl)-1,3-cyclopentanedione) was earlier reported to be prepared by the reaction of the dione and methyl vinyl ketone as a solid, m.p. 117–118° (C.B.C. Boycl and J.S. Whitehurst, J. Chem. Soc., 1959, 2022). However, the compound was not the required product (Z.G. Hajos and D. R. Parrish, J. Org Chem., 1974, 1612).

3.1.13 Hetero Diels-Alder Adduct

Hetero Diels–Alder reaction is useful for the synthesis of heterocyclic compounds with nitrogen or oxygen-containing dienophiles. In this reaction (Note 1), an iminium salt (generated in situ under mannic like conditions) reacted with dienes in water give aza–Diel–Alder reaction products.

$$C_6H_5CH_2NH_2 \cdot HCl \xrightarrow[H_2O]{HCHO} \left[C_6H_5CH_2 \overset{+}{N} = CH_2Cl^- \right]$$

aza-Diel–Alder adduct

Materials

Benzylamine hydrochloride	11.7 g
Formalin 40%	25 ml
Cyclopentadiene	13 g

Procedure

Cyclopentadiene (13 g) was added to a stirred mixture of benzylamine hydrochloride (11.7 g) formalin (25 ml). The heterogeneous reaction mixture was vigorously stirred for 3 hr. at room temp. The separated addect was obtained in quantitatine yield (bicyclic amine). Record the NMR spectre of the product.

Note

1. Scot D. Larnsen and Paul A. Grieco, J. Am. Chem. Soc. 1985, 107, 1768.
2. This proceduse can be used for large number of bicyclic compounds.

3.2 SOLID STATE (SOLVENTLESS) REACTIONS

3.2.1 3-Pyridyl-4(3H)quinazolone

It is obtained by the reaction of anthranilic acid with formic acid and 2-aminopyridine under microwave irradiation.

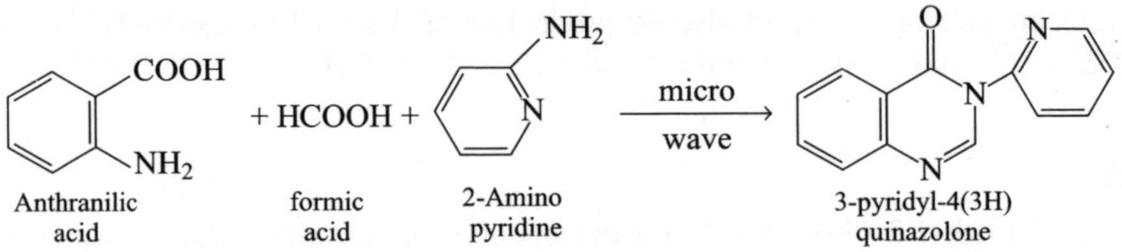

| Anthranilic acid | formic acid | 2-Amino pyridine | 3-pyridyl-4(3H) quinazolone |

Materials

Anthranilic acid	1.26 g
Formic acid	5 g
2-Aminopyridinie	0.92 g

Procedure

A mixture of anthranilic acid (1.26 g), formic acid (5 g) and 2-amino pyridinie (0.92 g) was heated in a microwave oven for 4 min. The product obtained melted at 156–157° (92 % yield).

Notes

1. The procedure adopted is the one described by M. Kidwai, S. Rastogi, R. Mohan and Ruby, Croatica, Chemica Acta, 2003, *76 (4)*, 365.
2. The methology in environmentally benign and completely eliminates the need of solvent from the reaction.

3.2.2 Diphenylcarbinol

It is obtained by reduction of benzophenone with sodium borohydride in solid state.

$$Ph_2CO + NaBH_4 \longrightarrow Ph_2CHOH$$

Benzopherone   Sodium borohydride   diphenylcarbinol

Materials

Benzophenone	1.8 g
Sodium borohydride	4.8 g

Procedure

A mixture of powdered benzophenone (1.8 g) and sodium borohydride (4.8 g) were kept in a dry box at room temperature with occasional mixing and grinding using an agate mortar and pestle for 5 days. Diphenyl carbinol was obtained in 100 % yield.

Note

1. The procedure is that described by K. Tanlka and F. Toda, Chem. Rev., 2000, *100*, 1028.

3.2.3 Phenyl benzoate

It is obtained by **Baeyer–Villiger oxidation** of benzophenone with m–chloroperbenzoic acid in solid state.

$$C_6H_5CO\,C_6H_5 \xrightarrow{\text{m–chloroperbenzoic acid}} C_6H_5COOC_6H_5$$

benzophenone $\qquad\qquad\qquad$ phenylbenzoate

Materials

Benzophenone	1.8 g
m–chloroperbenzoic acid	3.5 g

Procedure

Keep a powdered mixture of benzophenone (1.8 g) and m–chloroperbenzoic acid (3.5 g) at room temperature for 24 hr. Macerate the product with sodium bicarbonate solution and extract with ether. Distillation of the dried ether extract gave phenyl benzoate in 85 % yield.

Notes

1. The procedure is that described by K. Tanka and F. Toda, Chem. Rev., 2000, *100*, 1028.

2. Using phenyl benzyl ketone (PhCOCH$_2$Ph), the product, benzylbenzoate, PhCOOCH$_2$ Ph, is obtained in 97% yield.

3. The conventional Baeyer–Villiger oxidation in chloroform gives low yields (40–50%) of the product.

3.3 PHOTOCHEMICAL REACTIONS

3.3.1 Benzopinacol

Benzopinacol is obtained by photo reduction of benzophenone. It is a dimerization reaction brought about by exposing a solution of benzophenone in isopropyl alcohol to natural light.

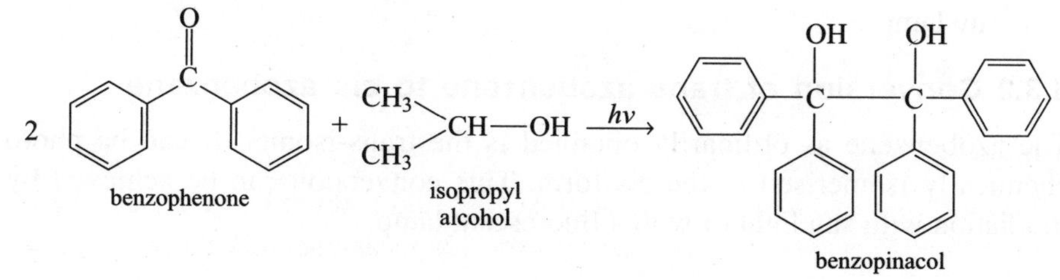

Mechanism

$$Ph_2C = 0 \xrightarrow{h\upsilon} Ph_2\overset{\cdot}{C} - O \cdot (S_1)$$

$$Ph_2\overset{\cdot}{C} - O \cdot (S_1) \xrightarrow{isc} Ph_2\overset{\cdot}{C} - O \cdot (T_1)$$

$$Ph_2\overset{\cdot}{C}—O \overset{\frown}{} H—\underset{\underset{CH_3}{|}}{\overset{\overset{CH_3}{|}}{C}}—OH \longrightarrow Ph_2\overset{\cdot}{C}—OH + \cdot\underset{\underset{CH_3}{|}}{\overset{\overset{CH_3}{|}}{C}}—OH$$
(T_1)

$$Ph_2\overset{\cdot}{C}—O \overset{\frown}{} HO—\underset{\underset{CH_3}{|}}{\overset{\overset{CH_3}{|}}{C}}\cdot \longrightarrow Ph_2\overset{\cdot}{C}—OH + O{=}C\overset{CH_3}{\underset{CH_3}{\big\langle}}$$
(T_1)

$$2\,Ph_2\overset{\cdot}{C}—OH \longrightarrow Ph—\underset{\underset{Ph}{|}}{\overset{\overset{OH}{|}}{C}}{-}{-}\underset{\underset{Ph}{|}}{\overset{\overset{OH}{|}}{C}}—Ph$$

Materials

Benzophenone	2.5 g
Isopropyl alcohol	10 ml

Procedure

Dissolve benzophenone (2.5 g) in isopropyl alcohol (10 ml) in a round bottomed flask (25 ml capacity). Add glacial acetic acid (0.2 ml) and fill the flask with isopropyl alcohol. Stopper the flask tightly and place the flask in bright sun light. The reaction takes about a week for completion (3 hrs. with a sun lamp). The product gets separated. It is filtered and crystallised from glacial acetic acid. Yield 98% m.p. 185–186.

Notes

1. It is a green reaction since it is carried out photochemically and there are no by-products formed.

2. The reaction can be completed in a much shorter time (3–4 hr) by using a uv lamp.

3.3.2 Conversion of trans azobenzene to cis azobenzene

The azobenzene as ordinarily obtained is the trans-isomer. It can be photo chemically isomerised to the cis form. This conversion can be achieved by irradiation with sun light or with a fluorescent lamp.

trans azobenzene   cis azobenzene

Materials

trans azobenzene	1 g	(see note 2)
Petroleum ether	50 ml	
b.p. 40–60°		

Procedure

Dissove trans azobenzene (1 g) in petroleum ether (50 ml) in a round bottomed flask (100 ml capacity). Place the flask in bright sun light. The reaction takes 4–5 days for completion (3 hr. with uv lamp irradiation). In the meanwhile prepare a chromatographic column from activated acid alumina grade I (50 g) as a slurry in petroleum ether. After the column is ready, place a circular filter paper on the top of the column (so that the alumina is not disturbed). Slowly pour the irradiated solution in to the column taking care not to disturb the alumina in the column Elute the column with petroleum ether (b.p. 40–60°). A sharp coloured band (about 2 cm) of cis azobenzene appears at the top of the column while the diffuse coloured region of the trans form moves down the column. Protect the upper band by covering with a carbon paper (so that the cis form is not reconverted into trans). When the chromatography is complete, extrude the column, cut the upper band and shake with petroleum ether (100 ml) containing methanol (1 ml). Filter the solution, wash the filtrate with water (2 × 15 ml), dry the solution over anhydrous sodium sulphate and evaporate the solution. The residual coloured product (m.p. 71.5°) is pure cis azobenzene.

Notes

1. The puriy of cis azobenzene can be confirmed by recording the uv absorption spectrum in ethanol solution as soon as possible after its isolation, cis-azobenzene has λ_{max} 281 nm (ε 5260); trans azo-benzene has λ_{max} 320 nm (ε 21300) in ethanol solution.

2. The azobenzene (trans) required is prepared as follows:
   To a suspension of magnesium turnings (1 g), nitrobenzene (1.8 ml), methanol (35 ml) add a small crystal of iodine in a round bottomed flask. Fit the flask with a reflux condenser. If the reaction does not commence in 2–3 minutes, warm the mixture (water-bath) to start the reaction. In case the reaction becomes too vigorous, use a cold water bath for few seconds. When most the magnesium has reacted, add more magnesium turnings (1 g). Let the reaction proceed (as above). Finally heat the reaction mixture on a water-bath. Cool the mixture, pour into water (65 ml), acidify with glacial acetic acid (until the mixture is acidic to congo red). Cool the mixture (ice-bath) filter the separated azobenzene and crytallise from ethanol (90%). Yield 1 g (34 %), m.p. 67–68°.

3.3.3 Conversionn of trans stilbene into cis stilbene

The stilbene as ordinarily obtained is the trans isomer. It can be photochemically isomerised to the cis form. This can be achieved by irradiation with sun light or with a uv fluorescent lamp.

trans-stilbene   cis-stilbene

The trans form is readily available by a variety of reactions and is much more stable than the cis isomer because it is less sterically hindered. However, it is possible to produce a mixture containing mostly the cis isomer by irradiation a solution of the trans isomer in presence of a suitable photosensitizer. (e. g. benzophenone). The stilbene required is prepared by wittig reaction of benzaldehyde with triphenyl phosphonium salt (section 3.1.11).

3.4 PTC CATALYSED REACTIONS

3.4.1 Phenylisocyanide ($C_6H_5N \equiv C$)

It is prepared by the reaction of dichlorocarbene (generated in situ) by the PTC method from aniline (1° amine).

$$C_6H_5NH_2 + CHCl_3 + NaOH \xrightarrow{C_6H_5CH_2 \overset{+}{N}Et_3 \ Cl^-} C_6H_5N \equiv C$$
$$\text{aq} \qquad\qquad\qquad\qquad\qquad\qquad\qquad 55\%$$

Materials

Aniline	4 ml
Chloroform	12 ml
aq. NaOH (50% solution)	20 ml
benzyltriethylammonium chloride	0.25 g

Procedure

Add a mixture of aniline (4 ml), chloroform (12 ml) and benzyltriethyl ammonium chloride (0.25 g) to a vigrorously stirred solution of sodium hydroxide solution (50%, 20 ml). Use a reflux condenser, since the mixture starts refluxing. The reaction subsides in 4–5 min and the stirring continued for 1 hr. After the reaction is over, add cold water. Extract the mixture with methylene chloride, wash the organic layer with aqueous sodium chloride (5 %, 30 ml) and dry over anhyd-sodium sulphate. Distil the clear solution under nitrogen. Phenyl isocyanide is collected at 50–52°/ 11 mm. Yield 57%.

Notes

1. The required PTC, viz. benzyl triethylammonium chloride is obtained by refluxing a solution of triethylamine (3.3 g) and benzyl cholride (5 g) in absolute ethanol for about 50 hrs. Cool the solution to room temperature and add ether. Filter the separated benzyltriethylammonium chloride. Purify it by dissolving in hot acetone and reprecipitation with ether. Yield 8.5 g (90%).

2. The usual method of preparation of phenyl isocyanide without the PTC gives very poor yields (5–10%).

3.4.2 1-Cyano Octane ($CH_3(CH_2)_6CH_2CN$)

We know that alkyl halides do not react with sodium cyanide under a variety of conditions, e.g, stirring and heating for long time. However, if a small quantity of a phase transfer catalyst is used, the reaction goes to completion in about 2 hrs time. Thus, 1-cyanoctane is obtained as follow.

$$CH_3(CH_2)_6Cl \ + \ NaCN \ \xrightarrow[C_{16}H_{33} \ P^+ \ (C_4H_9)_3 \ Br^-]{PTC} \ CH_3 \ (CH_2)_6 \ CH_2 \ CN$$

| 1-chlorooctane | or KCN | 94% yield |

Materials

1-Chloro octane	5 g	
Sodium cyanide	5 g	
Water	5 ml	
Hexadecyl tributyl Phosphoniun bromide	0.5 g.	(see note 1)

Procedure[2]

Heat a mixture of 1-chloro octane (5 g), sodium cyanide (5 g), water (5 ml) and hexadecyl tributyl–phosphonium bromide (0.5 g) with stirring at 105° for 2 hrs in a round bottomed flask (250 ml capacity), cool the reaction mix, add water (100 ml) and extract with dichloromethane. Wash the organic extract with water, dry over anhydrous sodium sulphate and distil. 1-Cyano octane gets collected at 140°/1mm. Yield 94% (97% pure).

Note

1. The PTC, hexadecyl tributyl phosphonium bromide [$n-C_{16}H_{33} P^+ (n-C_4H_9)_3$ Br^-] is prepared (CM. Starks J. Am. Chem. Soc., 1971, *93*, 195) by heating a mixture of 1-bromohexadecane (1.5 g) and tributyl phosphine (1 g) at 70–90° for 3–4 days. Cool the mixture, filter the separated PTC and crystallise from hexane. Yield 1.7 g (60%), m.p. 54°.

2. C.M. starks, J. Am. Chem Soc, 1971, 93, 195

3. PTC methalogy can also be used for the preparation of following alkyl or acyl halides.

(a)
$$C_6H_5CH_2CH_2Cl \xrightarrow[\text{PTC}]{\text{NaCN/H}_2\text{O}} C_6H_5CH_2CH_2CN$$

$$N^+ (CH_3)_3 \, CH_2 \, C_6H_5Cl^- \qquad 91\%$$
3 hr. 90–95°

Ref. N. Sugemol, T. Fujita, N. Shigematsu and A. Ayadha, Chem. Pharm. Bull., 1962, *10*, 427; Japanese Patent 1961/63.

(b)
$$C_6H_5COCl \xrightarrow[\text{PTC Bu}_4\text{N}^+\text{X}^-]{\text{Na CN/H}_2\text{O}} C_6H_5COCN$$

60–70%

Ref. K.E. Koening and W.P. Weber, Tetrahedron Lett., 1974, 2274.

3. Benzonitrite can conveniently prepared from benzamide by the reaction with dichlorocarbene [generated in situ].

3.4.3 1-Oxaspiro-[2,5]-octane-2-carbonitrile

It is obtained by the PTC catalysed **Darzen condensation** of cyclohexanone with chloroacetonitrile (A. Jonczyk, M. Fedorynski and M. Makosza, Tetrdhedron Lett., 1972, 2395).

cyclohexanone   chloroaceto nitrile   1-oxaspiro-[2,5]-octane-2-carbonitrile

Materials

Cyclohexanone	5.4 g
Sodium hydroxide Solution, 50%	10 ml
Benzyltriethyl ammonium chloride	0.2 g
Chloroacetonitrile	3.8 g

Procedure

To a stirred mixture of cyclohexanone (5.4 g), sodium hydroxide solution (50%, 10 ml) and benzyltriethyl-ammonium chloride (0.2 g) at 15–20°, was added dropwise chloroacetonitrile (3.8 g). Stir the mixture at 15–20° for 30 min. Extract the mixture with dichloro methane, wash the organic phase with water, dry over anhyd sodium sulphate and distil. 1-Oxaspiro-[2,5] octane-2-carbonitrile is obtained at 87°/5 mm. Yield 70%.

Note

1. Darzen reaction was earlier performed under anhydrous conditions. It has now been possible in aqueous phase using PTC.

3.4.4 3,4-Diphenyl-7-hydroxycoumarin

3,4-Diphenyl coumarins, known for their antifertility activity, were prepared earlier in low yields and required anhydrous conditions. These are now obtained in excellent yield and purity by the use of a phase transfer catalyst in presence of aqueous potassium carbonate by the reaction of o-hydroxy benzophenones with phenyl acetyl chloride.

2-Hydroxy-4-methoxy
benzophenone

phenyl acetyl
chloride

3,4-Diphenyl
-7-hydroxycoumarin

Materials

2-hydroxy-4-methoxy benzophenone	2.45 g
Phenyl acetyl chloride	1.5 ml
Tetrabutyl ammonium hydrogen sulphate	0.1 g
Aqueous potassium carbonate 20%	50 ml

Procedure[1]

Add dropwise a solution of phenyl acetyl chloride (1.5 ml) in benzene (10 ml) to a stirred mixture of 2-hydroxy-4-methoxybenzophenone (2.45 g) in benzene (50 ml), tetrabutylammonium hydrogen sulphate (100 mg) and aqueous potassium carbonate (20 %, 50 ml). Stir the mixture for 5 hr, separate the organic layer, wash with water, dry (anhyd. Na_2SO_4) and distil the solvent. Crystallise the residue from ethanol to give 3,4-diphenyl-7-hydroxycoumarin, m.p. 168–69°. Yield 80%.

Note

1. V.K. Ahluwalia and C.H. Khanduri, Indian J. Clem., 1989, *28B*, 599.

2. The PTC, tetrabutyl ammonium hydrogen sulphate is obtained (W.T. Ford and R.J. Haurt, J. Am. Chem. Soc., 1973, *95*, 7381) as follows.
   Add dimethyl sulphate (4.6 g) is a stirred mixture of tetrabutyl ammonium bromide (9.8 g) and chlorobenzene (15 ml) at 80–85° in a two neck round bottomed flask fitted with a short distillation column and a dropping funnel. Collect the formed methyl bromide as a distillate using a trap cooled in acetone–dry ice mixture. After the distillation of methyl bromide ceases, increase the heating until the temperature at the top of the distillation column starts to rise rapidly. Add cautiously a solution of concentrated sulphuric acid (0.75 ml) in water (300 ml). Reflux the mixture for 48 hr. Evaporate the solution to almost dryness under reduced pressure. Dissolve the residue in dichloromethane (250 ml), wash the solution with water (2 × 30 ml), dry (anhyd-sodium sulphate) and distil the solvent. The PTC,

tetrabutylammonium hydrogen sulphate (10 g) separated. It is almost pure and can be crystallised from isobutyl methyl ketone.

3.4.5 Flavone

Flavones, a class of natural products were synthesised by a number of methods and in most of the methods, the yield is low and the work-up procedure difficult. These can be obtained in excellent yield (90%) by the reaction of appropriate o–hydroxyacetophenone with appropriately substituted benzoyl chloride in benzene solution with a phase transfter catalyst in presence of sodium hydroxide or sodium carbonate followed by cyclisation the formed o-hydroxydibenzoylmethane with p-toluene sulphonic acid. The last step is known as **Baker–Venkataraman synthesis**. Simple flavone is synthesied as shown below.

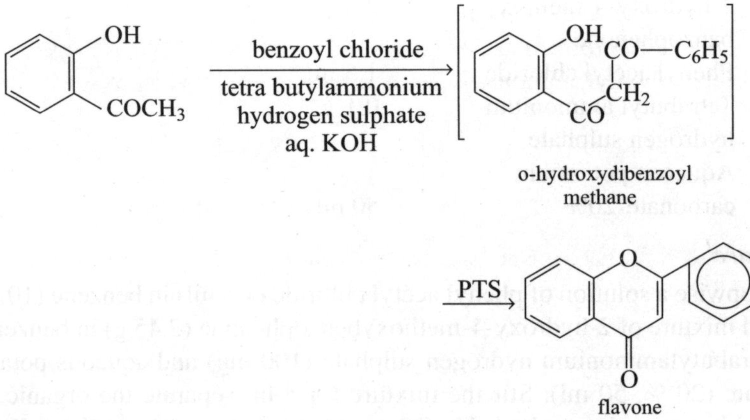

o-hydroxydibenzoyl methane

flavone

Materials

o–Hydroxyacetophenone	0.43 g
Benzoyl chloride	0.42 g
Tetrabutyl ammonium hydrogen sulphate	0.2 g
aq. potassium hydroxide 10%	20 ml
p-toluene sulphonic acid	0.1 g

Procedure[1]

Add a solution benzoyl chloride (0.42 g) in benzene (20 ml) to a stirred mixture of o-hydroxyacetophenone (0.43 g), tetrabutyl ammonium hydrogen sulphate (0.2 g) and aq potassium hydroxide (10%, 20 ml). Continue stirring for 2–3 hr until the starting ketone disappeared (TLC). Separate the benzene solution (separatory funnel), wash with water (2 × 15 ml), dry (anhydrous sodium sulphate) and distil. The residual product consisting of o-hydroxydibenzoylmethane was dissolved in benzene (50 ml) and refluxed with p-toluene sulphonic acid (0.1 g). Remove the water formed by distillation using a Dean–Stark apparatus. Reflux the mixture for 45–60 min, extract the benzene solution with sodium bicarbonate solution (5 %, 25 ml) to remove p-toluene sulphonic acid. Remove the solvent by distillation and crystallise the residue of flavone from ethyl acetate-petroleum ether. Yield 95%, m.p. 117°.

Notes

1. V. K. Ahluwalia *et al.* unpublished results.

2. The PTC, tetrabutylammonium hydrogen sulphate is obtained as described in note 2 in the preparation at 3,4-diphenyl-7-hydroxycoumarin (section 3.4.4).

3.4.6 Dichloronorcarane [2,2-Dichlorobicyclo (4.1.0) heptane]

Dichloronorcarene or 2,2-dichlorobicyclo [4.1.0] heptane is prepared by addition of dichlorocarbene to cyclohexene. The dichlorocarbene is generated in situ by the reaction of chloroform and sodium hydroxide in presence of phase transfer catalyst (tetra-n-butyl ammonium bromide or benzyltriethylammonium chloride).

cyclohexene   dichloro
                                                   norcarane
                                    (2,2-Dichlorobicyclo (4·1·0)heptane)

Materials

Chlroform	12 ml
Cyclohexene	4.1 g (5.1 ml)
Tetra-n-butyl ammonium bromide	0.25 g
sodium hydroxide	20 ml (50 %)

Procedure

Add chloroform (12 ml) to a round bottomed flask (100 ml capacity) followed by addition of cyclohexene (4.1 g or 5.1 ml) and tetra-n-butyl ammonium bromide (0.25 g). Fit the flask with a reflux condenser, continue stirring with the help of a magnetic needle. Pour in one lot sodium hydroxide solution (50%, 20 ml) and water (15 ml). Heat the flask gently and stir the mixture vigorously for 30 min. Heating should be done so that chloroform refluxed slowly as seen by an occasional drop falling from the reflux condenser.

After the heating is complete, add ice to cool the mixture to room temperature. Then extract with ether (2 × 30 ml) by using a separatory funnel. Dry the ether solution over anhydrous magnesium sulphate and distil. Collect the fraction distilling between 192–197°C. Yield 5.7 g (about 65%).

Notes

1. Using styrene (5.2 g) in place of cyclohexene, you can prepare 1,1-dichloro-2-phenyl cyclopropane (b.p. 103°C/10 Torr).

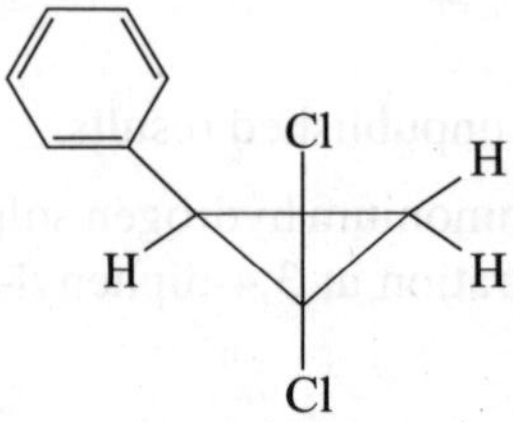

2. Using 1,5-cyclooctadiene (3.1 ml) in place of cyclohexene in the above experiment you can get the bis-adduct.

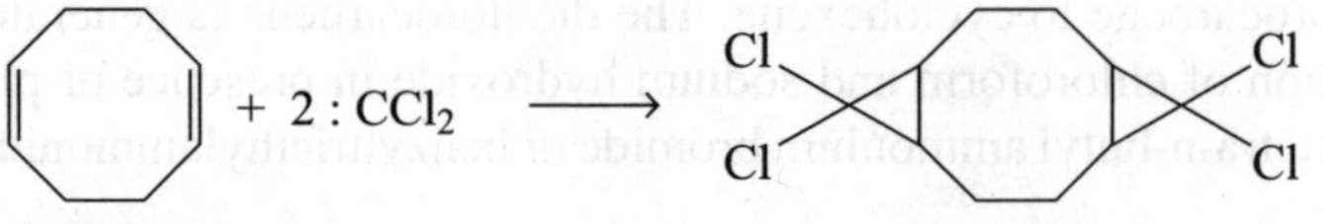

1,5-cyclooctadiene   bis-adduct

3. Dichlorocarbene can also be generated by direct sonication (S.L Regen, A.K. Singh, J. Org. Chem, 1982, *47*, 1587) between powdered sodium hydroxide and chloroform. This procedure is simple and efficient and avoids the use of phase transfer catalyst. The generated dichlorocarbene in situ undergoes addition reaction to alkenes. One example is

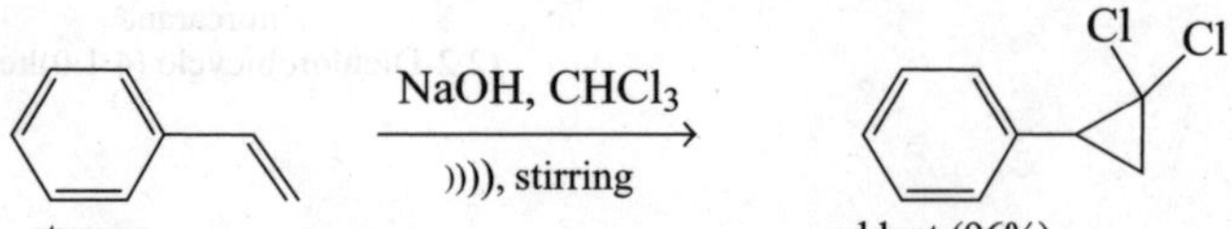

styrene   adduct (96%)

4. Reimer–Tiemann reaction can also be performed using the same procesdure (section 3.4.10).

3.4.7 Oxidation of toluene to benzoic acid

Normally toluene is oxidised to benzoic acid by alkaline $KMnO_4$ solution. However, even after prolonged refluxing, the yield is only 40–50%. Use of a phase transfer catalyst (e.g. cetyltrimethyl ammonium chloride) or a crown ether (*eg.* [18] crown 6) gives much better yield (80–90%) in shorter time.

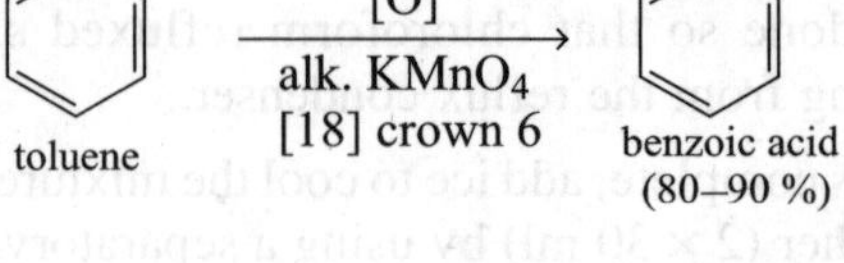

toluene   benzoic acid (80–90 %)

Materials

Toluene	2.5 ml
Potassium permanganate	4 g

Sodium carborate (2N)   1.6 ml
[18] crown 6.
> or   0.1 g
Cetyl trimethyl ammonium chloride

Procedure[1]

Place a mixture of toluene (2.5 ml), sodium carbonate solution (2N, 1.6 ml) and [18]-crown-6 (0.1 g) in a round bottomed flask (100 ml capacity) filled with a reflux condenser. Add potassium permanganate solution through the reflux condenser while the mixture is kept gently refluxing on a wire guage. Reflux for a total of 3 hr. Cool the reaction mixture. Filter the alkaline solution to remove precipitated MnO_2 and pass sulphur dioxide gas or add of saturated solution of sodium sulphite and dilute sulphuric acid till the solution become colourless. Extract the solution with ether (2 × 10 ml) and evaporate the ether (caution). Crystallise the residual product from hot water. Yield 2.2 g. m.p. 122°.

Note

1. Crown ether forms a complex with $KMnO_4$ which is soluble in organic phase and thus $KMnO_4$ becomes more effective for oxidation of toluene.

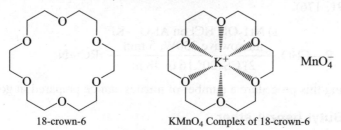

18-crown-6 $KMnO_4$ Complex of 18-crown-6

3.4.8 Benzonitrile from benzamide

Benzamide on reaction with dichloro carbene generated in situ by the reaction of chloroform and sodium hydroxide in presence of a PTC gives benzonitrile.

$$C_6H_5CONH_2 + CHCl_3 + NaOH \xrightarrow[\text{aq}]{C_6H_5CH_2 \overset{+}{N} Et_3\ Cl^-} C_6H_5 CN$$

Materials

Benzamide	6.05 g
Benzyl triethyl ammonium chloride	0.34 g
Chloroform	6 ml
aq. sodium hydroxide (50%)	25 g NaOH in 50 ml H_2O

Procedure[1]

Stir a mixture of benzamide (6.05 g), benzyltriethyl ammonium chloride (0.34 g), chloroform (6 ml) and aq. sodium hydroxide (50 % solution, 25 g NaOH in 50 ml water) at room temperature for 2 hr. Extract the reaction mixture with chloroform, wash organic layer with water, dry (anhyd Na₂ SO₄) and distil to give benzonitrile (84 % yield).

Notes

1. T. Saraie, K. Ishiguno, K. Kawashima and K. Morita, Tetrahedron Lett., 1971, 2121.

2. For the Preparation of PTC (benzyl triethyl ammonium chloride see note 1 of preparation of phenyl isocyanide (section 3.4.1).

3. A number of methods are available for the synthesis of nitriles. In most of the cases the aldoxime is initially prepared and then dehydrated by a wide variety of reagents. The method described for the synthesis of benzonitrile from benzamide using dichlorcarbene in aqueous phase in presence PTC is a convenient procedure.

4. In an efficient procedure, the aldehyde is converted into adsorbed oximate by reaction with hydroxylamine hydrochloride and potassium fluoride on alumina under microwave activation and without a solvent. The absorbed oximate is transformed into nitrile by treatment with carbon disulphide at room temperature (D. Villemin, M. Lalaoui, A. B. Alloum, Chem. Ind., 1991, 176).

$$R\text{—}CHO \xrightarrow[\text{2) CS}_2\,20°,16\text{ or }48\text{ hr}]{\substack{\text{1) NH}_2\text{OH HCl on Al}_2\text{O}_3\text{–KF} \\ \text{microwave 350W, 5 min}}} RC\equiv N$$

Using this procedure a number of nitriles can be prepared in good yield.

3.4.9 n-Butyl benzyl ether

n-Butyl benzyl ether, an unsymmetrical ether is obtained by an improved **Williamson ether synthesis**[1] using phase transfer catalyst. Thus the reaction of butyl alcohol with benzyl chloride in presence of sodium hydroxide and tetrabutylammonium bisulphate (TBAB) as catalyst gives n-butyl benzyl ether.

$$CH_3CH_2CH_2CH_2OH + C_6H_5CH_2Cl \xrightarrow[\text{TBAB}]{50\% \text{ Na OH}} CH_3CH_2CH_2CH_2O\,CH_2C_6H_5$$

n-butyl alcohol   benzyl chloride   n-butyl benzyl ether
     + NaCl + H₂O

Materials

n-Butyl alcohol	6.2 g
Benzyl chloride	6.3 g
Tetrabutylammonium Bisulphate (TBAB)	1 g
Sodium hydroxide 50%	10 g in 20 ml H₂O.

Procedure

A mixture of n-butyl alcohol 6.2 g), benzyl chloride (6.3 g), sodium hydroxide solution (10 g sodium hydroxide in 20 ml water) and TBAB (1 g) is stirred at 35–

$40°$ for 1.5 hr. Extract the mixture with THF, wash THF solution with NaCl sat'd with 50% aqueous sodium hydroxide. Distillation of the dried THF solution gives the required n-butyl benzyl ether in 92% yield.

Notes

1. The method adopted is that described by H.H. Freedman and R.A. Dubois, Tetrahedron Lett., 1975, 3251.

2. This is the best procedure for the preparation of mixed ethers. The usual Williamson method gives a mixture of products.

3. In this procedure, only a minor amount of symmetrical ether is formed.

4. Primary alcohol are complelely alkylated by aliphatic chlorides; secondary alcohols require longer time or greater amount of catalyst.

3.4.10 Salicylaldehyde

It is obtained by the reaction of phenol with dichlorocarbene, which is generated in situ from chloroform and sodium hydroxide solution in presence of a PTC catalyst[1]. The reaction is known as **Reimer–Tiemann Reaction**.

$$CHCl_3 \xrightarrow[\text{PTC}]{^-OH} :\bar{C}Cl_3 + H_2O$$

$$:\bar{C}Cl_3 \xrightarrow{-Cl^-} :CCl_2$$
dichlorocarbene

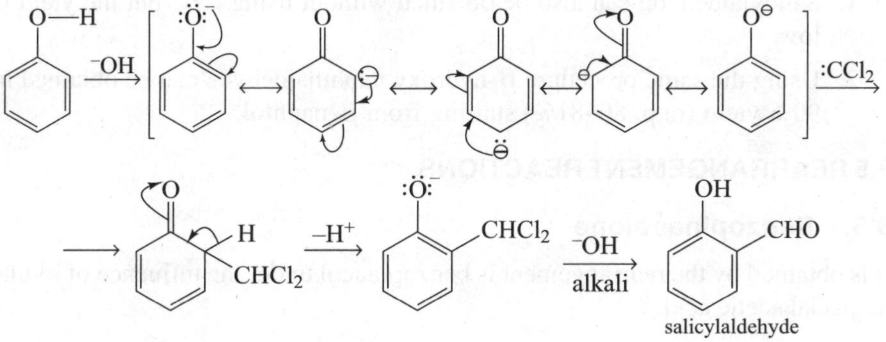

salicylaldehyde

Materials

Phenol	16.3 g
Chloroform	27.3 ml
Sodium hydroxide	50% 20 ml
Tetra-n-butyl ammonium bromide	0.25 g

Procedure

Add chloroform (16.3 g) to a round bottomed flask (100 ml capacity) followed by the addition of phenol (16.3 g) and tetra-n-butylammonium bromide (0.25 g).

Fit the flask with a reflux condenser, continue stirring with the help of a magnetic needle. Pour in one lot sodium hydroxide solution (50 %, 20 ml) and water (15 ml). Heat the flask gently and stir the mixture vigorously for 30 min.

Heating should be done so that chloroform refluxed slowly as seen by an occasional drop falling from the condenser.

After the reaction is comple, remove excess chloroform by steam distillation. Acidify the remaining aqueous alkaline solution cautiously with dilute sulphuric acid and then steam distil till no more oily drops are collected. Extract the distillate containing salicylaldehyde with ether (2×15 ml). Remove the ether by distillation, shake vigorously the residual product containing phenol and salicylaldehyde with a saturated solution of sodium bisulphate and allow to stand for 1 hr. Filter the bisulphite adduct, wash with water and a little alcohol and finally with ether. Decompose the adduct by warming on a water bath with dilute sulphuric acid. Extract the cooled mixture with ether, dry the ether solution over anhydrous sodium sulphate and evaporate the ether. Salicylaldehyde is collected by distillation, b.p. 195–197° yield 7.1 g (40 %).

The p-hydroxybenzaldehyde obtained as a by product is isolated from the residue left after steam distillalion. Filter the residual solution, cool and extract with ether. Remove the solvent. Crytallise the yellow solid from aqueous sulphurous acid. Yield 1.5 g (7 %), m.p. 116–117°.

Notes

1. Salicylaldehyde can also be obtained without using PTC but the yield is low.

2. Using the same procedure β–hydroxy naphthaldehyde can be obtained in 90% yield (m.p. 80–81%) starting from β–naphtol.

3.5 REARRANGEMENT REACTIONS

3.5.1 Benzopinacolone

It is obtained by the rearrangement is benzopinacol under the influence of iodine in glacial acetic acid.

benzopinacol   benzopinacolone

The above acid catalyst rearrangement is called the **pinacol rearrangement**.

Mechanism

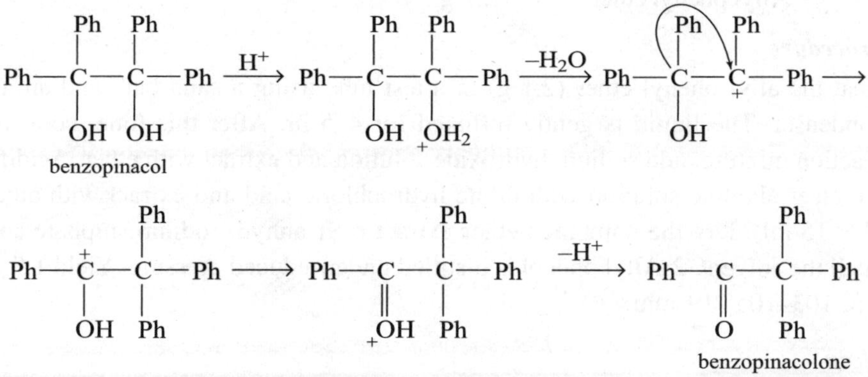

benzopinacol

benzopinacolone

Materials

        Benzopinacol   2.5 g
        Iodine solution   12.5 ml
  Iodine solution is 0.015 m solution of iodine dissolved in glacial acetic acid

Procedure

In a round bottomed flask take 12.5 ml solution of iodine dissolved in glacial acetic acid. Add 2.5 g benzopinacol. Reflux the solution for 5 min. Crystals appear from the solution. Cool the solution, filter the separated benzopinacolone, wash with cold glacial acetic acid (2ml) and dry. Record the yield. M.p. 182° C.

Note

Since it is a rearrangement reaction so there is 100 % atom economy and so is a green reaction.

3.5.2 2-Allyl phenol

It is obtained by **claisen rearrangement** of allyl phenyl ether.

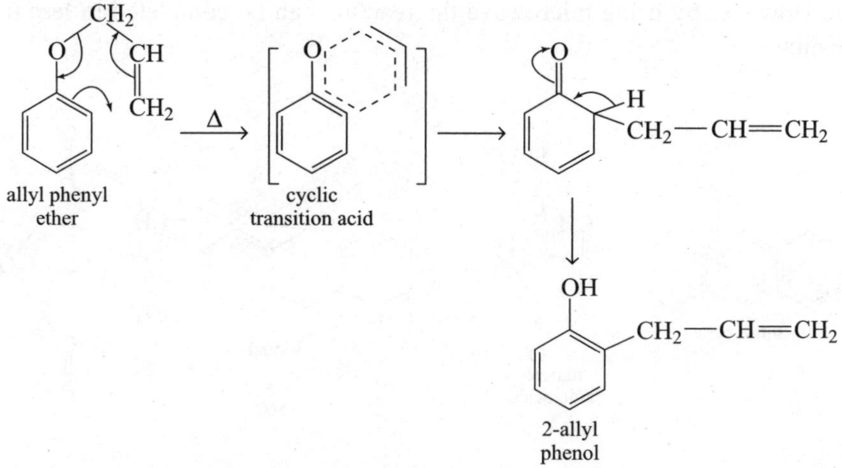

allyl phenyl
ether

cyclic
transition acid

2-allyl
phenol

Materials

        Allyl phenyl ether   2.1 g

Procedure

Heat the allyl phenyl ether (2.1 g) in a test tube using a sand bath and an air condenser. The liquid is gently refluxed for 4–5 hr. After this time, cool the reaction mixture, add sodium hydroxide solution and extract with ether. Acidify the clear alkaline solution with dilute hydrochloric acid and extract with ether (2 × 15 ml). Dry the combined ether extract over anhyd. sodium sulphate and distil the solvent. 2-Allyl phenol is distilled under reduced pressure. Yield 1.5 g, b.p. 103–105°/19 mm.

Note

1. The required allyl phenyl ether is obtained as follows: Reflux a mixture of phenol (1.9 g), allyl bromide (2.5 g), anhydrous potassium carbonate (3 g) and dry acetone (20 ml) on a water bath using a round bottomed flask fitted with a reflux condenser and calcium chloride guard tube for 6 hr. Distil the acetone (as much as possible) on a boiling water bath and add water (15 ml) to the residue. Extract the mixture with ether (2 × 20 ml). Wash the ether extract with sodium hydroxide solution (10% 2 × 10 ml) and finally with water (2 × 10 ml). Dry the ether solution over anhydrous sodium sulphate and distil. The allyl phenyl ether is obtained by distillation under reduced pressure b.p. 65°/19 mm, yield 2.1 g (78%).

3.6 MICROWAVE INDUCED REACTIONS

3.6.1 9,10-Dihydroanthracene-endo-α,β-succinic anhydride (Anthracene-maleic anhydride adduct)

It is obtained by **Diels–Alder reaction** of anthracene with maleic anhydride. Normally the reaction is carried out by refluxing the mixture in xylene for 15–20 min. However by using microwave the reaction can be completed in less than one minute.

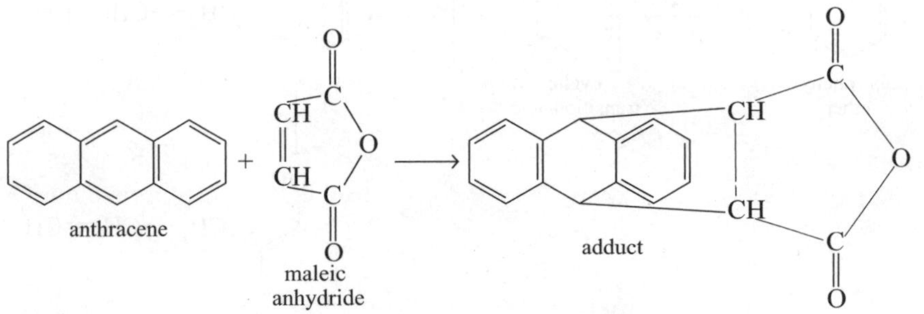

anthracene

maleic anhydride

adduct

Materials

Anthracene	3 g
Maleic anhydride	1.15 g
Diglyme	5 ml

Procedure

Place the grinded mixture of anthracene (3 g) and maleic anhydride (1.15 g) in a beaker (250 ml capacity). Add diglyme (5 ml), shake the mixture gently and cover the beaker with a watch glass irradiate the mixture in a microwave oven for 90 sec. at a medium power level. Remove the beaker from the oven, allow it cool to room temperature. Filter the separated adduct by suction, wash the product with methanol, yield 3.3 g. m.p. 262–263° (decom.).

Note

1. It is possible to conduct Diels–Alder reaction in aqueous phase also (R. Breslow (a review) Acc. Chem. Research, 1991, *24*, 159). Also see Diels–Alder reaction of furan with maleic acid (Sce. 3.1.10).

3.6.2 3-methyl-1-phenyl-5-pyrazolone

It is prepared by the condensation of ethyl acetoacetate with phenyl hydrazine by heating in oil bath at 110–120° for 4 hr. However, in microwave oven, the reaction takes only 10 minutes for completion.

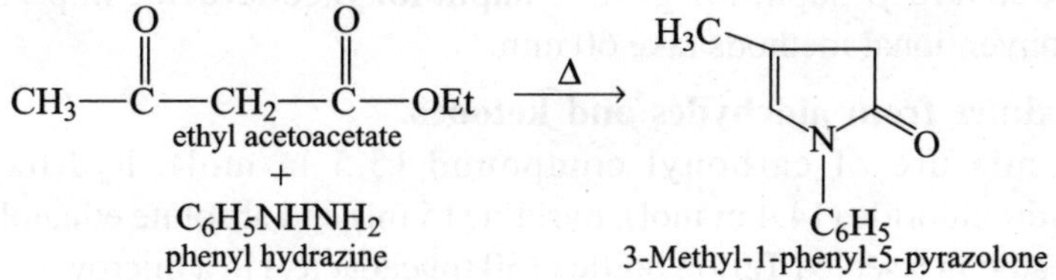

$$CH_3-\overset{O}{\underset{}{C}}-CH_2-\overset{O}{\underset{}{C}}-OEt \quad \xrightarrow{\Delta}$$

ethyl acetoacetate

+

$C_6H_5NHNH_2$

phenyl hydrazine

3-Methyl-1-phenyl-5-pyrazolone

Materials

Ethyl acetoacetate	2.9 g
Phenyl hydrazine	2.7 g

Procedure[1]

Heat a mixture of ethylacetoacetate (2.9 g) and phenyl hydrazine (2.7 g) in a conical flask (50 ml capacity) in microwave oven (280 W) for 10 min. Cool the reaction mixture, filter the separated product and recrystallise from alcohol–water (1:1) to get quantitative yield of 3-methyl-1 phenyl-5-pyrazolone. M.p. 129°.

Note

1. D. Villenmin, B. Labiad, Synthetic Commun, 1990, *20 (20)*, 3213.

3.6.3 Preparation of derivatives of some organic compounds

Microwave irradiation has been used for the preparation of following derivatives. p-Nitrobenzyl esters from carboxylic acids, Aryloxyacetic acids from phenols, Oximes from aldehydes and ketones.

(a) **p-Nitrobenzyl esters from carboxylic acids.**
A mixture of carboxylic acid (10 m mol) and water (1 ml) is neutralized with sodium hydroxide solution (10%). A solution of p-nitrobenzyl chloride (9 m mole) in ethanol (5 ml) is added and the mixture is heated in microwave oven for 2 min. in a sealed 150 ml. Teflon bottle. After cooling, water (2 ml) is added. The separated ester is filtered, washed with sodium carbonate solution (5%) and crysttllised from alcohol. Using cinnamic acid, corresponding p-nitrobenzyl cinnamate, m.p. 115–16° is obtained in 40 % yield.

(b) **Aryloxyacetic acids from phenols.**
A mixture of phenol (1.4 m mol) in sodium hydroxide (6 M, 3 ml) and aqueoues chloroacetic acid (50%, 0.5 ml) is heated in a sealed Teflon tube (150 ml capacity) in a microwave oven for 2 min. After cooling, water (2 ml) is added and the solution acidified with dilute hydrochloric acid. The mixture is extracted with ether, ether extract washed with water and extracted with sodium carbonate solution (5%). The sodium carbonate extract on acidification gives the aryloxyacetic acid derivative. Using this procedure β-naphthol gave 2-naphthoxyacetic acid, m.p. 154–56°. Conventional methods take 60 min.

(c) **Oximes from aldehydes and ketones.**
A mixture of carbonyl compound (5.5 m mol), hydroxylamine hydrochloride (14.4 m mol), pyridine (5 ml) and absolute ethanol (5 ml) is heated in a sealed Teflon bottle (150 ml capacity) in a microwave oven for 2 min. After cooling, the solvent is removed under reduced pressure. The formed oxime is stirred with cold water and crystallised from alcohol. Using this procedure, benzophenone give the oxime, m.p. 140–41° in 70% yield.

(d) **Carboxylic acids from ester (saponfication).** A mixute to the ester (7.4 m mol) and aqueous sodium hydroxide (25%, 10 ml) is heated in a sealed Teflon tube (150 ml capacity) for 2.5 min. in a microwave oven. The solution is cooled and acidified with hydrochloric acid and extracted with ether. The dried ether extract is evaporated to give the acid. Using this procedure methyl benzoate gave benzoic acid, m.p. 121–22° in 84 % yield.

Note

The procedure followed is that described by. R.N. Gedye, F.E. Smith, K.C. Westaway, Can. J. Chem. 1988, *66*, 17.

3.7 ENZYMATIC TRANSFORMATIONS

3.7.1 Ethanol

It is obtained by the fermentation of sucrose. It is not possible to obtain more than 10–15% ethanol by this method (since fermentation is inhibited if the concentration of alcohol exceeds 15%). More concentrated alcohol is isolated by fractional distillation. The fermentation of sucrose is represented as shown below.

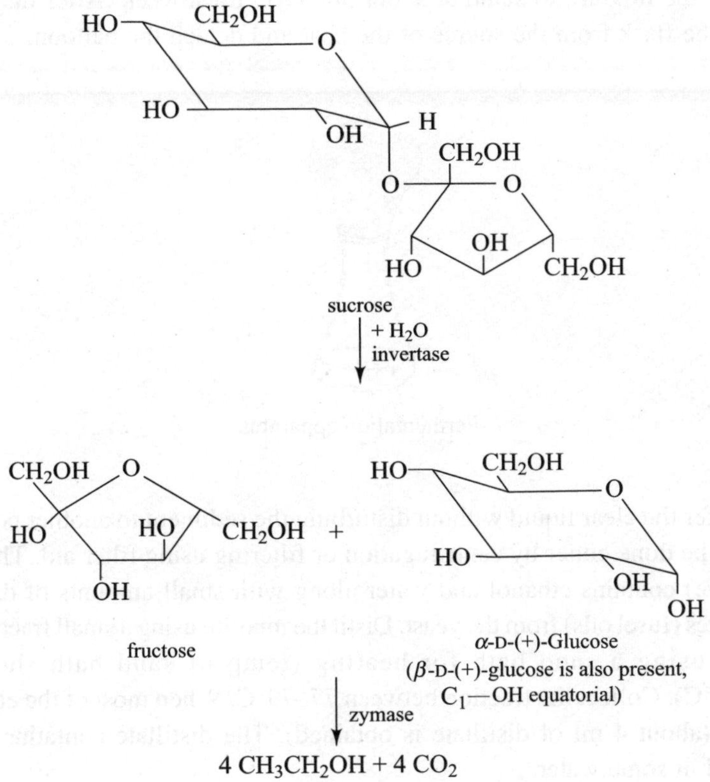
HO
CH_2OH
O
HO
OH
H
CH_2OH
O
O
OH
HO
CH_2OH
sucrose
$+ H_2O$
invertase
CH_2OH
O
HO
HO
CH_2OH +
OH
HO
CH_2OH
O
HO
OH
OH
fructose
α-D-(+)-Glucose
(β-D-(+)-glucose is also present,
C_1—OH equatorial)
zymase
$4\ CH_3CH_2OH + 4\ CO_2$

Materials

Sucrose	10 g
Baker's yeast	1 g
Pasteur's salts solution	10 ml

Note

A solution of Pasteur's salt consist of potassium dihydrogen phosphate (1 g), calcium phosphate (monobasic) (0.1 g), magnesium sulphate (0.1 g) and ammonium tartrate (diammanium salt) (5 g) dissolved in water (430 ml).

Procedure

Place sucrose (10 g) in a Erlenmeyer flask (500 ml capacity). Add water (70 ml) warmed to 25–30°C, pasteur's salt solution (10 ml) and dried baker's yeast (1 g). Shake the contents to mix them thoroghly and then attach a balloon directly to the flask (sec fig. below). The gas will make the balloon expand as fermentation continues. In this way contact with atmospheric oxygen is excluded, otherwise ethanol formed would be converted into acetic acid. As long as CO_2 continues to be liberated, alcohol is being formed.

Allow the mixture to stand at about 30–35°C for a week. After this period, remove the flask from the source of the heat and detach the balloon.

Fermentation apparatus

Transfer the clear liquid without disturbing the sediment to another container. This can be done either by centrifugation or filtering using filter aid. The liquid (or filtrate) contains ethanol and water along with small amounts of dissolved metabolites (fusel oils) from the yeast. Distil the mixture using a small fractionaling column using a sand bath for heating (temp of sand bath should be 150–200°C). Collect the fraction between 77–79°C. When most of the ethanol is distilled (about 4 ml of distillate is obtained). The distillate contatins ethanol dissolved in some water.

Analysis of the distillate

Determine the total weight of the distillate to determine the approximate yield. (Water and alchol forms an azeotrope, which boils at 78°C.)

Determine the density of the distillate by transferring a known volume of the distillate (using a pipette) to a weighted bottle, closing the mouth and reweighing. This procedure gives good values up to two significant figures. Using the following table, determine the percentage composition by weigth of alcohol in the distillate from the density as determined.

% Ethanol by weight	Density at 20°C (g.ml)
75	0.856
80	0.843
85	0.831
90	0.818
95	0.804
100	0.789

This preparation is green synthesis since no waste which may cause pollution is obtained as a by product.

3.7.2 (S)-(+)-Ethyl 3-hydroxybutanoate

It is obtained by chiral reduction of ethyl acetoacetate using baker's yeast.

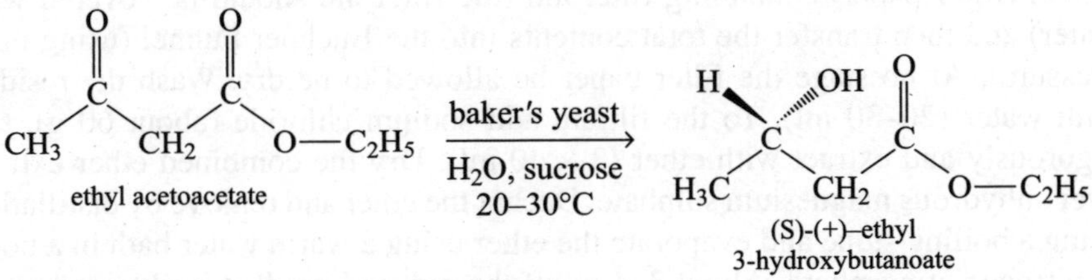

ethyl acetoacetate

baker's yeast
$\xrightarrow{\text{H}_2\text{O, sucrose} \atop 20\text{–}30°\text{C}}$

(S)-(+)–ethyl
3-hydroxybutanoate

A small amount (<10%) of the enantiomer with (R) configuration is also obtained.

Meterials

Ethyl acetoacetate	6.0 g
Sucrose	90 g
Dry bakers yeast	10.4 g

Procedure

Fit up the apparatus as shown to the fig. below. Place a magnetic stirring bar and a one hole rubber stopper with a glass tubing leading to a beaker containing a solution of barium hydroxide. Add some mineral oil on the top layer in the beaker to protect the barium hydroxide from air.

Add 150 ml water, 45 g sucrose and 5.2 g bakers yeast to the flask in the order given while stirring. Continue stirring for 1 hr (about 20–30°C). Add 6 g of ethyl acetoacetate, allow the fermentating mixture to stand at room temperature while stirring for 24 hrs. Add a solution sucrose (45 g) in water (150 ml) solution (prepared at 40°C) along with 5.2 g bakers yeast to the fermenting liquid. Allow the mixture to stand for 48 hr (with the trap attached at room temperature).

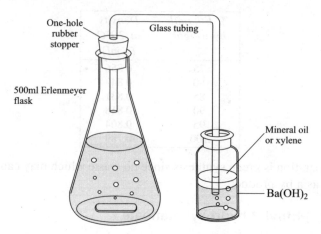

Fig. 3.1 Set up of apparatus for the reduction of ethyl acetoacetate

After the reaction is complete, filter it first by decantation through a buchner funnel (filter paper) containing filter aid (the filter aid should be covered with water) and then transfer the total contents into the Buchner Funnel (using mild pressure). At no stage the filter paper be allowed to be dry. Wash the residue with water (20–30 ml). To the filtrate add sodium chloride (about 60 g), stir vigorously and extract with ether (3 × 40 ml). Dry the combined ether extract over anhydrous magnesium sulphate. Decant the ether and remove by distillation using a boiling stone and evaporate the ether using a warm water bath in a hood in nitrogen atmosphere. About 3–4 ml of the reduced product is obtained.

Purify the crude hydroxy ester by column chromatography on alumina column using methylene chloride (to pour the ester in the column) and eluting with methylene chloride. Evaporation of the eluate give the pure hydroxy ester. Find its weight.

Take the 1R spectrum of the hydroxy ester. Make sure that a stretching peak at 3200–3500 cm^{-1} observed for O–H stretching peak and a stretching peak at about 1715 cm^{-1} has disappeared (there will still a C=O stretching peak from the ester functional group around 1735 cm^{-1}). In case, C=O stretching peak (at about 1715 cm^{-1}) remains or if the O—H stretching peak (at about 3200–3500 cm^{-1}) remains, the reduction has not taken place and the experiment has to be repeated.

Calculation of optical purity or enantiomeric excess

With the help of a polarimeter determine the spectric rotation of the sample prepared. Then % optical purity = % enantiomeric excess

$$= \frac{\text{observed specific rotation}}{\text{specific rotation of pure enantiomer}} \times 100$$

The specific rotation of the pure enantiomer is known from standard tables.

The published specific rotation of (+)-ethyl-3-hydroxybutanoate is $[\alpha]_D^{25} = +43.5°$

Let us suppose that optical purity as determined from the above equation comes to be 60%

$$\therefore \ \% \ (+) \text{ enantiomer} = 60 + \frac{(100 - 60)}{2} = 80\%$$

$$\text{and } \% \ (-) \text{ enantiamer} = \left(\frac{100 - 60}{2}\right) = 20 \ \%$$

It should be understood that the difference between these two calculated value is equal to the optical purity or enantiomeric excess (ee).

Note

The specific rotation is defined as the number of degrees of rotation observed when light is passed through 1 decimeter (10 centimeter) of its solution having a concentration of 1 gram per millilitre. The specific rotation in calculated by using the equation.

$$[\alpha]_D^{t^0} = \frac{\alpha_{aba}}{l \times c}$$

Where $[\alpha]_D^{t^0}$ stands for specific rotation determined at t°C using D–line of sodium light, α_{abc} is the observed angle of rotation; l is length of the solution is decimetrers; and c is the cencentration of the active compound in grams per milliliter.

Note

This is green synthesis, since no by-products which may cause pollution are formel.

3.7.3 Benzoin

It is well known that benzoin is obtained from benzaldehyde using cyanide ion, an inorganic reagent as catalyst.

A green synthesis of benzoin consist in the reaction of benzaldehyde with a biological coenzyme, thiamine hydrochloride as the catalyst (R. Breslow, J. Am. Chem. Soc., 1958, *80*, 3719).

Mechanism

Thiamine loses a proton to the solvent (or to the enzyme) to give conjugate base of thiamine, which adds to benzaldehyde to give Thiamine-benzaldehyde. The Thiamine-benzaldehyde loses a proton to give a resonance-stabilized equivalent of acyl carbanion, which adds to a second molecule of benzaldehyde to give a product that eliminates thiamine to give benzoin and to regenerate the catalyst.

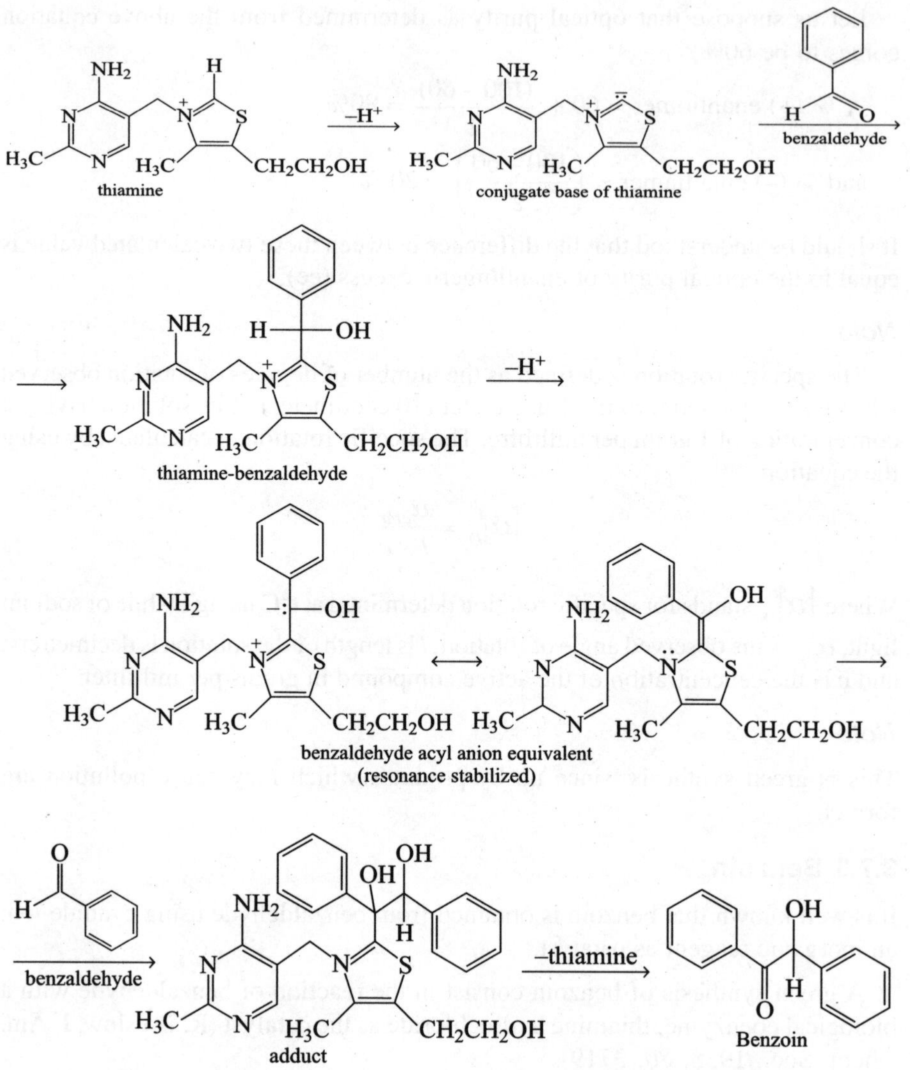

thiamine

conjugate base of thiamine

benzaldehyde

thiamine-benzaldehyde

benzaldehyde acyl anion equivalent
(resonance stabilized)

benzaldehyde

adduct

Benzoin

Materials

Benzaldehyde	0.9 ml
Thiamine hydrochloride	0.3 g
Ethanol (95%)	3 ml
Sodium hydroxide solution	0.9 ml
(obtained by dissolving 2 g. NaOH is 25 ml H$_2$O)	

Procedure

In an Erlenmeyer flask (25 ml capacity) dissove thiamine hydrochloride (0.3 g) in water (0.5 ml). Add ethanol (3 ml), shake the container till all solid dissolves. To this, add sodium hydroxide solution (0.9 ml, 8%) and swirl the flask until the bright yellow colour fades to a pale yellow colour. Add benzaldehyde (0.9 ml).

Swirl the flask until homogeneous solution is obtained. Stopper the flask and let it stay in dark place for 48 hrs.

Cool the flask (ice-bath), scratch the sides of the flask (to induce crystallistation) and filter the formed benzoin. Crystallise from alcohol (using 0.8 ml alcohol/0.1 g of crude benzoin). Record the yield, m.p. 134–35°C.

Notes

1. Pure benzaldehyde should be used. For use it is best to shake benzaldehyde with an equal volume of 5% aq. Na_2CO_3 solution. Remove the lower layer of Na_2CO_3 solution, wash the aldehyde layer with H_2O and dry over anhyd $CaCl_2$. The resulting purified benzaldehyde is suitable for the above preparation.

2. It is equally important to use pure thiamine hydrochloride, which should be stored in a refrigerator.

3. Thiamine serves as coenzyme for the following three types of enzymatic reactions

   (a) Non-oxidative decarboxylation of α-ketoacids.

$$R-\overset{O}{\overset{\|}{C}}-COOH \xrightarrow{B_1} R-\overset{O}{\overset{\|}{C}}-H + CO_2$$

   (b) Oxidative decarboxylation of α–ketoacids.

$$R-\overset{O}{\overset{\|}{C}}-COOH \xrightarrow{B_1, O_2} R-\overset{O}{\overset{\|}{C}}-OH + CO_2$$

   (c) Formation of acyloins (α–hydroxy ketones).

$$R-\overset{O}{\overset{\|}{C}}-COOH + R-\overset{O}{\overset{\|}{C}}-H \xrightarrow{B_1} R-\overset{O}{\overset{\|}{C}}-\overset{OH}{\overset{|}{C}H}-R + CO_2$$

   or

$$2\,R-\overset{O}{\overset{\|}{C}}-COOH \xrightarrow{B_1} R-\overset{O}{\overset{\|}{C}}-\overset{OH}{\overset{|}{C}H}-R + 2\,CO_2$$

   or

$$2\,R-\overset{O}{\overset{\|}{C}}-H \xrightarrow{B_1} R-\overset{O}{\overset{\|}{C}}-\overset{OH}{\overset{|}{C}H}-R$$

4. It is reported that benzoin condensations of aldehydes are strongly catalysed by quaternary ammonium cyanide in a two phase system (J. Soludar, Tetradhedron Lett., 1971, 287).

3.7.4 1-Phenyl-(1S) ethan-1-ol from acetophenone

It is obtained by the reduction of acetophenone with Daucus carota root.

$$
\underset{\text{acetophenone}}{C_6H_5-\overset{\overset{\displaystyle O}{\|}}{C}-CH_3} \xrightarrow{\text{D. Carota}} \underset{\text{1-Phenyl -(1S)-ethan-1-ol}}{C_6H_5-\overset{\overset{\displaystyle OH}{|}}{CH}-CH_3}
$$

Procedure

Remove the external layer of carrot and cut the remaining into small thin pieces (approx. 1 cm long slice). To acetophenone (1 ml) and water (20 ml) add slices of carrots (10 g). Shake the reaction mixture occasionally. The reaction takes about 40 hrs for completion. Isolate the product by ether extraction and purifying by chromatography over silica gel (200 mesh). Elution was done with ether-petroleum ether mixture (1:2). Yield 73 %. % ee 92. The alcohol obtained was

1-phenyl -(1S)-ethan-1-ol. Record its NMR spectra and rotation $[\alpha]_D^{25} = -39.1$ (C = 3.5, CHCl$_3$).

Notes

1. The procedure adopted for asymmetric reduction is that of J.S. Yadav, S. Nanda, P. Thirupathi Reddy and A. Bhaskar Rau, J. Org. Chem., 2000, *67*, 3900.

2. The procedure is better than that uses Baker's yeast as the recovery of the product is not straight forward (O.P. Ward and C.S. Young, Enzyme Microbiol Technol., 1998, *12*, 4822).

3. The Chiral alcohol obtained had S configuration, which is in perfect agreement with Prelog's rule.

4. Using the above procedure a number of ketones (viz, p–Cl, p–Br–, p–F–, p–NO$_2$–, p–CH$_3$, p–OCH$_3$ –, p–OH–acetophenones could be reduced to the corresponding S alcohol with ee 90–98% and in 70–82% yield. The reduction was monitored by TLC carried out on 0.25 mm silica gel plates with uv light and 2.5% ethanolic anisaldehyde (with 1% CH$_3$COOH and 3% conc. H$_2$SO$_4$)–heat as developing agent.

5. The procedure can also be used for the reduction of β–ketoesters, cyclic ketones, azido ketones and open chain ketones like 2-butanone, 2-pentanone etc.

3.7.5 Deoximation of oximes by ultrasonically stimulated Bakers yeast

The enzymatic conversion of oximes by ultrasonically stimulated bakers yeast yields the corresponding aldehydes and ketones.

Materials

Oxime of p-methoxy benzaldehyde	0.7 g
Ethanol	15 ml
Baker's yeast	15 g

Procdeure

To ultrasonically pretreated baker's yeast (15 g) in phosphate buffer (pH 7.2; 150 ml) add the oxime (0.7 g) in ethanol (15 ml).

The mixture was incubated at 37°C for 2-3-days. It was filtered, filtrate extracted with ethylacetate. The organic phase was dried and evaporated under reduced pressure. The residue obtained was purified was column chromatogrraphy to give p-methoxy benzaldehyde. b.p. 247°. Yield 98%.

Note

1. The procedure followed was that described by A. Kamal, M.V. Rao and H.M. Meshram, J. Chem. Soc. Perkin Trans I., 1991, 2056.

2. Oxime of p-methoxy benzaldehyde was prepared by the usual procedure.

3.8 SONICATION REACTIONS

3.8.1 Butyraldehyde

It is obtained by **Bouveault reaction**[1] of 1-chlorobutane by ultrasonic irradiation with lithium and dimethyl formamide

$$CH_3CH_2CH_2CH_2Cl \xrightarrow{Li} CH_3CH_2CH_2CH_2 Li \xrightarrow{DMF}$$

1–chlorobutane

$$\longrightarrow \left[CH_3CH_2CH_2CH \begin{array}{c} OLi \\ \diagup \\ \diagdown \\ NMe_2 \end{array} \right] \xrightarrow{H_3O^+} CH_3CH_2CH_2CHO + HNMe_2$$

Materials

1-chlorobutane	1.85 g
Lithium	0.1 g
Dimethyl formamide	1.5 g

Procedure

A mixture of 1-chlorobutane (1.85 g) dry dimethyl fomamide (1.5 g) and lithium sand (0.1 g in 4 ml dry THF) is sonicated in an ultrasound cleaner (96 w/l, 40 KHz). The reaction is complete at 10–20° in 15 min. Extract the mixture with ether (50 ml), wash ether with dil. HCl and sodium chloride solution and dry (anhydrous Na_2SO_4). Distillation of the ether solution gave n-butyraldehyde in 78% yield.

Notes

1. The procedure followed is that of C. Petrier, A.L. Gemal and J. L. Luche, Tetrahedron Lett., 1982, *23*, 3361.

2. Lithium metal is obtained as suspension in mineral oil. For using, it is washed with anhydrous THF under inert atmosphere.

3. Using appropriate halides (Chlorides or bromides) a number of aldehydes can be prepared. For example, cyclohexyl halides give the corresponding aldehyde.

3.8.2 2-Chloro-N-aryl anthranilic acid

2-Chloro-N-arylanthranilic acid is prepared by the **Ullmann condensation** of 2-chlorobenzoic acid with 2-Choloraniline in presence of copper powder and cuprous iodide in boiling DMF with ultrasonic irradiation.

Reaction scheme:

2-chloro benzoic acid (COOH, Cl) + 2-chloro aniline (Cl, H₂N) →[Cu powder / cuprous iodide / ultrasonic irradiation / 20 min]→ 2-chloro-N-aryl anthranilic acid (CO₂H, Cl, NH)

Materials

2-chlorobenzoic acid	1.5 g
2-chloroaniline	1.0 g
Copper powder	0.1 g
Cuprous iodide	0.1 g
Potassium carbonate	0.55 g

Procedure

Heat a mixture of 2-chlorobenzoic acid (1.5 g), 2-chloroaniline (1.0 g), potassium carbonate (0.55 g), copper powder (0.1 g) and cuprous iodide (0.1 g) in boiling DMF (5 ml) for 20 min. with ultrasonic irradiation (virsonic 300 cell disrupter at 20 kHz). Pour the reaction mixture on to water (20 ml). Filter the separated solid, treat with conc. K_2CO_3 solution. Acidify the clear filtrate with dilute acetic acid (1:3). The product separates at pH 6–8. Crystallise the product, m.p. 196–8° (75 % yield).

Notes

1. The procedure described is that of R. Carrasco, R.F. Pellon, Jose' Elguero, Pilar Goya and Juan Antorio Paez, Synthetic communications, 1989, *19 (11 and 12)*, 2077–2080.

2. The procedure using ultrasound is very convenient, shortens the reaction time, i.e., 20 minutes in comparison to 4–6 hrs by the usual procedure. Also the reaction proceeds in a much cleaner way and affords higher purity of crude products.

3.9 ESTERIFICATION

3.9.1 Benzocaine (Ethyl p-aminobenzoate)

It is a local anesthetic and is prepared by direct esterification of p-aminobenzoic acid with ethanol.

$$H_2N-\langle\text{benzene ring}\rangle-\overset{\overset{O}{\|}}{C}-OH + CH_3CH_2OH \overset{H^+}{\rightleftharpoons} H_2N-\langle\text{benzene ring}\rangle-\overset{\overset{O}{\|}}{C}-OC_2H_5$$

p-amino benzoic acid

ethyl p-amino-
benzoate
(benzocane)

Materials

p-amino benzoic acid	4 g
absolute ethanol	40 ml
conc. sulphuric acid	3 ml

Stir a mixture of p-aminobenzoic acid (4 g) and absolute alcohol (40 ml) in a round bottomed flask till complete solution results. To the stirred solution add conc. sulphuric acid dropwise (3 ml). A precipitate forms but it dissolves on stirring and subsequent refluxing. Reflux the mixture using a reflux condenser for 1–1.5 hr. at about 105°C. Stir the mixture during this period.

Allow the reaction mixture to cool down to room temperature. Add water (30 ml) and then add sodium carbonate solution (10 %) dropwise to neutralise the mixture. As the pH increase, a white precipitate of benzocaine is produced. Add more sodium carbonate solution until pH is about 8. Filter the separated benzocaine, wash with water and crystallise from methanol–water. Record the yield and m.p. The pure benzocaine melts at 92°C.

Notes

1. You may test benzocaine for its anesthetic action on frog's muscle.

2. Carboxylic acids can be conveniently estirified [H.E. Hennis, L.R. Thompson and J. P. Long, Ind. Eng. Chem. Prod. Res. Dev., 1968, 7, 96. H.E. Hennis, J. P. Esterly, L.R. Collins and L. R. Thompson, Ind. Eng. Chem. Prod Res. Dev., 1967, 6, 143] with alkyl halide and triethyl amine. In this case, PTC is generated in situ by the reaction of triethylamine and alkyl halide.

$$Et_3N + RX \longrightarrow Et_3\overset{+}{N} RCl^-$$

Et₃N — ; RX alkyl halide ; Et₃NRCl⁻ PTC

$$R'COONa + RX \overset{PTC}{\longrightarrow} R'COOR + NaX$$

Sod. salt of carboxylic and (aq. solution) ; alkyl halide ; ester

In the above case alkyl halide must be highly reactive, e. g., benzylchloride. Alternatively quaterary ammonium or phosphonium salts can be directly used for the esterification of carboxylic acid with alkyl halides (R. Holmbery and S. Hansen, Tetrahedron Lett., 1975, 2307).

3. Esterification can be very conveniently carried out using a microwave in better yields and in much shorter time (5–10 min.) (R.N. Gedye, F.E. Smith, K.C. Westaway, Can. J. Chem., 1988, *66*, 17; R.N. Gedye, W. Ramnk, K.C. Westaway, Can. J. Chem., 1991, *69* 706).

3.9.2 Isopentyl acetate (Banana oil)

The ester, isopentyl acetate is referred to as banana oil, because it has familiar odour of this fruit. It is prepared by esterification of acetic acid with isopentyl alcohol. Since the equilibrium does not favour the formation of the ester, it must be shifted to the right in order to get better yield of the ester. For this purpose, one of the starting material is used in excess. Acetic acid being less expensive than isopentyl alcohol and more easily removed from the reaction mixture is used.

$$CH_3-\underset{\underset{acetic\ acid}{}}{\overset{\overset{O}{\parallel}}{C}}-OH + CH_3-\underset{\underset{isopentyl\ alcohol}{}}{\overset{\overset{CH_3}{|}}{CH}}-CH_2CH_2OH \underset{}{\overset{H^+}{\rightleftharpoons}} CH_3-\overset{\overset{O}{\parallel}}{C}-O-CH_2CH_2-\underset{\underset{isopentyl\ acetate}{}}{\overset{\overset{CH_3}{|}}{CH}}-CH_3 + H_2O$$

The mechanism of the reaction is similar to that described for the preparation of methyl salicylate (see 3.9.3).

Materials

Isopentyl alcohol (isoamyl alcohol)	12.5 ml
Acetic acid	18 ml
Conc. sulphuric acid	2.5 ml

Procedure

In a round bottomed flask take isopentyl alcohol (12.5 ml) and glacial acetic acid (18 ml). To the clear solution add dropwise conc. sulphuric acid (2.5 ml) during shaking. Reflux the mixture for 1 hr. To the cooled solution add water (25 ml). Remove the lower aqueous layer and discard (use a small separatory funnel). Extract the organic layer with aqueous sodium bicarbonate (2 × 5 ml) and then with water. Dry the organic layer over anhydrous sodium sulphate, filter and distil. Yield 10 g. b. p. 142 °C.

Notes

1. See note 2 in the preparation of benzocaine (3.9.1)

2. See note 3 in the preparation is Benzocaine (3.9.1).

3.9.3 Methyl salicylate (oil of wintergreen)

Methyl salicylate was isolated as a familiar–smelling organic ester–oil of wintergreen from the wintergreen plant (gaul theria) in 1843. It was found to have analgesic and antipyretic character almost identical to that of salicylic acid when taken internally. This medicinal characteristic is due to the ease with which methyl salicylate is hydrolysed to salicylic acid under alkaline conditions in the intestinal tract. Salicylic acid is known to have analgesic and antipyretic properties. Methyl salicylate can be taken internally or absorbed through the skin. It finds application in liniment preparations. On applying to skin, it produces a mild tingling and soothing sensation. The ester is also used to a small extent as a flavouring principle due to its pleasant odour.

Methyl salicylate is obtained by the esterification of salicylic acid with methanol in presence of acid. It is an equilibrium reaction. The equilibrium can shift to right, i.e., more product can be obtained by increasing the concentration of one of the reactant (methyl alcohol being cheaper is used in excess).

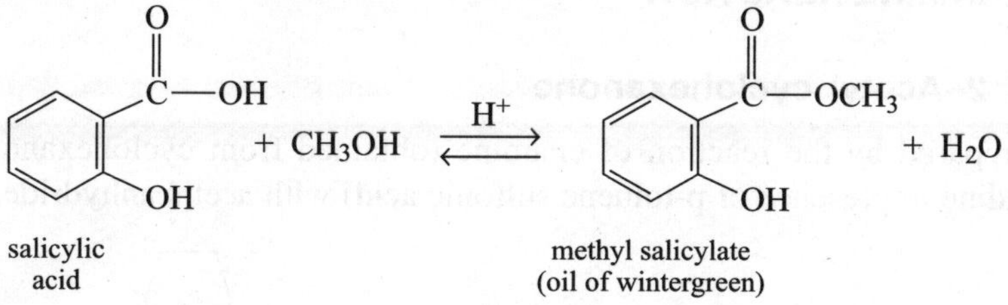

salicylic
acid

methyl salicylate
(oil of wintergreen)

Mechanism

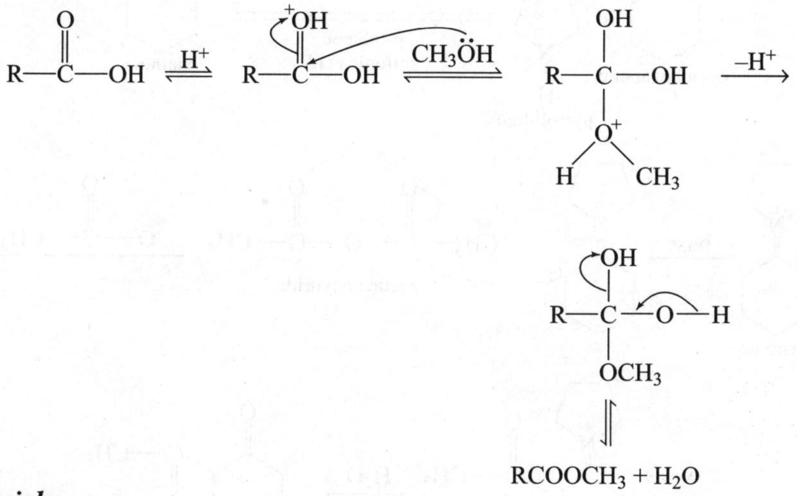

$$RCOOCH_3 + H_2O$$

Materials

Salicylic acid	6.5 g
Methyl alcohol	20 ml
Conc. Sulphuric acid	3 ml

Procedure

In a round bottomed flask dissolve salicylic acid (6.5 g) into methanol (20 ml). To the clear solution add conc. H_2SO_4 dropwise (3 ml) during shaking. Reflux the solution at 80° for 2.5 hrs. Remove excess methyl alcohol by distillation (steam bath) and pour the residual product into water (40 ml). Extract with ether (2 × 50 ml), wash ether extract with sodium bicarbonate solution (till free of acid) and finally with water. Dry ether extract over anhyd. sodium sulphate. Methyl salicylate is obtained as a colourless liquid with fragrant smell. B.P. 223–25°. Yield 6.5 g (85 %).

Note

1. In a similar way methyl benzoate (b.p. 190°C), known as oil of Niobe can be prepared.

2. See notes 2 and 3 in the preparation of benzocaine (3.9.1).

3.10 ENAMINE REACTION

3.10.1 2–Acetyl cyclohexanone

It is prepared by the reaction of enamine (obtained from cyclohexanone and pyrrolidine in presence of p-toluene sulfonic acid) with acetic anhydride.

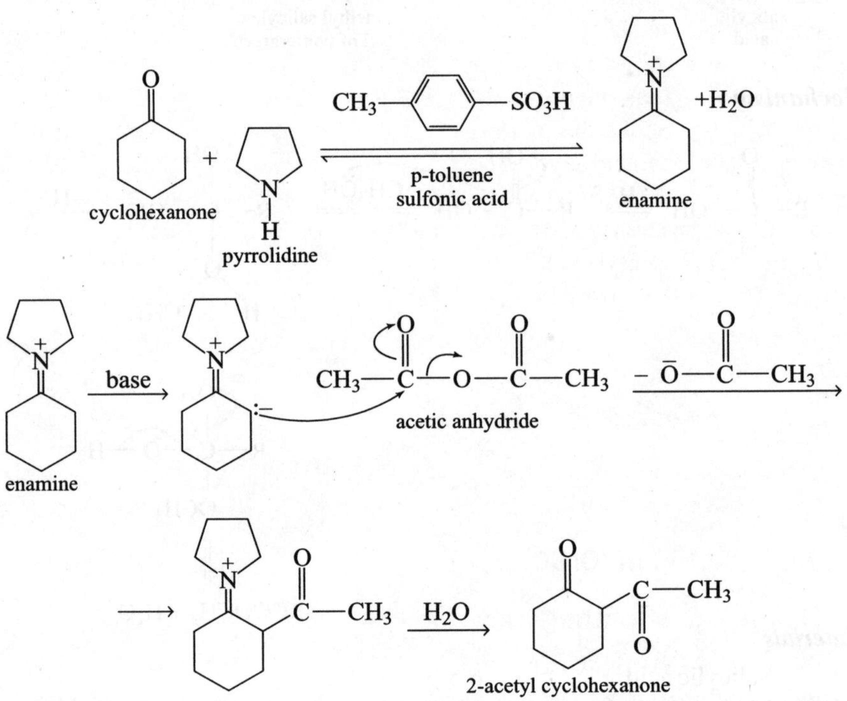

Step 1 Preparation of enamine

Materials

Cyclohexanone	1.92 ml
p-toluene sulfonic acid monohydrate	55 mg
Pyrrolidine	1.62 ml
Toluene	12 ml

Procedure

Take cyclohexanone (1.92 ml) in a round bottomed flask (25 ml capacity). Add Toluene (12 ml) and p-toluene sulfonic acid monohydrate (55 mg). To the mixture add pyrrolidine (1.62 ml). Stir the mixture using a magnetic needle for 10–15 min under anhydrous conditions. While stirring set up the distillation apparatus. Heat the reaction mixture to 140–45° (sand bath). The rate of distillation should be controlled so that the distillate 11–12 ml gets collected in about 30 minutes. By this process, the water formed in the reaction is removed as an azeotrope. The flask contains the enamine and is used as such for the next step.

Step 2 2-Acetyl cyclohexanone

To the enamine obtained in the step 1, add acetic anhydride (1.92 ml) dissolved in toluene (3 ml). Swirl the flask (which has been corked) for about 5 min. at room temperature and allow the mixture to stand for 48 hrs.

After this period, add water (3 ml). Reflux the mixture (using a water cooled condenser) at about 120°C for 30 minutes. Cool the flask to room temperature, add water (3 ml) and separate the upper toluene layer using a small separatory funnel. Add to toluene layer (containing 2-acetyl cyclohexanone) hydrochloric acid (6 M, 6 ml). Shake the mixture. By this procedure pyrrolidine is extracted by HCl. Discard the acid layer. Shake the organic layer with water (3 ml). Dry the solution (anhyd. Na_2SO_4) and evaporate the dried toluene solution in a water bath at 70° using a stream of dry air or nitrogen. When all the toluene is removed, the dry product (2–acetyl cyclohexanone) is purified by column chromatrography using alumina as the absorbent and methylene chloride as the eluant.

2-Acetyl cyclohexanone is obtained as a yellow liquid. Determine its yield and record the nmr spectra.

Notes

1. The above procedure can also be used for the preparation of 2-alkyl cyclohexanone by using alkyl halide in place of acetic anhydride.

3.11 REACTIONS IN IONIC LIQUIDS

3.11.1 1-Acetylnaphthalene

It is obtained by **Friedel–Crafts reaction** of naphthalene with acetyl chloride in presence of ionic liquid such as the [emin] Cl–AlCl$_3$ [emin = 1-methyl 3-ethylimidazolium cation].

naphthalene + CH$_3$COCl $\xrightarrow[\text{r.t}]{\text{[emin] Cl-AlCl}_3}$ 1-acetyl naphthalene

Materials

Naphthalene	6.4 g
Acetyl chloride	3.9 g
[emin] Cl-AlCl$_3$	20 g

Step (i) 1-Methyl-3-ethylimidazolium chloride. It is prepared by the procedure described by R.S. Verma, V.V. Namboodiri, Chem. Commun., 2001, 643.

A mixture of ethyl chloride (7.8 g) and 1-methyl-3-ethyl imidazole (11.2 g) was mixed thoroughly and the mixture heated in unmodified household microwave oven (240 W) for 1 min., till a clear single phase is obtained. Cool the resulting ionic liquid and wash with ether (3 × 20 ml) to remove unreacted starting material and the product, 1-methyl-3-ethylimidazolium chloride is obtained.

Step (ii) 1-Acetylnaphthalene

A mixture of naphthalene (6.4 g), acetyl chloride (3.9 g) and the ionic liquid [emin]Cl–AlCl$_3$ [emin-1-methyl-3-ethyl-imidazolium cation] (20 g) [obtained by mixing equimolar amounts of 1-methyl-3-ethyl imidazolium chloride with anhyd. aluminium chloride] was allowed to react at 0° at room temperature for 1 hr. The product obtained (after extraction with ether) was crystallised from alcohol to give 1-acetyl naphthalene in 89 % yield. Record its nmr spectra.

Notes

1. The procedure deseribde is adopted from C.J. Adams, M.J. Earle, G. Roberts and R. Seddon Chem. Common, 1998, 2097.

2. Friedel–Crafts acylation of naphthalene in ionic liquid gives the thermodynamically unfavourable 1-isomer. On the other hand conventional Friedel–Crafts reaction gives the 2-isomer.

3. Using the above procedure, toluene, chlorobenzene and anisole give the corresponding 4-acetyl compounds in 97–99% yields.

4. The ionic liquid can be used again.

3.11.2 Ethyl 4-methyl-3-cyclohexene carboxylate

It is prepared by **Diels–Alder reaction** of isoprene (dienophile) with ethyl acrylate (diene) in neutral ionic liquids, viz 1-ethyl-3-methyl imidazolium tetrafluorobarate, [emin] BF$_4$.

isoprene   ethyl acrylate   ethyl 4-methyl-
                                                        3-cyclohexene carboxylate

Step (i)   1-Ethyl-3 methyl imidiazolium tetrafluoruborate [emin] BF_4.

It is prepared by the procedure described by J.D. Holbreg and K.R. Seddon, J. Chem. Soc. Dalton Trans., 1999, 2133–2139.

Tetrafluoroboric acid (12.2 ml, 0.116 mol, 48% solution in water) was added slowly to a rapidlly stirred slurry of silver (I) oxide (13.49 g, 0.058 mol) in water (50 ml) over 15 min. The reaction mixture was covered with aluminium foil to prevent photodegradation. The stirring was continued (1 hour) until all the Ag(I) oxide had completely reacted to give a clear solution. A solution of 1-ethyl-3 methyl imidazolium bromide (22.24 g, 0.116 mol) in water (200 ml) was added to the reaction mixture and stirred at room temperature for 2 hr. The resulting yellow precipitate of silver bromide was removed by filtration. The solvent was removed from the supernatant liquor by heating at 70° initially under reduced pressure and finally in vacuo to yield the tetrafluoro borate salt as a pale yellow liquid, yield 21.36 g, 93%.

Step (ii)   Ethyl 4-methyl-3-cyclohexene carboxylate

The reaction of diene (isoprene) with dienophile (ethyl acrylate) in the binary liquid [bmin] BF_4] was perfoumed in 1:5 : 1.0 :1.0 molar ratio mixture of diene: dienophite : solvent at 70° for 24 hr. to give 97% yield of the required product (endo: exo ratio 4.0:1).

Notes

1. The procedure described is that of M.J. Earle, P. B. McCormac and K.R. Seddon, Green, Chemistry, 1999, 23.

2. Using isoprene (diene) and but-2-en-3-one (dienophile) in [bmin] [BF_4] at 20°, for 2 hr. gave 90% yield of 4-acetyl-1-methyl cyclohexene.

3. The addition of 5 mole % of zinc (II) iodide dramatically increase the rate of the reaction.

4. The final product was isolated by extraction with ether.

5. The binary liquid can be used again.

Index

A

1-Acetylnaphthalene 201
2–Acetyl cyclohexanone 200
p-Acetamidophenol 161
2-Allyl phenol 183
Anthracene-maleic anhydride adduct 184
Acetaldehyde 98
Acetoin 53
Acid-catalysed aldol condensation 23
Aconitic acid 121
Acyloin condensation 17
Acyloin condensation using coenzyme, thiamine 20
Addition reactions 3, 11
Aldol condensation 22, 28, 70, 159
Aldol condensation 70, 159
Aldol condensation in solid phase 29
Aldol condensation reaction 158
Aldol type condensations 27
Aldosterone-21-acetate 51
Aldosteroneacetate oxime 51
Aliphatic claisen rearrangement 65
Allyrethrone 122
Anthanthrone 139
Aqueous phase claisen rearrangement 67
Aqueous phase reaction 14, 155
Arndt-eistert synthesis 33
Aryl cinnamonitriles 109
Aryloxyacetic acids from phenols 186

Azobenzene (trans) 170, 171

B

Baeyer–villiger oxidation 77, 169
Baeyer–villiger oxidation in aqueous phase 41
Baeyer–villiger oxidation in solid state 42
Baeyer–villiger oxidation 38
Baker–venkataraman rearrangement 56
Baker–venkataraman synthesis 176
Baker's yeast 189
Banana oil 198
n-Butyl benzyl ether 180
Barbier reaction 47
Barbier reaction under sonication 47
Barton reaction1 50
Benzil 55
Benzil-benzilic acid rearrangement 61
Benzilic acid 55, 61
Benzocaine 196
Benzoin 191
Benzoin condensation 52
Benzoin condensation under catalytic conditions 53
Benzonitrile 179
Benzopinacol 169
Benzopinacolone 182
Benzylacetate 43
Benzyl triethylammonium chloride 173
Benzyl triphenyl phosphonium chloride 166

Benzylacetate 43
Benzylidene acetone 72
Benzylidene acetophenone 72
Betaine 144
cis-Bicyclo [3·3·0] octane-3,7-dione-
   2,4,6,8-tetracarboxylate 140
Bicyclo [4.1.0] heptane 130, 132
Bicyclobutene carboxylic acid 38
Biochemical baeyer–villiger oxidations
   45
1-Butene 100
2-Butene 100
Butylbenzene 153
Blaise reaction 126
Bouveault reaction 57, 195
Bouveault reactions under sonication
   58
2-Butanol 94
t-Butylacetate 44
Butylbenzene 74
Butylacohol 97
Butyraldehyde 195

C

ε-Caprolactone 43
α- Campholide 44
Cannizzaro Reaction 58
Cannizzaro reactions under sonication
   61
1-Cyano octane 173
2-Carbethyoxycyclopentanone 82
2-carbethoxycyclohexanone 82
β-carotene 150
Endo-cis-1,4-endoxo- Δ^5-cyclohexene-
   2,3-dicarboxylic acid 164
Chloro-N-aryl anthranilic acid 196
Carboxylic acids from ester 186
Caronic acid 121
Catalytic reagents 8
Catechol 76
Chalcone 156, 157
Chromones 57
Cinnamaldehyde 71

[18] Crown-6 178
Citral 30, 128
Citric acid 129
Claisen rearrangement 63, 183
Claisen schmidt reaction in aqueous
   phase 70
Claisen-schmidt reaction 26, 69, 162

Clemmensen reduction 73
Cope rearrangement 63
Cope-knoevenagel reaction 107
Crossed aldol condensation 24, 26, 69,
   156
Crossed cannizzaro reaction 30, 60
3-Cyanocumarins 108
2-Cyano-2,3-epoxytetrahydro benzo
   thiophene 80
Cycloaddition reactions 13
Cycloburane 100
Cyclohexane 76
Cyclohexanone 84
Cyclohexanone oxygenase, 43
Cyclopentane 76
Cyclopentanone 47, 84
Cylindrocarpon radicicola 45

D

2,4-Dihydroxybenzoic acid 163
9,10-Dihydroanthracene-endo-α,
   β-succinic anhydride 184
Dakin reaction 39, 46, 76
Darzen condensation 174
Darzen reaction 79
Darzen reaction in presence of phase
   transfer cata 79
Daucus carota root 193
Deoximation of oximes 194
Derivatization 7
Desoxybenzion 55
p,p´-Diaminobiphenyl 138
Dibenzo-18-crown-6 80
Dichloronorcarane 177
Dieckmann condensation 82

Dieckmann condensation in solid
state 83
2,2-Dimethyldecane 154
3,4-Dimethly hexane 154
cis-1,2-Diethylcyclopropane 131
trans-1,2-Diethylcyclopropane 131
2,2-Dimethyl-3-nitropropanenitrile 121
5,7-Dimethyl flavonone 116
(–)N-methyl-N-benzylephedrinium
bromide 118
2,2′-Dinitrobiphenyl 136, 137
4,4′-Diphenic acid 138
Dieckmann condensation under
sonication 83
Dieckmann condensation using
Polymer support techn 84
Diels–alder reaction 63, 86, 164,
184, 202
Diels–alder reaction in ionic liquids 89
Diels–alder reactions in aqueous phase
89
Diels–alder reactions under microwave
irradiation 89
2,7-Dimethoxy-9,10-dihydropenanth-
rene 138
Dihydrosterculic acid 130
3,4-Dimethoxyphenol 46
3,4-Dimethoxy phenylacetic acid 36
3,5-Dimethyl catechol 78
cis-Dimethyl 1,2,3,6-tetrahydropht-
halate 87
trans-Dimethyl 1,2,3,6-trahydrophtha-
late 88
Dimidone 119
Diphenyl amine 136, 139
Diphenyl ether 136, 139
Diphenylcarbinol 168
1,2-Divinyl cyclohexanol 102
3,4-Diphenyl-7-hydroxycoumarin 174
(2,2-Dichlorobicyclo (4·1·0)heptane
177
Dithioacetic acid 101
Dl-Tyrosine 135

Doebner modification 106
DSIDA 135

E

Electrocyclic reaction 13
Elimination reactions 4, 12
Enamine reaction 200
1-Ethyl-3 methyl imidiazolium tetraf-
luoruborate [emin] BF$_4$ 203
Endo-anthracene maleic anhydride 87
Endo-cis-1,4-endoxo–D5-
cyclohexene-2,3-dicarboxyli 164
Energy requirement 6
Enzymatic Baeyer–Villiger Oxidation 42
Enzymatic Transformations 187
epoxy-N,N-diethylpropanamide 81
Esterification 196
2-ethyl-1-methyl-piperdine 75
1-ethyl-4-piperidone 85
Ethanol 187
Ethyl cinnamate 72
Ethyl p-aminobenzoate 196
Ethyl 3-methyl-4-nitrobutyrate 121
Ethyl 2,3-dimethoxy cinnamate 110
Ethyl 2-methyl-3-p-toly2-butenoate
128
Ethyl 3-phenyl-3-hydroxy-propionate
127
Ethyl 4-methyl-3-cyclohexane carbox-
ylate 202
1-Ethyl cyclohexene 96
6-Ethoxycarbonyl-3,5-diphenyl-2-
cyclohexenone 157
p-Ethoxyacetanilide 160
(S)-(+)-Ethyl 3-hydroxybutanoate 189
Ethyl benzene 74
ethyl p-amino-benzoate 197
ethyl proply ether 142

F

Flavones 56, 57, 176
Furyl alcohol 62

G

Glycolic acid 60
Green chemistry 1
Green reaction 17
Green synthesis 155
Grignard reaction 90
Grignard reaction in solid state 93
Grignard reaction under sonication 90

H

Haloform reaction 164
Heck Reaction 103
Heck reaction in aqueous phase 103
Heck reaction in ionic liquids 105
Heck Reaction under PTC conditions 104
Henry reaction 27
n-Heptane 74
Hetero diels-alder adduct 167
hexadecyl tributyl phosphonium bromide 173
homoveratroyl chloride 36
Horner-wadsworth-emmons modification 146
Huang-minlon modification of the wolff-kishner red 73
Hydrobenzoin 55
Hydroxymethyldimethyl nitromethane 32

I

Intramolecular aldol condensation 32
Internal claisen condensation 56, 119
Intramolecular cannizzaro reaction 60
Intramolecular claisen condensation 82
Intramolecular darzen reaction 80
Iodoform 164
Iodomethylenezinc iodide 130
Ionic-liquid 14
β-ionone 30, 72
Isopentyl acetate 198
Isopropylidene cyanoaceticester 107

K

Knoevenagel condensation 106
Knoevenagel reaction in ionic liquids 109
Knoevenagel reaction in solid state 109
Knoevenagel type addition 53
Knoevenagel-stobbe condensation[3] 108

M

Maincone 123
Malic acid 110
Mescaline 36
Mesityl oxide 31
o-Methoxy benzoic acid 61
o-Methoxy benzyl alcohol 61
Methylbicyclo [2,1,1] hexane carboxy- late 37
4-Methoxy benzoion 54
Methyl cyanide 98
cis-9-Methyldecalin 75
Methyl δ-furanyl glutarate 37
3-Methyl-3-pentanol 96
4-Methyl-4-phenyl caproic acid 36
Methyl propyl ether 98
2-Methyl propyl amine 98
Methyl salicylate 199
Methylbicyclo [2, 1, 1] hexane carboxylate 37
Methylene cyclohexane 145, 149
Michael addition 111, 157, 159, 166
Michael addition in aqueous medium 113
Michael addition in ionic liquids 118
Michael addition in solid state 116
michael addition reaction 157
Michael addition under PTC conditions 113
Michael condensation 107
Microwave induced reactions 184
Modifications of wittig reagent 146
Mukaiyama reaction 70, 123
Mukaiyama reaction in aqueous phase 124

2-Methyl-2-(3-oxobutyl)-1,3-cyclopentanedione. 166
3-Methyl-1-phenyl-5-pyrazolone 185
1-Methyl-3-ethylimidazolium chloride 202

N

2-Nitrophenyl acetic acid 36
p-Nitrobenzyl esters from carboxylic acids 186
Need of green chemistry 1
Ntramolecular aldol condensation 32

O

[3.3.0] Octane-3,7-dione-2,4,6,8-tetracarboxylate 140
$\Delta^{1.9}$-2-Octalone 159
Oil of wintergreen 198
2-Oxo-1-phenyltetrahydropyrrole 139
Organic synthesis in solid state 14
Oximes from aldehydes and ketones 186
1-Oxaspiro-[2,5]-octane-2-carbonitrile 174

P

Penicillium chrysogenum 45
Penicillium lilacinum 45
Pentaerythritol 31, 62
Percentage atom utilization 2, 10
Percentage yield 2
Pericyclic reaction 13, 63
Perylene 139
Phase transfer catalysed Williamson ether synthesi 141
Phase transfer catalysed Wittig-Horner reaction 147
Phenacetin 160
Phenyl acetate 38, 44
Phenyl benzoate 169
γ-Phenylbutanoic acid 76
α-Phenyl-γ-fluorotetronic acid 127

3-Pyridyl-4(3H)quinazolone 168
1-Phenyl-(1S) ethan-1-ol from acetophenone 193
Phenyl p-nitrobenzoate 41
Phenylalanine 134
Phenylisocyanide 172
Phorone 31
Phosphorane 144
Photochemical reactions 169
Phynyl acetate 44
Photochemical arndt-eistert reaction 37
Planning a green synthesis 10
Prelog's rule 194
Principles of green chemistry 2
Propane 100
Propylalcohol 94
Pseudoionone 30, 72
PTC catalysed reactions 172

Q

Quinol 78

R

Reactions in ionic liquids 201
Reactions in ionic-liquid 14
Rearrangement reactions 3, 182
Rearrangements 11
β-Resorcylic acid 163
Reformatsky reaction 125
Reformatsky reaction in solid state 127
Reformatsky reaction using sonication 126
regiospecific 88
Reimer–Tiemann Reaction 181
Robinson annulation 119

S

Salicylaldehyde 181
Salicylic acid 155
Saponification 155
Sigmatropic rearrangements 13
Simmons–Smith Reaction 130
Simmons–Smith Reaction under

sonication 131
Solid state (solventless) reactions 168
Solid supported organic synthesis 15
sonication 27
Sonication reactions 195
stilbene 55, 102
Stoichiometric reagents 8
Strecker synthesis 133
Strecker synthesis under sonication 134
Substitution seactions 4, 11
Sulfinic acid 101

T

Testololactone 45
Tetrabutyl ammonium hydrogen
    sulphate 175, 177
Tetraethyl propane-1,1,3,3-
    tetracarboxylate 107
Thianine 191
Thiophene-2-carbinol 62
Trans stilbene 165
Trihydroxymethylenenitromethane 32
2,4,4-Trimethyl-cyclopentanone 75
5,6,7-Trihydroxy flavone 78
2,4,6-Trinitrobiphenyl 138
2,2',4,4'-Tetramethyl-biphenyl 138
Triphenyl phosphene oxide 143
Triptycene 87
2,4,6-Tristyrylpyrimidine 32
dl-Tyrosine 135
Tylenol 161

U

Ullmann coupling reaction 136

Ullmann coupling under sonication 137
Ullmann Reaction 136

V

δ-Valero lactone 39
Vanillideneactone 162
Vinylogous aldol addition 28
Vinylogous aldol reaction 28
Vitamin A 150
Vitamin A1 129

W

Weiss-cook reaction 140
Wieland-miescher ketone 114
Williamson ether synthesis1 180
Williamsons ether synthesis 140
Wittig reaction 143
Wittig reaction in Ionic liquids 149
Witttig reaction with aqueous sodium
    hydroxide 145
Wittig Reaction in Solid Phase 148
Wittig-Horner Reaction 147
Wolff rearrangement 34
Wurtz Reaction 152
Wurtz-Fittig reaction 153
Wurtz-Type coupling 154

Y

Ylide 144

Z

Zerewitinoff determination of active
    hydrogens 99